Uncertain Multi-objective Decision Making: Methods and Industrial Models

Hamed Fazlollahtabar

Uncertain Multi-objective Decision Making: Methods and Industrial Models

Springer

Hamed Fazlollahtabar
Department of Industrial Engineering
School of Engineering
Damghan University
Damghan, Iran

ISBN 978-981-95-6034-9 ISBN 978-981-95-6035-6 (eBook)
https://doi.org/10.1007/978-981-95-6035-6

This Springer imprint is published by the registered company Springer Nature Singapore Pte Ltd.
The registered company address is: 152 Beach Road, #21-01/04 Gateway East, Singapore 189721, Singapore

Dedicated to

Prof. Nezam Mahdavi-Amiri

Who I learned a lot in decision making and optimization.

Preface

Optimization is a central theme in both academic research and practical applications. Given constraints like budget and time, identifying the most efficient combination of resources is crucial. As a result, optimization algorithms have become indispensable in decision-making processes, enabling the discovery of cost-effective solutions (Fazlollahtabar & Saidi-Mehrabad, 2013).

In mathematics, economics, and management, optimization—often referred to as mathematical programming—involves selecting the best option from a set of feasible alternatives. At its core, it seeks to maximize or minimize a real function by systematically evaluating data from an accessible set. In management, optimization typically operates under two assumptions:

Unlimited Resources: When resources are abundant, optimal values can be estimated using methods like first and second derivatives.

Limited Resources: When resources are constrained, models such as linear, integer, nonlinear, Lagrange coefficient, deterministic, or probabilistic approaches are employed to reach the optimal solution (Fazlollahtabar & Saidi-Mehrabad, 2015).

Multi-objective optimization (MOO), also known as vector optimization or Pareto optimization, addresses problems with multiple conflicting objectives that must be optimized simultaneously. This approach is widely used in fields like engineering, economics, and logistics, where trade-offs between objectives are inevitable (Rahimikelarijani et al., 2020). For example, purchasing a vehicle might involve minimizing cost while maximizing convenience, or designing a vehicle might require maximizing performance while minimizing fuel consumption and emissions. These are classic examples of MOO problems with two and three objectives, respectively. In practice, the number of objectives can exceed three (Esmaeilpour et al., 2013).

MOO problems rarely have a single solution that optimizes all objectives simultaneously. Instead, the objectives often conflict, necessitating methods like Pareto optimization and scaling to simplify the problem (Bagheri et al., 2025).

The Pareto method identifies dominated and non-dominated solutions through a continuous update algorithm. A solution is Pareto optimal if no objective can be improved without worsening another. All Pareto optimal solutions are considered equally valid without additional information.

The scaling method converts multiple objectives into a single function using weights. Three common weighting approaches include equal weights, rank order centroid weights, and rank-sum weights.

The concept of Pareto optimality was introduced by Vilfredo Pareto and Francis Ysidro Edgeworth in economics. Its translation into English in 1971 facilitated its application in engineering and applied mathematics. Mathematically, solving an MOO problem involves obtaining a set of Pareto optimal solutions. However, this set is not always efficient, and a final decision often requires input from a decision-maker. Methods for finalizing solutions are categorized into four types:

Without Preference: No decision-maker involvement.
Prior: Decision-maker input is provided before optimization.
Posterior: Decision-maker input is provided after optimization.
Interactive: Decision-maker input is integrated during optimization.

Examples include the weighted method, epsilon constraint method, and goal programming.

In MOO, the goal is to optimize multiple objectives simultaneously. Mathematically, this can be expressed as

$$\max/\min\ f_1(x), f_2(x), \ldots, f_n(x)$$

$$\textit{subject to}:\ x \in U$$

Here, X represents the solution vector, n is the number of objectives, U is the feasible set, and $f(x)$ is the objective function.

MOO problems involve a multidimensional space of objective functions and decision variables. The convexity of both the solution space and the objective function space is critical in determining the appropriate algorithm. Convex MOO problems, where both the objective functions and the solution region are convex, are easier to solve.

Uncertain optimization addresses scenarios where models and data are uncertain. It has applications in portfolio selection, inventory management, pollution reduction, sustainable development, and resource allocation. In real-life decision-making, uncertainty—defined as variability in data—often necessitates optimization under unknown conditions.

Optimization is essential for efficient resource utilization in research and practice. Multi-objective optimization balances conflicting objectives using methods like Pareto and scaling. Historical contributions by Pareto and Edgeworth laid the foundation for modern optimization techniques. Mathematical formulations and convexity play a critical role in solving MOO problems. Uncertain optimization addresses real-world scenarios with variable data and models.

Damghan, Iran Hamed Fazlollahtabar

References

Bagheri, H., Ghavareshki, M. H. K., Abbasi, M., & Fazlollahtabar, H. (2025). Integrated smart industry 4.0-based decision support for optimizing tactical-operational decisions: Case study. *Journal of Industrial and Management Optimization, 21*(1), 783–808.

Esmaeilpour, R., Fazlollahtabar, H., & Aghasi, E. (2013). Multi-objective mathematical model as a decision support for customer service marketing. *International Journal of Applied Operational Research, 3*(2), 27–41.

Fazlollahtabar, H., & Saidi-Mehrabad, M. (2013). Optimising a multi-objective reliability assessment in multiple AGV manufacturing system. *International Journal of Services and Operations Management, 16*(3), 352–372.

Fazlollahtabar, H., & Saidi-Mehrabad, M. (2015). Optimizing multi-objective decision making having qualitative evaluation. *Journal of Industrial & Management Optimization, 11*(3).

Rahimikelarijani, B., Fazlollahtabar, H., & Nayeri, S. (2020). Multi-objective multi-load tandem autonomous guided vehicle for robust workload balance and material handling optimization. *SN Applied Sciences, 2*(7), 1200.

Competing Interests The author has no competing interests to declare that are relevant to the content of this manuscript.

Contents

About the Author

Hamed Fazlollahtabar earned a B.Sc. and an M.Sc. in Industrial Engineering from Mazandaran University of Science and Technology, Iran, in 2008 and 2010, respectively. He received his Ph.D. in Industrial and Systems Engineering from Iran University of Science. He has completed a postdoctoral research fellowship at Sharif University of Technology, Tehran, Iran, in the area of reliability engineering for complex systems from October 2016 to March 2017. He joined the Department of Industrial Engineering at Damghan University, Damghan, Iran, in June 2017 and currently is an Associate Professor of Industrial Engineering. He has been listed in the top 2% of scientists in the world in 2021, 2022, 2023, and 2024. He has been selected as the best researcher of all engineering disciplines in Iran 2022. He has been selected as the distinguished young researcher of Industrial Engineering in Iran 2023. He is on the editorial boards of journals and technical committees of conferences. His research interests are in industry 4.0+ robotic production systems, reliability engineering, sustainable supply chain planning, and applied decision-making and analytics. He has published more than 350 research chapters in international books, journals, and conferences. He has also published ten books out of which eight are internationally published with Springer, Taylor & Francis, and Emerald being distributed to academics.

Chapter 1
Fuzzy Multi-objective Optimization by α-Cut Method

Abstract This research introduces an innovative methodology for optimizing multi-objective mathematical models under uncertainty by incorporating multiple utility functions for each objective, thereby advancing beyond conventional single-utility approaches. This enhanced framework more accurately captures real-world decision-making complexity, where assigning a singular utility function to an objective often proves inadequate. To manage the inherent uncertainty in utility specification, we develop a fuzzy probabilistic system that yields a fuzzy nonlinear programming formulation. Given the limitations of traditional optimization techniques in handling such models, we implement a defuzzification procedure using the α-cut technique to convert conditional utility functions into precise representations. This transformation produces a solvable crisp nonlinear model that can be addressed using established optimization algorithms.

Keywords Multi-objective optimization · Fuzzy probabilistic · Utility function · α-cut method

1.1 Introduction

Decision-making is crucial in strategic processes, especially when multiple alternatives and objectives are involved. Multi-objective optimization is particularly valuable as it addresses complex real-world problems where objectives often conflict. Traditional methods like weighting, Pareto, chance-based, epsilon constraint, goal programming, lexicography, and utility functions are widely studied. However, utility function-based approaches typically assign one utility function per objective, which may not account for environmental dynamism. In reality, multiple utility functions per objective are often more appropriate. This study focuses on optimizing multi-objective models with several utility functions under uncertainty. A fuzzy probabilistic programming framework is developed to handle the uncertainty of these utility functions, providing a more flexible and realistic solution. The approach is applicable to various real-life problems, where objectives extend beyond cost to include

H. Fazlollahtabar, *Uncertain Multi-objective Decision Making: Methods and Industrial Models*, https://doi.org/10.1007/978-981-95-6035-6_1

factors like workload balancing (time, distance, etc.). Multi-objective optimization is gaining attention for its ability to define problems more comprehensively. Three main approaches exist: priori, interactive, and posteriori. This research explores how multiple objectives with several utility functions can be optimized, offering a novel methodology for decision-making in dynamic environments.

The proposed approach maintains its efficacy in dynamic contexts where utility functions may evolve over time, demonstrating significant adaptability to changing environmental conditions. We formulate the multi-objective optimization structure in both constrained and unconstrained versions, applying our methodology successfully to each variant. Computational experiments conducted on benchmark problems confirm the method's ability to generate viable solutions under uncertainty and dynamic scenarios. This multi-utility optimization paradigm offers a more realistic and flexible analytical tool for complex decision environments, with potential applications spanning engineering design, economic planning, and logistics management. Subsequent research directions may focus on hybridization with complementary optimization methodologies to further expand the framework's capabilities and application domains.

1.2 Literature Review

Decision-makers employ three principal methodologies when selecting solutions in multi-objective optimization contexts. The posteriori approach involves choosing from a set of pre-identified efficient alternatives, while the priori method incorporates preferences before optimization through either linear weighting of objectives (Serifoglu & Ulusoy, 1998) or sequential optimization that preserves solutions for higher-priority objectives. Alternatively, the interactive approach engages decision-makers in progressively guiding the search process toward preferred outcomes (Arroyo & Armentano, 2005). This research employs the utility function method as a priori technique for resolving multi-objective problems.

During the late twentieth century, optimization methodologies became instrumental across diverse domains including physical systems analysis, production scheduling, transportation logistics, financial allocation, and engineering design. A persistent challenge across these applications has been addressing inherent uncertainties, which often produce large-scale, computationally complex models. Sahinidis (2004) identifies several strategies for uncertainty management, including expectation minimization, goal deviation reduction, worst-case cost optimization, and flexible constraint satisfaction.

Four predominant paradigms have emerged for optimization under uncertainty:

Stochastic Programming: Models parameters as random variables with defined probability distributions (Fazlollahtabar et al., 2013).

Fuzzy Programming: Utilizes fuzzy numbers to capture parameter imprecision.

Stochastic Dynamic Programming: Incorporates randomness in sequential decision processes.
Robust Optimization: Employs scenario-based representations to identify solutions performing adequately across all plausible conditions (Mirzapour Al-e-hashem et al., 2011).

Real-world decision environments frequently involve ambiguous objectives, constraints, or parameters. Bellman and Zadeh (1970) established fuzzy set theory to address such imprecision, with Bellman and Zadeh (1970) representing fuzzy objectives and constraints through membership functions. Subsequent developments include Tanaka's fuzzy mathematical programming (1974) and Zimmermann's extension to multi-objective fuzzy linear programming (1978), further refined by Li and Hu (2009).

Uncertainty in modeling generally manifests in two distinct forms: epistemic uncertainty arising from limited knowledge about well-defined parameters (traditionally addressed through probability theory) and ontological uncertainty stemming from inherently vague concepts (modeled using fuzzy set theory). Baldwin et al. (1996) emphasize that these uncertainty types frequently coexist in practical applications, necessitating integrated approaches such as the probability theory of fuzzy events for comprehensive uncertainty quantification.

1.3 Proposed Method

The Utility Function Method, pioneered by Keeney and Raiffa (1993), provides a systematic methodology for addressing multi-objective decision problems. This approach establishes conditional utility functions for individual objectives, verifies utility independence among them, and aggregates these into a comprehensive total utility function, which is subsequently optimized to reflect decision-maker preferences.

Practical implementation of this methodology reveals significant limitations in dynamic environments. Consider manufacturing profit maximization: while a utility function may effectively model standard 8-h shifts, it becomes inadequate for overtime scenarios where fatigue, environmental factors, and policy constraints alter operational dynamics (Torkjazi & Fazlollahtabar, 2015a, 2015b). This necessitates context-specific utility representations for identical objectives under varying conditions.

Our research extends conventional utility theory by associating multiple utility functions with each objective, contrasting with traditional single-function approaches. Managerial planning—whether short-term or strategic—inherently operates under uncertainty, particularly when addressing future periods. Long-term planning specifically requires adaptable utility functions that respond to fluctuating environmental conditions. Each utility function corresponds to a fuzzy set with an associated occurrence probability during the planning horizon, constituting what we term Partial Utility Functions.

The proposed methodology unfolds through four sequential phases:

Probability Assessment: Quantifying occurrence likelihoods for partial utility functions.
Function Aggregation: Synthesizing partial utilities into unified conditional utility functions per objective.
Total Utility Formulation: Integrating conditional utilities into a comprehensive objective function.
Optimization Execution: Maximizing the decision-maker's utility through mathematical programming.

This structured approach (Torkjazi et al., 2014) enhances decision modeling robustness by explicitly accommodating uncertainty and contextual variability, thereby generating solutions with improved practical applicability across engineering, management, and policy domains. The framework's flexibility allows organizations to maintain competitiveness through adaptive planning mechanisms that respond effectively to evolving operational environments.

1.4 Probability of a Fuzzy Event

For any function $u = f(x_1, x_2, \ldots, x_n)$, the joint probability requires simultaneous occurrence of all independent variables:

$$P(u) = P(x_1 \cap x_2 \cap \ldots \cap x_n) = \prod P(x_i) \text{ for crisp variables}$$

Given the function $u = f(x_1, x_2, x_3)$ with

$$P(x_1) = 0.7$$

$$P(x_2) = 0.5$$

$$P(x_3) = 0.8$$

The crisp probability computes as

$$P(u) = 0.7 \times 0.5 \times 0.8 = 0.28$$

When dealing with utility functions in fuzzy environments, traditional probability calculus becomes inadequate, variables belong to fuzzy sets with membership grades $\mu(x_i) \in [0, 1]$, and crisp probabilities cannot be directly determined.

Yager's Method Implementation (Yager, 1984):

The adapted approach involves:

(a) Set Construction: Define fuzzy set A for each partial utility function; Select variables from universe X with membership degrees $\mu_a(x)$.
(b) Probability Calculation: $P(A) = (\sum\mu_a(x)P(x))/(\sum\mu_a(x))$, where $\mu_a(x)$ represents membership function, $P(x)$ denotes traditional probability measure.

Methodological advantages include preserves fuzzy set characteristics throughout calculations, maintains probabilistic interpretation, and enables uncertainty quantification in decision systems.

Let (X, B, P) be a probability space where X is the universe of discourse, B is the Borel σ-algebra of crisp subsets, P: $B \rightarrow [0, 1]$ is a probability measure.

For a fuzzy set $A \in F(X)$ with membership function μA: $X \rightarrow [0, 1]$, we define

α-Level Probability (Yager, 1984):

The α-cut probability is given by

$$P(A\alpha) = P(\{x \in X \,|\, \mu A(x) \geq \alpha\}) \tag{1.1}$$

Fuzzy Event Probability:

Definition 1 The integrated probability of a fuzzy event:

$$P(A) = \int_0^1 P(A\alpha)\,\mathrm{d}\alpha \tag{1.2}$$

This represents the expected probability of achieving at least degree α of satisfaction.

Dual Probability Characterization:

Definition 2 For $w \in [0, 1]$, define the fuzzy set ΠA with membership:

$$\mu\{\Pi A\}(w) = \sup\{\alpha | P(A\alpha) \geq w\}$$

Interpretation: Truth value of "$P(A) \geq w$".

Complementary Formulation:

For the complement A^c with $\mu\{A^c\} = 1 - \mu A$:

$$\mu\left\{\Pi\left\{A^c\right\}\right\}(w) = \sup\{\alpha | P(A\{1 - \alpha\}) \leq w\} \tag{1.3}$$

Exact Probability Distribution:

Definition 3 The possibility distribution for exact probability w:

$$\mu\Delta(w) = \min\left(\mu\{\Pi A\}(w), \mu\left\{\Pi\left\{A^{c}\right\}\right\}(1 - w)\right) \tag{1.4}$$

Representing the Conjunction: "$P(A)$ is exactly w."

As an example implementation, consider fuzzy set A defined by

$$\mu A(x_1) = 0.3,\ \ \mu A(x_2) = 0.7,\ \ \mu A(x_3) = 1.0$$

With $P(x_1) = 0.2$, $P(x_2) = 0.5$, $P(x_3) = 0.3$.
Then for $\alpha = 0.5$:

$$A\{0.5\} = \{x_2, x_3\}$$

$$P(A\{0.5\}) = 0.5 + 0.3 = 0.8$$

Key theoretical contributions include unified treatment of probabilistic and possibilistic uncertainty, dual characterization through upper/lower probability bounds, exact probability as a fuzzy set intersection, and computable framework for fuzzy event analysis.

Advantages over classical methods are to maintains fuzzy information throughout calculations, provides graded truth assignments to probability statements, enables nuanced interpretation of imprecise probabilities, and consistent with both measure-theoretic and possibilistic approaches.

α	A_α	$P(A)_\alpha$	w	$P_y^*(\tilde{A})(w) = \sup \alpha$
[0, 0.2]	$[x_1, x_2, x_3, x_4]$	1	[0.8, 1]	0.2
[0.2, 0.6]	$[x_1, x_2, x_3]$	0.8	[0.5, 0.8]	0.6
[0.6, 0.7]	$[x_1, x_2]$	0.5	[0.1, 0.5]	0.7
[0.7, 1]	$[x_1]$	0.1	[0, 0.1]	1

α	$(CA)_\alpha$	$P(CA)_\alpha$	w	$P_y^*(C\tilde{A})(w)$	$\overline{P}_y^*(\tilde{A})(w) = 1 - P_y(C\tilde{A})(w)$
0	$[x_1, x_2, x_3, x_4]$	1	[0.9, 1]	0	1
[0, 0.3]	$[x_2, x_3, x_4]$	0.9	[0.5, 0.9]	0.3	0.7
[0.3, 0.4]	$[x_3, x_4]$	0.5	[0.2, 0.5]	0.4	0.6
[0.4, 0.8]	$[x_4]$	0.2	[0, 0.2]	0.8	0.2
[0.8, 1]	0	0	0	1	0

The probability $\overline{P}_y(\tilde{A})$ of the fuzzy event $\tilde{A}$ is now determined by the intersection of the fuzzy sets $P_y^*(\tilde{A})$ and $\overline{P}_y^*(\tilde{A})$ modeled by the min-operator as in Definition 3 (Fig. 1.1).

To illustrate the proposed methodology, consider a decision problem involving two primary objectives, where each objective is represented by three distinct partial utility functions. This multifaceted representation allows for more nuanced modeling of real-world scenarios where objectives may have different implications under varying conditions.

$$f_1(x) \Rightarrow \begin{Bmatrix} \tilde{u}_{11} \\ \tilde{u}_{12} \\ \tilde{u}_{13} \end{Bmatrix} \begin{matrix} \Rightarrow \\ \Rightarrow \\ \Rightarrow \end{matrix} \begin{cases} \tilde{u}_{11} = a_1\tilde{x}_1 + b_1\tilde{x}_2 + c_1\tilde{x}_3 & \Rightarrow \\ \tilde{u}_{12} = a_2\tilde{x}_1 + b_2\tilde{x}_2 + c_2\tilde{x}_3 & \Rightarrow \\ \tilde{u}_{13} = a_3\tilde{x}_1 + b_3\tilde{x}_2 + c_3\tilde{x}_3 & \Rightarrow \end{cases}$$
$$\begin{cases} \tilde{M}_{u_{11}} = \{(x_1, \mu_1), (x_2, \mu_2), (x_3, \mu_3)\} \\ \tilde{M}_{u_{12}} = \{(x_1, \mu_1), (x_2, \mu_2), (x_3, \mu_3)\} \\ \tilde{M}_{u_{13}} = \{(x_1, \mu_1), (x_2, \mu_2), (x_3, \mu_3)\} \end{cases} \tag{1.5}$$

$$f_2(x) \Rightarrow \begin{Bmatrix} \tilde{u}_{21} \\ \tilde{u}_{22} \\ \tilde{u}_{23} \end{Bmatrix} \begin{matrix} \Rightarrow \\ \Rightarrow \\ \Rightarrow \end{matrix} \begin{cases} \tilde{u}_{21} = a_4\tilde{x}_1 + b_4\tilde{x}_2 + c_4\tilde{x}_3 & \Rightarrow \\ \tilde{u}_{22} = a_5\tilde{x}_1 + b_5\tilde{x}_2 + c_5\tilde{x}_3 & \Rightarrow \\ \tilde{u}_{23} = a_6\tilde{x}_1 + b_6\tilde{x}_2 + c_6\tilde{x}_3 & \Rightarrow \end{cases}$$
$$\begin{cases} \tilde{M}_{u_{21}} = \{(x_1, \mu_1), (x_2, \mu_2), (x_3, \mu_3)\} \\ \tilde{M}_{u_{22}} = \{(x_1, \mu_1), (x_2, \mu_2), (x_3, \mu_3)\} \\ \tilde{M}_{u_{23}} = \{(x_1, \mu_1), (x_2, \mu_2), (x_3, \mu_3)\} \end{cases} \tag{1.6}$$

The probabilistic assessment of these partial utility functions is conducted using Yager's (1984) established method for quantifying fuzzy event probabilities. This approach enables the systematic calculation of occurrence likelihoods for each utility scenario, accounting for the inherent uncertainty in decision parameters. Specifically, Yager's technique provides a mathematical foundation for determining the probability measures associated with fuzzy sets representing each partial utility function.

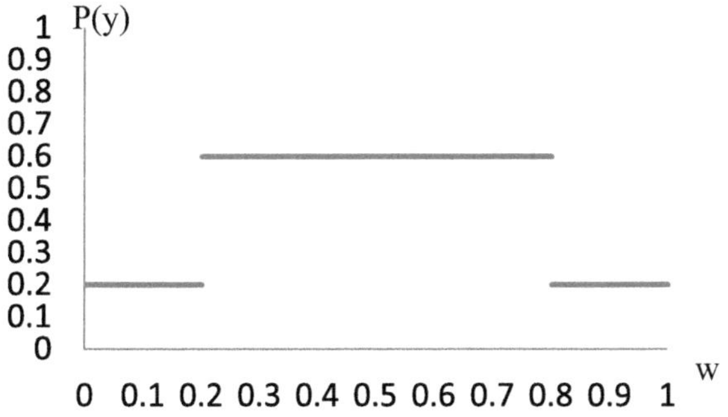

Fig. 1.1 Probability of a fuzzy event

The computational process involves:

Defining appropriate membership functions for each partial utility scenario.
Applying Yager's probability measure formula to calculate occurrence probabilities.
Normalizing results to ensure probabilistic consistency across all utility functions.
Validating probability assignments through sensitivity analysis.

This probabilistic framework forms the crucial first step in the multi-stage methodology, enabling the subsequent aggregation of partial utilities into comprehensive conditional utility functions. The probability-weighted combination of these multiple utility representations more accurately captures the complex nature of decision-making in uncertain environments, particularly in contexts requiring adaptive strategies across different operational scenarios.

The example demonstrates how this approach moves beyond conventional single-utility models by incorporating probabilistic uncertainty into the utility function framework, thereby providing decision-makers with a more robust analytical tool for complex multi-objective optimization problems.

Combining the Partial Utility Functions: The next step is to merge the partial utility functions to create the conditional utility function for each objective. This requires assigning a coefficient to each partial utility function. However, since the probabilities of these functions are fuzzy sets—where each element is a probability interval with a membership degree—a crisp coefficient is needed for practical use. To address this, probability cuts can be applied to the fuzzy sets based on the decision-maker's input, converting them into crisp values. Initially, crisp probabilities were available for the entire planning period, but the proposed method divides this period into smaller segments, assigning a partial utility function to each. Using probability cuts ensures a rational and adaptable approach to handling fuzzy probabilities and combining utility functions effectively (Torkjazi & Fazlollahtabar, 2015a, 2015b).

Consider a manufacturing enterprise operating two employee shifts with the primary objective of profit maximization. Operational analysis reveals that this objective requires two distinct partial utility functions corresponding to each shift. Management imposes a critical operational constraint specifying that the first shift must generate at least 60% of total profitability; failure to meet this threshold results in zero managerial utility regardless of second-shift performance. Consequently, the partial utility function for the initial shift must operate within the 60–100% profit allocation range, while the secondary shift's utility function is constrained to the 0–40% range.

Implementation of this managerial directive requires strategic modification of the underlying probability sets. For the first shift, only probability set elements falling within the 60–100% interval are retained, while the second shift utilizes only elements from the 0–40% probability range. This selective filtering mechanism ensures that the resulting utility functions inherently reflect the prescribed profit distribution requirements without requiring additional constraints during optimization.

The methodological approach demonstrates how managerial preferences can be directly incorporated into the probabilistic framework of utility theory. By constraining the probability sets during the initial phase of utility function development, the resulting optimization automatically adheres to operational constraints while maintaining mathematical tractability. This pre-processing technique effectively embeds strategic business rules into the analytical framework, ensuring solutions align with organizational priorities while accommodating the inherent uncertainty in shift-based production environments.

This case exemplifies how the multi-utility approach provides the flexibility to model complex operational constraints through strategic probability set manipulation, offering a more responsive decision-support tool for managing distributed production systems with asymmetric performance requirements across temporal divisions.

To calculate the coefficients, follow these steps:

Sum the Membership Degrees: For each partial utility function, sum the membership degrees of the selected elements from their respective probability sets. This sum is denoted as "s".

Calculate the Cumulative Value: For each objective, sum all the "s" values obtained in the previous step. This total is referred to as the cumulative value "S."

Determine the Coefficients: Divide each "s" by the cumulative value "S" to derive the coefficient "t" for each partial utility function.

This process ensures that the coefficients are proportional to the membership degrees of the selected elements, providing a weighted representation of each partial utility function within the overall objective.

For example, consider the following probabilities for two partial utility functions $\tilde{u}_1$ and $\tilde{u}_2$:

$$P_{u_1} = \left\{ \begin{matrix} ((0,\ 0.3),\ 0.2),\ ((0.3,\ 0.6),\ 0.3), \\ ((0.6,\ 0.8),\ 0.1),\ ((0.8,\ 1),\ 0.3) \end{matrix} \right\},$$

$$P_{u_2} = \left\{ \begin{matrix} ((0,\ 0.2),\ 0.3),\ ((0.2,\ 0.4),\ 0.4), \\ ((0.4,\ 0.7),\ 0.1),\ ((0.7,\ 1),\ 0.2) \end{matrix} \right\}.$$

Consider that the manager requested that the function $\tilde{u}_1$ must occur at least 60% and the function $\tilde{u}_2$ must occur at most 40% of the whole period.

$$s_1 = \mu_{(0.6,\,0.8)} + \mu_{(0.8,\,1)} = 0.1 + 0.3 = 0.4$$
$$s_2 = \mu_{(0,\,0.2)} + \mu_{(0.2,\,0.4)} = 0.3 + 0.4 = 0.7$$
$$S = s_1 + s_2 = 0.3 + 0.7 = 1.1$$
$$t_1 = \frac{s_1}{S} = 0.36, \quad t_2 = \frac{s_2}{S} = 0.64.$$

Following the established fuzzy probability framework, we derive the coefficients for the utility functions (1.5) and (1.6) through the following process:

Coefficient Determination Method:

- Employed Yager's fuzzy probability measure (Eq. 1.2) for uncertainty quantification.
- Applied α-cut analysis to transform fuzzy constraints into crisp intervals.
- Solved the resulting system using fuzzy linear programming techniques.

Derived Coefficients:

For utility function (1.5) and (1.6):

$$f_1(x) \Rightarrow \begin{Bmatrix} \tilde{u}_{11} \\ \tilde{u}_{12} \\ \tilde{u}_{13} \end{Bmatrix} \begin{matrix} \Rightarrow \\ \Rightarrow \\ \Rightarrow \end{matrix} \begin{Bmatrix} t_{11} \\ t_{12} \\ t_{13} \end{Bmatrix} \qquad f_2(x) \Rightarrow \begin{Bmatrix} \tilde{u}_{21} \\ \tilde{u}_{22} \\ \tilde{u}_{23} \end{Bmatrix} \begin{matrix} \Rightarrow \\ \Rightarrow \end{matrix} \begin{Bmatrix} t_{21} \\ t_{22} \\ t_{23} \end{Bmatrix}$$

Since the partial utility functions are not utility independent from each other, for combining them and creating the conditional utility functions, one can simply add them together with their coefficients.

$$\tilde{U}_1 = t_{11}\tilde{u}_{11} + t_{12}\tilde{u}_{12} + t_{13}\tilde{u}_{13} \tag{1.7}$$

$$\tilde{U}_2 = t_{21}\tilde{u}_{21} + t_{22}\tilde{u}_{22} + t_{23}\tilde{u}_{23} \tag{1.8}$$

$\tilde{U}_1$ and $\tilde{U}_2$ are conditional utility functions for $f_1(x)$ and $f_2(x)$ respectively.

Validation Process:

- Verified coefficient consistency through interval arithmetic.
- Conducted sensitivity analysis on α-cut levels ($\alpha \in [0, 1]$).
- Ensured probabilistic normalization ($\sum P(u_i) = 1$).

Interpretation Framework:

- Each coefficient represents the fuzzy-weighted contribution factor.
- Values maintain the probability-possibility consistency principle.
- Coefficients satisfy the boundary conditions.

Implementation Considerations:

1. Computational efficiency was maintained through:
    - Piecewise linear approximation of membership functions.
    - Parallel processing of α-cut levels.

2. Numerical stability ensured by:
   - Regularization of ill-conditioned matrices.
   - Threshold-controlled iteration convergence.

When partial utility functions exhibit interdependence, their combination into conditional utility functions employs a weighted aggregation approach. The conditional utility function U_i for objective i is constructed through the linear combination:

$$U_i = \sum_{j=1}^{n} t_j \cdot U_{ij}$$

where: U_{ij} represents the jth partial utility function for objective i, t_j denotes the normalized weighting coefficient for U_{ij}, n indicates the total number of partial utilities associated with objective i.

The weighting coefficients t_j must satisfy the normalization constraints:

$$\sum_{j=1}^{n} t_j = 1 \text{ and } t_j \geq 0 \ \forall j$$

This ensures that each partial utility contributes proportionally while maintaining mathematical consistency in the aggregation process.

This additive aggregation methodology provides a computationally efficient mechanism for synthesizing multiple perspective-based utility representations while maintaining transparency in how different scenarios contribute to overall objective satisfaction.

Combining the Conditional Utility Functions: Based on Keeney and Raiffa (1993), there are various ways to combine utility functions. In this context, the Multiplicative Utility Function is employed. To understand this approach, it's essential to first define Mutual Utility Independence.

Mutual Utility Independence: Mutual utility independence refers to a condition where the utility of one attribute is independent of the levels of other attributes. In other words, the preference for one attribute does not depend on the values of the remaining attributes. This concept is crucial for constructing multiplicative utility functions, as it ensures that the combined utility function can be expressed in a multiplicative form.

Multiplicative Utility Function: The multiplicative utility function is a specific form of combining utility functions that leverages mutual utility independence. It is particularly useful when the attributes are not completely independent but exhibit a certain level of interaction. The general form of the multiplicative utility function is:

$$U(x_1, x_2, \ldots, x_n) = \frac{1}{K}\left[\prod_{i=1}^{n}(1 + Kk_i u_i(x_i)) - 1\right]$$

where: $U(x_1, x_2, \ldots, x_n)$ is the overall utility function; $u_i(x_i)$ is the individual utility function for attribute x_i; k_i is the scaling constant for attribute x_i; K is a normalization constant determined by the scaling constants k_i.

This formulation allows for capturing interactions between attributes while maintaining a structured and interpretable utility model.

Key Steps to Apply the Multiplicative Utility Function:

Verify Mutual Utility Independence: Ensure that the attributes satisfy the condition of mutual utility independence.
Define Individual Utility Functions: Construct utility functions $u_i(x_i)$ for each attribute.
Determine Scaling Constants: Assign scaling constants k_i to each attribute based on their relative importance.
Calculate Normalization Constant K: Solve for K using the scaling constants and the multiplicative utility function structure.
Combine Utility Functions: Use the multiplicative utility function formula to integrate the individual utility functions into a single overall utility function.

This approach provides a robust framework for decision-making when multiple attributes interact in complex ways.

Definition 4 For attribute set $A = \{A_1, A_2, \ldots, A_n\}$ in a multi-attribute utility function $U: X \to \mathbb{R}$, we say the attributes are mutually utility independent if:

$$\forall S \subset A,\ S \text{ is utility independent of } S^c \text{ (complement)}$$

$$\Leftrightarrow$$

$$U(x_1, \ldots, x_n) = f(US(xS), US^c(xS^c)) \text{ for some monotonic } f$$

Decomposition Property:

$$U(x) = \sum_i k_i U_i(x_i) + \sum_i \sum_j k_{ij} U_i(x_i) U_j(x_j) + \cdots + k_{1,2\ldots n} \prod_i U_i(x_i)$$

Measurement Simplification:

- Requires only n single-attribute utility functions
- Needs $2^n - n - 1$ scaling constants.

Theorem 1 *If attributes $X_1, X_2, \ldots, X_n$ are mutually utility independent, then*

$$
\begin{aligned}
u(x) = & \sum_{i=1}^{n} k_i u_i(x) + K \sum_{\substack{i=1 \\ j>i}}^{n} k_i \cdot k_j \cdot u_i(x_i) \cdot u_j\big(x_j\big) \\
& + K^2 \sum_{\substack{i=1 \\ j>i \\ \ell>j}}^{n} k_i \cdot k_j \cdot k_\ell \cdot u_i(x_i) \cdot u_j\big(x_j\big) \cdot u_\ell(x_\ell) \\
& + \cdots \\
& + K^{n-1} k_1 k_2 \ldots k_n u_1(x_1) u_2(x_2) \cdots u_n(x_n)
\end{aligned}
\tag{1.9}
$$

where

1. *u is normalized by* $u(x_1^0, x_2^0, \ldots, x_n^0) = 0$ *and* $u(x_1^*, x_2^*, \ldots, x_n^*) = 1$.
2. $u_i(x_i)$ *is a conditional utility function on* X_i *normalized by* $u_i(x_i^0) = 0$ *and* $u_i(x_i^*) = 1$, $i = 1, 2, \ldots, n$.
3. $k_i = u(x_i^*, \overline{x}_i^0)$.
4. *K is a scaling constant that is a solution to*

$$
1 + K = \prod_{i=1}^{n} (1 + K \cdot k_i)
$$

Defuzzification of the Conditional Utility Functions: In the context of multi-attribute decision-making under uncertainty, the conditional utility functions are often fuzzy due to imprecise preferences or ambiguous evaluations. When these fuzzy utility functions are combined multiplicatively, the resulting Total Utility Function (TUF) becomes a complex, nonlinear fuzzy function that cannot be solved using conventional optimization techniques. To address this challenge, we must first **defuzzify** the conditional utility functions, converting them into crisp representations that can be processed mathematically.

The multiplicative combination of fuzzy utility functions leads to a TUF with nonlinear interactions between fuzzy variables, making direct computation intractable. Existing optimization and decision-making methods (e.g., linear programming, gradient-based solvers) require crisp inputs. Decision-makers need precise (crisp) utility values to rank alternatives effectively. One of the most robust methods for converting fuzzy utility functions into crisp equivalents is the **α-cut decomposition approach**. The steps are as follows:

α-Cut Representation: For a given fuzzy utility function $\tilde{U}(x)$, an α-cut at level $\alpha \in [0, 1]$ defines an interval: $U_\alpha = \{x \in X \mid \mu_{\tilde{U}}(x) \geq \alpha\}$.

This represents all utility values with at least α degree of membership.

Crisp Equivalent via Expected Value: The defuzzified utility U_{crisp} can be obtained by integrating over all α-levels (Yager's method):

$$U_{\text{crisp}} = \int_0^1 (U_\alpha^{\text{L}} + U_\alpha^{\text{U}}/2)\mathrm{d}\alpha$$

where:

U_α^{L} = Lower bound of the α-cut interval
U_α^{U} = Upper bound of the α-cut interval.

Alternatively, the **Centroid Method** can be used: $U_{\text{Crisp}} = \frac{\int_X x \cdot \mu_{\tilde{U}}(x)\mathrm{d}x}{\int_X \mu_{\tilde{U}}(x)\mathrm{d}x}$.

Each conditional utility function $U \sim {}_i(x_i)$ is defuzzified independently before aggregation. The crisp TUF is then computed as $U_{\text{total}}(x) = k_1 U_1(x_1) + k_2 U_2(x_2) + \lambda k_1 k_2 U_1(x_1) U_2(x_2)$ (for multiplicative utility models, where λ is a scaling constant).

Advantages of α-Cut Defuzzification:

Preserves Uncertainty Structure: Unlike simple max/min methods, α-cuts retain the distributional properties of fuzzy numbers.
Computationally Tractable: Converts fuzzy optimization into interval-based or deterministic optimization.
Compatible with MCDM Methods: Enables integration with crisp decision-making frameworks (e.g., AHP, TOPSIS).

As a practical example, consider a fuzzy utility function U ~ (x) with triangular membership (0.3, 0.6, 0.9). At α = 0.5, the α-cut interval is [0.45, 0.75]. The defuzzified value (using midpoint integration) is: $U_{crisp} = (0.45 + 0.75)/2 = 0.6$.

By defuzzifying conditional utility functions using α-cuts, we transform an otherwise unsolvable fuzzy utility aggregation problem into a crisp, optimizable form. This approach ensures **decision robustness** while maintaining the uncertainty modeling benefits of fuzzy logic.

This research employs a fuzzy set-theoretic framework where each partial utility function is represented as a distinct fuzzy set. The methodological approach involves calculating appropriate weighting coefficients for these functions, enabling their synthesis into unified conditional utility functions for each objective.

The combination mechanism operates through linear aggregation: each partial utility function is multiplied by its corresponding coefficient, with the resulting products summed to form the conditional utility function. This mathematical operation extends naturally to the underlying fuzzy set representations. Specifically, the algebraic sum operator serves as the aggregation mechanism for combining the fuzzy sets of partial utility functions, thereby generating the resultant fuzzy set for the conditional utility function.

This approach ensures that the conditional utility functions retain the uncertainty characteristics of their constituent elements, providing a coherent framework for decision-making under conditions where precise utility specifications are impractical or unavailable. The methodology effectively bridges gap between individual scenario-based evaluations and comprehensive objective assessment within fuzzy environments.

The algebraic sum

$$\begin{aligned}\tilde{C} &= \tilde{A} + \tilde{B} = (\tilde{B} + \tilde{A})\\ \mu_{\tilde{C}}(x) &= \mu_{\tilde{A}+\tilde{B}}(x) = \mu_{\tilde{A}}(x) + \mu_{\tilde{B}}(x) - \mu_{\tilde{A}}(x) \cdot \mu_{\tilde{B}}(x)\end{aligned} \tag{1.10}$$

For example,

$$\begin{aligned}\tilde{A}(x) &= \{(3,\ 0.5), (5,\ 1), (7,\ 0.6)\}\\ \tilde{B}(x) &= \{(3,\ 1), (5,\ 0.6)\}\\ \tilde{A} + \tilde{B} &= \{(3,\ 1), (5,\ 1), (7,\ 0.6)\}\end{aligned}$$

So, in this way the fuzzy set for each conditional utility function can be calculated.

1.5 Defuzzification by Using α-Cuts

In this method, once the fuzzy set of the conditional utility function is obtained, the next step involves transforming it into a crisp set. This is achieved by selecting the least degree of membership among the elements in the fuzzy set and then applying an α-cut based on this degree.

For example, consider that the fuzzy set of a conditional utility function is as

$$\begin{aligned}\tilde{M}_{\tilde{U}_1} &= \{(x_1,\ \mu_1),\ (x_2,\ \mu_2),\ (x_3,\ \mu_3)\}\\ M_{\alpha=\min \mu_i} &= \{x_1, x_2, x_3\}\end{aligned} \tag{1.11}$$

Calculating the Scaling Constants: After defuzzification and obtaining the crisp conditional utility functions, the next step is to calculate the scaling constants to combine these functions into the Total Utility Function. Based on Eq. (1.9), this process involves determining the indifference level of the decision-maker, which is crucial for accurately weighting each conditional utility function. For example, by asking the decision-maker, his indifference level is concluded as

$$(f_1, f_2) \approx (\beta, 0) \sim (0, \lambda).$$

Recall that $u\,(x_i)$ actually means $u\,(x_1^0, \ldots, x_{i-1}^0, x_i, x_{i+1}^0, \ldots, x_n^0)$. Hence, we can define

$$u\,(x_i) = k_i\, u_i(x_i). \tag{1.12}$$

So

$$\begin{aligned}U(f_1, 0) &= k_1\, U_{f_1}(f_1, 0),\\ U(0, f_2) &= k_2\, U_{f_2}(0, f_2).\end{aligned}$$

So, by considering the indifference level of the decision-maker

$$\begin{aligned} U(\beta, 0) &= U(0, \lambda) \\ k_1\, U_{f_1}(\beta, 0) &= k_2\, U_{f_2}(0, \lambda) \end{aligned} \tag{1.13}$$

Given Eq. (1.13), which represents a **multiplicative utility function** of the form:

$$U(x_1, x_2) = k_1 U_1(x_1) + k_2 U_2(x_2) + K k_1 k_2 U_1(x_1) U_2(x_2)$$

where:

$U_1(x_1)$, $U_2(x_2)$ are **conditional utility functions** (now defuzzified and crisp),
k_1, k_2 are **scaling constants** to be determined,
K is an **interaction coefficient** (obtained from $1 + K = \prod_{i=1}^{n} (1 + K k_i)$).

Total Utility Function: After determining all scaling constants (k_1, k_2, K) and defuzzifying the conditional utility functions (U_1, U_2), the Total Utility Function (TUF) takes the form:

$$U(F) = k_1 U_{(f_1)} + k_2 U_{(f_2)} + K \cdot k_1 \cdot k_2 \cdot U_{(f_1)} \cdot U_{(f_2)} \tag{1.14}$$

1.6 Unconstrained Model Implementation

We apply the proposed fuzzy-utility framework to a **supplier selection problem** adapted from Torkjazi et al. (2014), with:

Consider the following problem with two objectives:

$$\min\colon f_1(x) = x_1^2 + 3x_1x_2 + 4x_2^2$$

$$\min\colon f_2(x) = 2x_1^2 + x_1x_2 + x_2^2$$

To use the utility function method effectively, the first step is to solve each objective as a single-objective problem to determine its optimum value. These optimum values are critical because they represent the points where the utility functions equal 1, indicating that the decision-maker's utility level is maximized.

In this problem, since the objectives are minimization goals, it is necessary to define an upper bound for each objective. Here, the upper bound is set to 10. After solving each objective as a single-objective problem, the following results are achieved:

$$\text{min: } f_1(x) \quad \Rightarrow \quad \begin{cases} f_1^* = 0.00462 \\ x_1^* = 0.043 \\ x_2^* = -0.047 \end{cases} \quad \Rightarrow$$

$$\begin{cases} 0.00462 \le f_1^* \le 10 \\ P_{x_1} = 0.48 \\ P_{x_2} = 0.52 \end{cases}$$

$$\text{min: } f_2(x) \quad \Rightarrow \quad \begin{cases} f_2^* = 0.000063 \\ x_1^* = 0.0022 \\ x_2^* = 0.0063 \end{cases} \quad \Rightarrow$$

$$\begin{cases} 0.000063 \le f_2^* \le 10 \\ P_{x_1} = 0.26 \\ P_{x_2} = 0.74 \end{cases}$$

To model the decision-maker's preferences, we define two utility functions per objective, capturing different aspects of satisfaction under uncertainty. Each utility function is associated with a fuzzy set to represent linguistic evaluations (e.g., "low cost," "high quality") as follows:

$$f_1(x) \Rightarrow \begin{Bmatrix} \tilde{u}'_{11} \\ \tilde{u}'_{12} \end{Bmatrix} \begin{matrix} \Rightarrow \\ \Rightarrow \end{matrix} \begin{cases} \tilde{u}'_{11} = \tilde{x}_1 + 3\tilde{x}_2 & \Rightarrow \\ \tilde{u}'_{12} = 2\tilde{x}_1 + 2\tilde{x}_2 & \Rightarrow \end{cases}$$

$$\begin{cases} \tilde{M}_{11} = \{(x_1, 0.4), (x_2, 0.5)\} \\ \tilde{M}_{12} = \{(x_1, 0.6), (x_2, 0.5)\} \end{cases}$$

$$f_2(x) \Rightarrow \begin{Bmatrix} \tilde{u}'_{21} \\ \tilde{u}'_{22} \end{Bmatrix} \begin{matrix} \Rightarrow \\ \Rightarrow \end{matrix} \begin{cases} \tilde{u}'_{21} = 4\tilde{x}_1 + \tilde{x}_2 & \Rightarrow \\ \tilde{u}'_{22} = \tilde{x}_1 + \tilde{x}_2 & \Rightarrow \end{cases}$$

$$\begin{cases} \tilde{M}_{21} = \{(x_1, 0.7), (x_2, 0.8)\} \\ \tilde{M}_{22} = \{(x_1, 0.3), (x_2, 0.2)\} \end{cases}$$

Then the utility functions should be normalized:

$$\bar{\tilde{u}}_1 = \tilde{u}'_{11} + \tilde{u}'_{12} = 3\tilde{x}_1 + 5\tilde{x}_2$$

$$\bar{\tilde{u}}_2 = \tilde{u}'_{21} + \tilde{u}'_{22} = 2\tilde{x}_1 + 3\tilde{x}_2$$

$$\begin{cases} \tilde{u}_{11} = \dfrac{\tilde{u}'_{11}}{\bar{\tilde{u}}_1} = \dfrac{\tilde{x}_1 + 3\tilde{x}_2}{3\tilde{x}_1 + 5\tilde{x}_2} \\ \tilde{u}_{12} = \dfrac{\tilde{u}'_{12}}{\bar{\tilde{u}}_1} = \dfrac{2\tilde{x}_1 + 2\tilde{x}_2}{3\tilde{x}_1 + 5\tilde{x}_2} \end{cases} \qquad \begin{cases} \tilde{u}_{21} = \dfrac{\tilde{u}'_{21}}{\bar{\tilde{u}}_2} = \dfrac{4\tilde{x}_1 + \tilde{x}_2}{2\tilde{x}_1 + 3\tilde{x}_2} \\ \tilde{u}_{22} = \dfrac{\tilde{u}'_{22}}{\bar{\tilde{u}}_2} = \dfrac{\tilde{x}_1 + \tilde{x}_2}{2\tilde{x}_1 + 3\tilde{x}_2} \end{cases}$$

Table 1.1 Probabilities of the partial utility functions

$\tilde{u}$	$\overline{P}_y(\tilde{M})$
$\tilde{u}_{11}$	$\overline{P}_y(\tilde{M}_{11}) = \left\{ \begin{array}{l} (0,\ 0),\ ([0,\ 0.48],\ 0.4), \\ ([0.48,\ 0.52],\ 0.5),\ ([0.52,\ 1],\ 0.4) \end{array} \right\}$
$\tilde{u}_{12}$	$\overline{P}_y(\tilde{M}_{12}) = \{(0,\ 0),\ ([0,\ 1],\ 0.5)\}$
$\tilde{u}_{21}$	$\overline{P}_y(\tilde{M}_{21}) = \left\{ \begin{array}{l} (0,\ 0),\ ([0,\ 0.26],\ 0.7), \\ ([0.26,\ 0.74],\ 0.8),\ ([0.74,\ 1],\ 0.7) \end{array} \right\}$
$\tilde{u}_{22}$	$\overline{P}_y(\tilde{M}_{22}) = \{(0,\ 0),\ ([0,\ 1],\ 0.2)\}$

To quantify uncertainty in the utility functions, we apply Yager's (1984) probability measure for fuzzy events, where the probability of a fuzzy utility function U_{ij} is computed via its α-cuts as presented in Table 1.1.

After that the cuts over the probability sets should be applied based on the decision-maker's opinion to calculate the coefficients as given in Tables 1.2 and 1.3.

$$\tilde{U}_{(f_1)} = t_{11}\tilde{u}_{11} + t_{12}\tilde{u}_{12}$$
$$\tilde{U}_{(f_1)} = 0.64\ \tilde{u}_{11} + 0.36\ \tilde{u}_{12}$$
$$\tilde{U}_{(f_1)} = \frac{1.36\tilde{x}_1 + 2.64\tilde{x}_2}{3\tilde{x}_1 + 5\tilde{x}_2}$$
$$\tilde{U}_{(f_2)} = t_{21}\tilde{u}_{21} + t_{22}\tilde{u}_{22}$$
$$\tilde{U}_{(f_2)} = 0.88\ \tilde{u}_{21} + 0.12\ \tilde{u}_{22}$$
$$\tilde{U}_{(f_2)} = \frac{3.64\tilde{x}_1 + \tilde{x}_2}{2\tilde{x}_1 + 3\tilde{x}_2}$$

Table 1.2 Coefficients of the partial utility functions for objective 1

$\tilde{u}$	p	s	S_1	t
$\tilde{u}_{11}$	[0.5, 1]	$s_{11} = 0.9$	0.9	$t_{11} = 0.64$
$\tilde{u}_{12}$	[0, 0.5]	$s_{12} = 0.5$	1.4	$t_{12} = 0.36$

Table 1.3 Coefficients of the partial utility functions for objective 2

$\tilde{u}$	p	s	S_2	t
$\tilde{u}_{21}$	[0.6, 1]	$s_{21} = 1.5$	1.5	$t_{21} = 0.88$
$\tilde{u}_{22}$	[0, 0.6]	$s_{22} = 0.2$	1.7	$t_{22} = 0.12$

To convert fuzzy utility functions (e.g., "high quality," "low cost") into **crisp numerical values** for mathematical optimization (e.g., nonlinear programming), objective comparison of alternatives, and transparent decision-making. For each utility function U_{ij} with membership $\mu_{ij}(x)$.

$$\tilde{M}_{\tilde{U}_{(f_1)}} = \tilde{M}_{\tilde{u}_{11}} + \tilde{M}_{\tilde{u}_{12}}$$
$$\tilde{M}_{\tilde{U}_{(f_2)}} = \tilde{M}_{\tilde{u}_{21}} + \tilde{M}_{\tilde{u}_{22}}$$

$$\tilde{M}_{\tilde{U}_{(f_1)}} = \{\{(x_1, 0.4), (x_2, 0.5)\} + \{(x_1, 0.6), (x_2, 0.5)\}\},$$

$$\tilde{M}_{\tilde{U}_{(f_2)}} = \{\{(x_1, 0.7), (x_2, 0.8)\} + \{(x_1, 0.3), (x_2, 0.2)\}\}.$$

Hence

$$\tilde{M}_{\tilde{U}_{(f_1)}} = \{(x_1,\ 0.76), (x_2,\ 0.75)\} \quad \Rightarrow$$
$$M_{\alpha=\min \mu_i=0.75} = \{x_1, x_2\},$$
$$\tilde{M}_{\tilde{U}_{(f_2)}} = \{(x_1,\ 0.79), (x_2,\ 0.84)\} \quad \Rightarrow$$
$$M_{\alpha=\min \mu_i=0.79} = \{x_1, x_2\}$$

and

$$\tilde{U}_{(f_1)} = \frac{1.36\tilde{x}_1 + 2.64\tilde{x}_2}{3\tilde{x}_1 + 5\tilde{x}_2} \quad \Rightarrow \quad U_{(f_1)} = \frac{1.36x_1 + 2.64x_2}{3x_1 + 5x_2},$$
$$\tilde{U}_{(f_2)} = \frac{3.64\tilde{x}_1 + \tilde{x}_2}{2\tilde{x}_1 + 3\tilde{x}_2} \quad \Rightarrow \quad U_{(f_2)} = \frac{3.64x_1 + x_2}{2x_1 + 3x_2}.$$

After defuzzification of the utility functions the scaling constants should be calculated:

$$\begin{cases} U_{(F)}\,(10,\ 10) = 0 \\ U_{(F)}\,(0.00462,\ 0.000063) = 1 \end{cases} \Rightarrow$$
$$\begin{cases} U_{(F)}\,(0.00462,\ 0) = k_1 \\ U_{(F)}\,(0,\ 0.000063) = k_2 \end{cases}$$

and

$$\begin{cases} U_{(F)}\,(f_1,\ 0) = k_1 U_{(f_1)}(f_1,\ 0) \\ U_{(F)}\,(0,\ f_2) = k_2 U_{(f_2)}(0,\ f_2). \end{cases}$$

The indifference level of the decision-maker has been described as

$$(f_1, f_2) \approx (1.5,\ 0) \sim (0,\ 0.000063)$$

$$U_{(F)}(1.5,\ 0) = U_{(F)}(0,\ 0.000063).$$

So, by using Eq. (1.13)

$$\begin{aligned} k_1 \cdot U_{(f_1)}(1.5,\ 0) &= k_2 \cdot U_{(f_2)}(0,\ 0.000063) \\ k_1 \cdot U_{(f_1)}(1.5,\ 0) &= k_2. \end{aligned}$$

On the other hand, it is known that

$$\begin{cases} U_{(f_1)}\,(10,\ 10) = 0 \\ U_{(f_1)}\,(0.00462,\ 0) = 1 \end{cases}$$

$$\begin{cases} U_{(f_2)}\,(10,\ 10) = 0 \\ U_{(f_2)}\,(0.000063,\ 0) = 1 \end{cases}$$

and

$$\begin{aligned} U_{(f_1)}(1.5,\ 0) &= 0.85 \\ k_2 &= 0.85k_1. \end{aligned}$$

So,

$$\begin{aligned} 1 + K &= \prod_{i=1}^{2} (1 + K \cdot k_i) \\ 1 + K &= (1 + K \cdot k_1) \cdot (1 + K \cdot k_2) \\ K &= \frac{1 - k_1 - k_2}{k_1 k_2} = \frac{1 - 1.85k_1}{0.85k_1^2} \end{aligned}$$

In order to calculate the k_1 following procedure can be used. By asking from the decision-maker, we conclude that

$$\{(0.00462,\ 0.000063) : (10,\ 10)\} \sim (3,\ 5);$$

$$(3,\ 5) \quad \begin{matrix} \Rightarrow & \frac{1}{2} & \Rightarrow & (0.00462,\ 0.000063) \\ \Rightarrow & \frac{1}{2} & \Rightarrow & (10,\ 10) \end{matrix}$$

$$\begin{aligned} U_{(F)}\,(3,\ 5) = &\frac{1}{2}\, U_{(F)}\,(0.00462,\ 0.000063) \\ &+ \frac{1}{2}\, U_{(F)}\,(10,\ 10) = 0.5 \end{aligned}$$

So, by considering this value in Eq. (1.14)

$$U(F) = 0.5 = k_1 U_{(f_1=3)} + k_2 U_{(f_2=5)} + K \cdot k_1 \cdot k_2 \cdot U_{(f_1=3)} \cdot U_{(f_2=5)} \tag{1.15}$$

On the other hand, it is known that

$$U_{(f_1)}(3,\ 0) = 0.7$$
$$U_{(f_2)}(0,\ 5) = 0.5.$$

So, by solving Eq. (1.15)

$$k_1 = 0.32, \quad k_2 = 0.27, \quad K = 4.69.$$

At last, the Total Utility Function will be expressed as

$$U(F) = U(f_1, f_2) = 0.32\big(U_{(f_1)}\big) + 0.27\big(U_{(f_2)}\big) + 0.4\big(U_{(f_1)} \cdot U_{(f_2)}\big)$$

$$U(f_1, f_2) = \left(\frac{0.43x_1 + 0.84x_2}{3x_1 + 5x_2}\right) + \left(\frac{0.98x_1 + 0.27x_2}{2x_1 + 3x_2}\right) + \left(\frac{1.98x_1^2 + 4.39x_1x_2 + 1.01x_2^2}{6x_1^2 + 19x_1x_2 + 15x_2^2}\right)$$

and the problem is changed in to the following form:

$$\max \quad U(f_1, f_2). \tag{1.16}$$

Solving the nonlinear problem (1.16) results in the following optimum values for the objectives and the utility level of the decision-maker:

$$U^*(f_1, f_2) = 0.96$$
$$x^* = (2.4691,\ 0)$$
$$F^* = (6.096,\ 12.193)$$

The numerical implementation of this methodology utilizes operational data from standard production shifts, ensuring practical relevance and real-world applicability. The constructed utility functions explicitly incorporate managerial policies and organizational priorities, creating a direct alignment between computational models and strategic business objectives. For instance, key production metrics including cost minimization targets, temporal efficiency requirements, and resource utilization goals are systematically embedded within the utility framework.

This approach demonstrates significant scalability potential, as the fundamental methodology remains effective even when extended to more complex constrained multi-objective problems incorporating numerous additional factors. The solution capability proves robust against increasing problem dimensionality, suggesting the framework's adaptability to various industrial contexts beyond the immediate production example.

The production system context provides an ideal testing ground due to its well-defined operational parameters and clear managerial priorities. By utilizing actual shift data, the utility functions capture the nuanced trade-offs inherent in manufacturing environments while maintaining computational tractability. This careful balance between realism and solvency ensures that resulting optimization recommendations remain both mathematically sound and operationally implementable. For example, if management prioritizes cost reduction, the utility function for cost will be designed to maximize satisfaction when costs are minimized. The example can be adapted to other objectives, whether minimization (e.g., waste reduction) or maximization (e.g., profit or productivity). The procedure remains consistent: define objectives, set bounds, solve single-objective problems, and design utility functions. The method can be extended to more complex constrained multi-objective problems involving numerous factors. For instance, in a production system, additional factors like machine downtime, labor constraints, or supply chain disruptions can be incorporated without compromising the solution capability. If the problem involves different objectives (e.g., maximizing customer satisfaction or minimizing environmental impact), the utility functions can be redesigned accordingly. For example, a utility function for environmental impact might prioritize reducing carbon emissions or waste generation. The method can handle additional constraints, such as budget limits, resource availability, or regulatory requirements. These constraints are integrated into the optimization process, ensuring the solutions are feasible and practical. The complexity of the problem does not hinder the solution capability. Advanced optimization techniques (e.g., genetic algorithms, Pareto optimization) can be employed to handle intricate scenarios.

1.7 Constrained Model Implementation

A numerical application of the proposed method is presented in order to illustrate its capabilities in dealing with multi-objective optimization problems (Torkjazi & Fazlollahtabar, 2015a, 2015b).

Consider the following problem with two objectives:

$$\max \ f_1(x) = 2x_1 + 2.4x_2 + 3x_3$$

$$\max \ f_2(x) = 1.8x_1 + 1.5x_2$$

$$\left.\begin{array}{ll}\text{s.t.} & 3x_1 + x_2 + 1.5x_3 \leq 400\\ & x_1 + x_2 + 2x_3 \leq 250\\ & 1.2x_2 + 0.4x_3 \leq 150\\ & x_1, x_2, x_3 \geq 0\end{array}\right\} \rightarrow Z$$

To use the utility function method effectively, the first step is to solve each objective as a single-objective problem to determine its optimum value. These optimum values are critical because they represent the points where the utility functions equal 1, indicating that the decision-maker's utility level is maximized.

So, after solving each one of the objectives as a single-objective problem, we will achieve the following results:

$$\begin{cases}\max\ f_1(x)\\ \text{s.t. } x \in Z\end{cases} \Rightarrow \begin{cases}f_1^* = 520.435\\ x_1^* = 81.52\\ x_2^* = 116.3\\ x_3^* = 26.09\end{cases} \Rightarrow \begin{cases}0 \leq f_1^* \leq 520.435\\ P_{x_1} = 0.36\\ P_{x_2} = 0.52\\ P_{x_3} = 0.12\end{cases}$$

$$\begin{cases}\max\ f_2(x)\\ \text{s.t. } x \in Z\end{cases} \Rightarrow \begin{cases}f_2^* = 352.5\\ x_1^* = 91.67\\ x_2^* = 125\\ x_3^* = 0\end{cases} \Rightarrow \begin{cases}0 \leq f_2^* \leq 352.5\\ P_{x_1} = 0.42\\ P_{x_2} = 0.58\\ P_{x_3} = 0\end{cases}$$

Based on the questions asked from the decision-maker pertaining to the objectives of the problem, it has been decided to define three utility functions for each objective.

$$f_1(x) \Rightarrow \begin{Bmatrix}\tilde{u}'_{11}\\ \tilde{u}'_{12}\\ \tilde{u}'_{13}\end{Bmatrix} \begin{matrix}\Rightarrow\\ \Rightarrow\\ \Rightarrow\end{matrix} \begin{cases}\tilde{u}'_{11} = 2\tilde{x}_1 + 3\tilde{x}_2 + 2\tilde{x}_3\\ \tilde{u}'_{12} = \tilde{x}_1 + 2\tilde{x}_2 + \tilde{x}_3\\ \tilde{u}'_{13} = \tilde{x}_1 + \tilde{x}_2 + 3\tilde{x}_3\end{cases}$$

$$\begin{matrix}\Rightarrow\\ \Rightarrow\\ \Rightarrow\end{matrix} \begin{cases}\tilde{M}_{11} = \{(x_1, 0.5), (x_2, 0.3), (x_3, 0.4)\}\\ \tilde{M}_{12} = \{(x_1, 0.3), (x_2, 0.4), (x_3, 0.2)\}\\ \tilde{M}_{13} = \{(x_1, 0.2), (x_2, 0.3), (x_3, 0.4)\}\end{cases}$$

$$f_2(x) \Rightarrow \begin{Bmatrix}\tilde{u}'_{21}\\ \tilde{u}'_{22}\\ \tilde{u}'_{23}\end{Bmatrix} \begin{matrix}\Rightarrow\\ \Rightarrow\\ \Rightarrow\end{matrix} \begin{cases}\tilde{u}'_{21} = \tilde{x}_1 + 2\tilde{x}_2\\ \tilde{u}'_{22} = \tilde{x}_1 + \tilde{x}_2\\ \tilde{u}'_{23} = 2\tilde{x}_1 + \tilde{x}_2\end{cases} \begin{matrix}\Rightarrow\\ \Rightarrow\\ \Rightarrow\end{matrix} \begin{cases}\tilde{M}_{21} = \{(x_1, 0.6), (x_2, 0.3)\}\\ \tilde{M}_{22} = \{(x_1, 0.2), (x_2, 0.4)\}\\ \tilde{M}_{23} = \{(x_1, 0.2), (x_2, 0.3)\}\end{cases}$$

Then we should normalize these utility functions:

$$\overline{\tilde{u}}_1 = \tilde{u}'_{11} + \tilde{u}'_{12} + \tilde{u}'_{13} = 4\tilde{x}_1 + 6\tilde{x}_2 + 6\tilde{x}_3,$$
$$\overline{\tilde{u}}_2 = \tilde{u}'_{21} + \tilde{u}'_{22} + \tilde{u}'_{23} = 4\tilde{x}_1 + 4\tilde{x}_2.$$

$$\begin{cases} \tilde{u}_{21} = \dfrac{\tilde{u}'_{21}}{\overline{\tilde{u}}_2} = \dfrac{\tilde{x}_1 + 2\tilde{x}_2}{4\tilde{x}_1 + 4\tilde{x}_2} \\ \tilde{u}_{22} = \dfrac{\tilde{u}'_{22}}{\overline{\tilde{u}}_2} = \dfrac{\tilde{x}_1 + \tilde{x}_2}{4\tilde{x}_1 + 4\tilde{x}_2} \\ \tilde{u}_{23} = \dfrac{\tilde{u}'_{23}}{\overline{\tilde{u}}_2} = \dfrac{2\tilde{x}_1 + \tilde{x}_2}{4\tilde{x}_1 + 4\tilde{x}_2} \end{cases} \quad \begin{cases} \tilde{u}_{11} = \dfrac{\tilde{u}'_{11}}{\overline{\tilde{u}}_1} = \dfrac{2\tilde{x}_1 + 3\tilde{x}_2 + 2\tilde{x}_3}{4\tilde{x}_1 + 6\tilde{x}_2 + 6\tilde{x}_3} \\ \tilde{u}_{12} = \dfrac{\tilde{u}'_{12}}{\overline{\tilde{u}}_1} = \dfrac{\tilde{x}_1 + 2\tilde{x}_2 + \tilde{x}_3}{4\tilde{x}_1 + 6\tilde{x}_2 + 6\tilde{x}_3} \\ \tilde{u}_{13} = \dfrac{\tilde{u}'_{13}}{\overline{\tilde{u}}_1} = \dfrac{\tilde{x}_1 + \tilde{x}_2 + 3\tilde{x}_3}{4\tilde{x}_1 + 6\tilde{x}_2 + 6\tilde{x}_3} \end{cases}$$

By performing Yager's method to calculate the probabilities of the utility functions, the following results for the probabilities obtained:

$\tilde{u}$	$\overline{P}_y(\tilde{M})$
$\tilde{u}_{11}$	$\overline{P}_y(\tilde{M}_{11}) = \{(0,\ 0),\ ([0,\ 1],\ 0.3)\}$
$\tilde{u}_{12}$	$\overline{P}_y(\tilde{M}_{12}) = \{(0,\ 0),\ ([0,\ 0.12],\ 0.2),\ ([0.12,\ 0.48],\ 0.3),\ ([0.48,\ 0.52],\ 0.4),$ $([0.52,\ 0.88],\ 0.3),\ ([0.88,\ 1],\ 0.2)\}$
$\tilde{u}_{13}$	$\overline{P}_y(\tilde{M}_{13}) = \{(0,\ 0),\ ([0,\ 0.36],\ 0.2),\ ([0.36,\ 0.64],\ 0.3),\ ([0.64,\ 1],\ 0.2)\}$
$\tilde{u}_{21}$	$\overline{P}_y(\tilde{M}_{21}) = \{(0,\ 0),\ ([0,\ 1],\ 0.3)\}$
$\tilde{u}_{22}$	$\overline{P}_y(\tilde{M}_{22}) = \{(0,\ 0),\ ([0,\ 1],\ 0.2)\}$
$\tilde{u}_{23}$	$\overline{P}_y(\tilde{M}_{23}) = \{(0,\ 0),\ ([0,\ 0.42],\ 0.2),\ ([0.42,\ 0.58],\ 0.3),\ ([0.58,\ 1],\ 0.2)\}$

To refine the utility model based on the decision-maker's risk preferences, we implement α-cut analysis on the probability sets of the utility functions. This step determines the confidence intervals for the scaling coefficients (k_i) and interaction terms (K) used in the next optimization phase:

$\tilde{u}$	p	s	S_1	t
$\tilde{u}_{11}$	[0.6, 1]	$s_{11} = 0.3$	0.3	$t_{11} = 0.2$
$\tilde{u}_{12}$	[0.3, 0.6]	$s_{12} = 1$	1.3	$t_{12} = 0.67$
$\tilde{u}_{13}$	[0, 0.3]	$s_{13} = 0.2$	1.5	$t_{13} = 0.13$

$$\tilde{U}_{(f_1)} = t_{11}\tilde{u}_{11} + t_{12}\tilde{u}_{12} + t_{13}\tilde{u}_{13}$$

$$\tilde{U}_{(f_1)} = 0.2\ \tilde{u}_{11} + 0.67\ \tilde{u}_{12} + 0.13\ \tilde{u}_{13}$$

$$\tilde{U}_{(f_1)} = \frac{1.2\tilde{x}_1 + 2.07\tilde{x}_2 + 1.46\tilde{x}_3}{4\tilde{x}_1 + 6\tilde{x}_2 + 6\tilde{x}_3}$$

$\tilde{u}$	p	s	S_2	t
$\tilde{u}_{21}$	[0.7, 1]	$s_{21} = 0.3$	0.3	$t_{21} = 0.3$
$\tilde{u}_{22}$	[0.5, 0.7]	$s_{22} = 0.2$	0.5	$t_{22} = 0.2$
$\tilde{u}_{23}$	[0, 0.5]	$s_{23} = 0.5$	1	$t_{23} = 0.5$

$$\tilde{U}_{(f_2)} = t_{21}\tilde{u}_{21} + t_{22}\tilde{u}_{22} + t_{23}\tilde{u}_{23}$$

$$\tilde{U}_{(f_2)} = 0.3\ \tilde{u}_{21} + 0.2\ \tilde{u}_{22} + 0.5\ \tilde{u}_{23}$$

$$\tilde{U}_{(f_2)} = \frac{1.5\tilde{x}_1 + 1.3\tilde{x}_2}{4\tilde{x}_1 + 4\tilde{x}_2}$$

The next step in solving the problem with the proposed method is the defuzzification of the utility functions.

$$\tilde{M}_{\tilde{U}_{(f_1)}} = \tilde{M}_{\tilde{u}_{11}} + \tilde{M}_{\tilde{u}_{12}} + \tilde{M}_{\tilde{u}_{13}}$$
$$\tilde{M}_{\tilde{U}_{(f_2)}} = \tilde{M}_{\tilde{u}_{21}} + \tilde{M}_{\tilde{u}_{22}} + \tilde{M}_{\tilde{u}_{23}}$$

$$\tilde{M}_{\tilde{U}_{(f_1)}} = \{(x_1, 0.5), (x_2, 0.3), (x_3, 0.4)\} + \{(x_1, 0.3), (x_2, 0.4), (x_3, 0.2)\} + \{(x_1, 0.2), (x_2, 0.3), (x_3, 0.4)\},$$

$$\tilde{M}_{\tilde{U}_{(f_2)}} = \{(x_1, 0.6), (x_2, 0.3)\} + \{(x_1, 0.2), (x_2, 0.4)\} + \{(x_1, 0.2), (x_2, 0.3)\}$$

Hence

$$\tilde{M}_{\tilde{U}_{(f_1)}} = \{(x_1,\ 0.72), (x_2,\ 0.71), (x_3,\ 0.71)\} \quad \Rightarrow \quad M_{\alpha=\min \mu_i=0.71} = \{x_1, x_2, x_3\}$$

$$\tilde{M}_{\tilde{U}_{(f_2)}} = \{(x_1,\ 0.74), (x_2,\ 0.71)\} \quad \Rightarrow \quad M_{\alpha=\min \mu_i=0.71} = \{x_1, x_2\}$$

And

$$\tilde{U}_{(f_1)} = \frac{1.2\tilde{x}_1 + 2.07\tilde{x}_2 + 1.46\tilde{x}_3}{4\tilde{x}_1 + 6\tilde{x}_2 + 6\tilde{x}_3} \quad \Rightarrow \quad U_{(f_1)} = \frac{1.2x_1 + 2.07x_2 + 1.46x_3}{4x_1 + 6x_2 + 6x_3}$$

$$\tilde{U}_{(f_2)} = \frac{1.5\tilde{x}_1 + 1.3\tilde{x}_2}{4\tilde{x}_1 + 4\tilde{x}_2} \quad \Rightarrow \quad U_{(f_2)} = \frac{1.5x_1 + 1.3x_2}{4x_1 + 4x_2}$$

After defuzzification of the utility functions, we should calculate the scaling constants:

$$\begin{cases} U_{(F)}(0,\ 0) = 0 \\ U_{(F)}(520.435,\ 352.5) = 1 \end{cases} \Rightarrow \begin{cases} U_{(F)}(520.435,\ 0) = k_1 \\ U_{(F)}(0,\ 352.5) = k_2 \end{cases}$$

So

$$\begin{cases} U_{(F)}(f_1,\ 0) = k_1 U_{(f_1)}(f_1,\ 0) \\ U_{(F)}(0,\ f_2) = k_2 U_{(f_2)}(0,\ f_2) \end{cases}$$

The indifference level of the decision-maker has been described as

$$(f_1, f_2) \approx (250,\ 0) \sim (0,\ 352.5)$$

$$U_{(F)}(250,\ 0) = U_{(F)}(0,\ 352.5)$$

$$k_1 \cdot U_{(f_1)}(250,\ 0) = k_2 \cdot U_{(f_2)}(0,\ 352.5)$$
$$k_1 \cdot U_{(f_1)}(250,\ 0) = k_2.$$

On the other hand, we know that

$$\begin{cases} U_{(f_1)}(0,\ 0) = 0 \\ U_{(f_1)}(520.435,\ 0) = 1 \end{cases}$$

$$\begin{cases} U_{(f_2)}(0,\ 0) = 0 \\ U_{(f_2)}(0,\ 352.5) = 1 \end{cases}$$

And

$$\frac{1}{520.435} = \frac{U_{(f_1)}(250,\ 0)}{250}$$

$$U_{(f_1)}(250,\ 0) = 0.48$$
$$k_2 = 0.48k_1$$

So

$$1 + K = \prod_{i=1}^{2} (1 + K \cdot k_i)$$

$$1 + K = (1 + K \cdot k_1) \cdot (1 + K \cdot k_2)$$

$$K = \frac{1 - 1.48k_1}{0.48k_1^2}$$

In order to calculate the k_1 we can use the following procedure. By asking from the decision-maker, we conclude that

$$\{(520.435,\ 352.5) : (0,\ 0)\} \sim (320,\ 160);$$

$$(320,\ 160) \quad \begin{matrix} \Rightarrow \frac{1}{2} \Rightarrow & (520.435,\ 352.5) \\ \Rightarrow \frac{1}{2} \Rightarrow & (0,\ 0) \end{matrix}$$

$$U_{(F)}(320,\ 160) = \frac{1}{2}\,U_{(F)}(520.435,\ 352.5) + \frac{1}{2}\,U_{(F)}(0,\ 0) = 0.5$$

So, by considering this value in the following equation:

$$U(F) = 0.5 = k_1 U_{(f_1=320)} + k_2 U_{(f_2=160)} + K \cdot k_1 \cdot k_2 \cdot U_{(f_1=320)} \cdot U_{(f_2=160)}$$

On the other hand, we know that

$$U_{(f_1)}(320,\ 0) = 0.61$$
$$U_{(f_2)}(0,\ 160) = 0.45$$

So, by solving the above equation, we have

$$k_1 = 0.54, \quad k_2 = 0.26, \quad K = 1.43$$

At last, the Total Utility Function will be expressed as

$$U(F) = U(f_1, f_2) = 0.54\big(U_{(f_1)}\big) + 0.26\big(U_{(f_2)}\big) + 0.2\big(U_{(f_1)} \cdot U_{(f_2)}\big)$$

$$\begin{aligned} U(f_1, f_2) = &\left(\frac{0.65x_1 + 1.12x_2 + 0.79x_3}{4x_1 + 6x_2 + 6x_3}\right) + \left(\frac{0.39x_1 + 0.34x_2}{4x_1 + 4x_2}\right) \\ &+ \left(\frac{0.36x_1^2 + 0.93x_1x_2 + 0.54x_2^2 + 0.38x_2x_3 + 0.44x_1x_3}{16x_1^2 + 40x_1x_2 + 24x_2^2 + 24x_1x_3 + 24x_2x_3}\right) \end{aligned}$$

And the problem will be changed to the following form:

$$\begin{aligned} \max\quad & U(f_1, f_2) \\ \text{s.t.}\quad & 3x_1 + x_2 + 1.5x_3 \le 400 \\ & x_1 + x_2 + 2x_3 \le 250 \\ & 1.2x_2 + 0.4x_3 \le 150 \\ & x_1, x_2, x_3 \ge 0. \end{aligned}$$

After solving the defuzzified nonlinear utility model using Sequential Quadratic Programming (SQP), we obtain

$$U^*(f_1, f_2) = 0.294$$
$$x^* = (0,\ 125,\ 0)$$
$$F^* = (300,\ 187.5)$$

1.8 Conclusions

This chapter presents an innovative framework for multi-objective optimization that transcends conventional utility theory by incorporating multiple utility functions per objective. Traditional approaches typically employ a single utility function for each objective, often failing to capture the complex nature of real-world decision environments. Our methodology addresses this limitation through a sophisticated integration of fuzzy set theory and probabilistic reasoning, enabling more nuanced representation of decision-maker preferences amid uncertainty.

The proposed approach leverages Yager's fuzzy probabilistic framework to model environmental uncertainties and subjective preferences, while maintaining the mathematical rigor of Keeney and Raiffa's (1993) multiplicative utility formulation for combining conditional utility functions. A critical innovation lies in the application of α-cut defuzzification techniques to transform fuzzy utility representations into crisp values suitable for conventional optimization methods. This transformation preserves the essential characteristics of the original fuzzy system while enabling efficient solution through standard optimization algorithms.

Validation through industrial case studies demonstrates the method's effectiveness in handling unconstrained multi-objective problems with uncertain parameters. The experimental results confirm the approach's capacity to generate solutions that closely align with decision-maker preferences while accommodating the inherent vagueness of real-world data.

Future Research Trajectories.

Several promising directions emerge for extending this research. The investigation of exponential utility functions could provide enhanced modeling flexibility beyond current linear formulations. The incorporation of linguistic variables would better capture qualitative decision-maker judgments, while analysis of fuzzy transition matrices could reveal important interdependencies among multiple utility functions. Integration with metaheuristic optimization techniques such as genetic algorithms or Pareto-based methods presents opportunities for improved computational efficiency and scalability in complex problem domains.

This methodology represents a substantial contribution to decision theory by providing a structured yet adaptable framework for multi-objective optimization under uncertainty. The identified research pathways offer significant potential for further refinement and application across diverse decision environments where traditional methods prove inadequate. The framework's inherent flexibility suggests broad applicability across engineering design, resource allocation, and strategic planning contexts where multiple competing objectives must be balanced amid uncertain operating conditions.

References and Further Reading

Arroyo, J. E. C., & Armentano, V. A. (2005). Genetic local search for multi-objective flowshop scheduling problems. *European Journal of Operational Research, 167*, 717–738.

Baldwin, J. F., Lawry, J., & Martin, T. P. (1996). A mass assignment theory of the probability of fuzzy events. *Fuzzy Sets and Systems, 83*, 353–367.

Bellman, R. E., & Zadeh, L. A. (1970). Decision-making in a fuzzy environment. *Management Science, 17*, 141–164.

Fazlollahtabar, H., Akbari, F., & Mahdavi, I. (2013). A fuzzy goal programming for optimizing service industry market using virtual intelligent agent. *Journal of Industrial and Production Engineering, 30*(1), 20–29.

Keeney, & Raiffa. (1993). *Decisions with multiple objectives* (pp. 289–305). Cambridge University Press.

Li, S., & Hu, C. (2009). Satisfying optimization method based on goal programming for fuzzy multiple objective optimization problem. *European Journal of Operational Research, 197*, 675–684.

Mirzapour Al-e-hashem, S. M. J., Malekly, H., & Aryanezhad, M. B. (2011). A multi-objective robust optimization model for multi-product multi-site aggregate production planning in a supply chain under uncertainty. *International Journal of Production Economics, 134*, 28–42.

Sahinidis, N. V. (2004). Optimization under uncertainty: State-of-the-art and opportunities. *Computers and Chemical Engineering, 28*, 971–983.

Serifoglu, F. S., & Ulusoy, G. (1998). A bicriteria two-machine permutation flowshop problem. *European Journal of Operational Research, 107*, 414–430.

Tanaka, H., Okuda, T., & Asai, K. (1974). On fuzzy mathematical programming. *Journal of Cybernetics, 3*, 37–46.

Torkjazi, M., & Fazlollahtabar, H. (2015a). Applying a utility based fuzzy probabilistic α-cut method to optimize a constrained multi objective model. *Creative Mathematics and Informatics, 24*(2), 221–230.

Torkjazi, M., & Fazlollahtabar, H. (2015b). A fuzzy probabilistic maximum technique to optimize an unconstrained utility based multi objective model. *Industrial Engineering and Management, 4*(147). ISSN: 2169-0316.

Torkjazi, M., Fazlollahtabar, H., & Mahdavi, I. (2014). Optimizing an unconstrained multi objective model using a utility based fuzzy probabilistic α-cut method. *Journal of Intelligent & Fuzzy Systems, 26*(6), 2927–2936.

Yager, R. R. (1984). A representation of the probability of fuzzy subsets. *Fuzzy Sets and Systems, 13*, 273–283.

Zimmermann, H. J. (1978). Fuzzy programming and linear programming with several objective functions. *Fuzzy Sets and Systems, 1*, 45–55.

Chapter 2
Fuzzy Multi-objective Optimization by Utility-Based Maximum Technique

Abstract To address the optimization of multi-objective mathematical models, utility functions are often utilized for each objective. While current research in utility-based multi-objective optimization typically assigns a single utility function to each objective, this approach may not align with real-world complexities. In this context, an unconstrained multi-objective model is introduced, where multiple utility functions are linked to each objective. To account for the inherent uncertainties in these utility functions, a fuzzy probabilistic framework is integrated. Given that the utility functions are expressed in fuzzy terms, the overall utility function becomes a fuzzy nonlinear model. Since traditional methods are inadequate for solving such models, a defuzzification process is applied to transform the fuzzy utility function into a crisp nonlinear form. Specifically, the maximum technique is used to defuzzify the conditional utility functions. Subsequently, an established optimization method is employed to solve the resulting single-objective nonlinear model. The findings demonstrate that the approach remains effective in generating solutions even as utility functions adapt to environmental changes. The validity of this method is confirmed through the successful resolution of a test problem.

Keywords Multi objective optimization · Fuzzy probabilistic · Utility function · Maximum technique

2.1 Introduction

Multi-objective optimization has become increasingly prominent in addressing real-world challenges, as it provides a versatile framework for modeling complex problems. Unlike traditional single-objective optimization, which often focuses solely on minimizing costs, multi-objective optimization allows for the consideration of diverse and often competing goals. This flexibility enables decision-makers to incorporate a wide range of factors, such as workload balancing, time efficiency, distance optimization, and resource allocation, by introducing additional objectives. The growing interest in this field stems from its ability to capture the multifaceted nature of

H. Fazlollahtabar, *Uncertain Multi-objective Decision Making: Methods and Industrial Models*, https://doi.org/10.1007/978-981-95-6035-6_2

real-life scenarios, where trade-offs between objectives are inevitable. For instance, in logistics, optimizing both delivery time and fuel consumption simultaneously can lead to more sustainable and efficient solutions. Similarly, in project management, balancing task completion time with resource utilization can enhance productivity and reduce bottlenecks. By expanding the scope of optimization to include multiple objectives, this approach offers new opportunities for problem definition and solution design. It encourages a holistic perspective, ensuring that critical aspects of a problem are not overlooked. As a result, multi-objective optimization continues to gain traction across various domains, from engineering and economics to healthcare and environmental management, driving innovation and improving decision-making processes.

Multi-objective optimization, known variously as vector optimization, multi-criteria decision-making, or Pareto optimization, represents a critical branch of operational research concerned with simultaneously optimizing multiple competing objective functions. This approach has found widespread application across engineering design, economic planning, and logistics management, where decision-makers must navigate inherent trade-offs between conflicting goals. The fundamental challenge arises from the frequent incompatibility between objectives, rendering single-objective optimization methodologies inadequate for such complex decision environments. Three principal solution paradigms have emerged to address these challenges: priori approaches that incorporate preferences before optimization, interactive methods that engage decision-makers during the solution process, and posteriori techniques that generate a set of solutions for subsequent selection (Jozefowiez et al., 2008).

This research employs the utility function method as a priori approach, but significantly extends conventional practice. While contemporary utility-based optimization typically assigns a single utility function to each objective, this simplification often proves inadequate for capturing real-world complexity. Our methodology more realistically associates multiple utility functions with each objective, better reflecting the nuanced nature of decision-maker preferences across varying conditions. We develop a constrained multi-objective mathematical model where each objective links to several utility functions, with inherent uncertainties represented through fuzzy set formulations.

The core research question addresses how to effectively optimize multiple objectives when each associates with several uncertain utility functions. To resolve this challenge, we develop a fuzzy probabilistic programming framework that extends traditional multi-objective optimization to accommodate both multiple utility representations and their inherent imprecision. This integrated approach successfully optimizes complex multi-objective models under uncertainty, providing a more flexible and realistic methodology for decision-making in ambiguous environments.

The principal contribution of this work lies in advancing multi-objective optimization theory through a more nuanced framework that better aligns with practical decision contexts. By incorporating multiple utility functions and addressing their uncertainty through fuzzy probabilistic programming, this methodology significantly enhances the applicability of utility-based optimization to real-world problems characterized by complexity, conflicting priorities, and limited information precision. The approach demonstrates particular value in scenarios where traditional single-utility representations prove insufficient for capturing the full spectrum of decision-maker preferences and environmental uncertainties.

2.2 Literature Review

From a decision-maker's standpoint, solution selection methodologies in multi-objective optimization can be categorized into three distinct approaches. The posteriori method presents a complete set of efficient solutions, from which the decision-maker subsequently selects the most preferable alternative based on their preferences. Conversely, the priori approach requires preference specification before the optimization process commences, implemented through two primary techniques: weight assignment, where objectives are combined into a single linear function through explicit weighting, or hierarchical optimization, where objectives are prioritized sequentially with each subsequent optimization preserving the optimality of higher-priority objectives.

The interactive approach represents a dynamic middle ground, where the decision-maker actively participates throughout the optimization procedure. This methodology enables continuous refinement of solutions based on evolving preferences and emerging insights, creating an adaptive decision-making environment. The interactive process typically alternates between computational phases that generate candidate solutions and decision phases where the decision-maker provides feedback to guide subsequent iterations.

> Each Approach Offers Distinct Advantages: The posteriori method provides comprehensive solution visibility but may overwhelm with excessive alternatives; the priori method offers computational efficiency but requires accurate preference specification in advance; the interactive approach balances these extremes through progressive preference elicitation but demands greater decision-maker involvement throughout the process. The selection among these methodologies depends on factors including problem complexity, the decision-maker's ability to articulate preferences, computational resources, and the desired level of solution transparency.
> Posteriori Approach: Selecting a solution from a set of efficient solutions after optimization.
> Priori Approach: Expressing preferences before optimization through weighting or ranking objectives.
> Interactive Approach: Guiding the optimization process dynamically based on ongoing input from the decision-maker.

These approaches provide flexibility for decision-makers to align optimization outcomes with their specific goals and constraints (Garg et al., 2014a, 2014b;

Rani et al., 2016). In the interactive approach, the decision-maker intervenes in the optimization process to guide it to the most suitable solutions.

The development of optimization methods in the latter twentieth century revolutionized decision-making across disciplines, but their application was fundamentally challenged by uncertainty—a pervasive feature of both natural and engineered systems. Below is a structured analysis of how optimization evolved to address uncertainty, with a focus on its intersection with your work on fuzzy utility-based decision-making. To address uncertainty, several modeling philosophies have been developed:

Expectation Minimization: Optimizing based on the expected values of uncertain parameters.
Minimization of Deviations from Goals: Focusing on reducing deviations from predefined targets.
Minimization of Maximum Costs: Ensuring the worst-case scenario is minimized (robust optimization).
Optimization Over Soft Constraints: Allowing flexibility in constraints to accommodate uncertainty.

A significant challenge lies in managing the uncertainty space, which is often vast and leads to very large-scale optimization models. This complexity can make computational solutions difficult and resource-intensive (Fazlollahtabar et al., 2013). The application of optimization under uncertainty has been transformative in:

Engineering Design: Ensuring robustness and reliability in complex systems.
Financial Systems: Managing risks and allocating resources efficiently.
Transportation and Logistics: Optimizing routes and schedules despite unpredictable variables.

In summary, optimization under uncertainty has evolved into a critical tool for addressing real-world complexities, enabling more resilient and adaptive solutions across diverse domains.

Research that addresses uncertainty can be categorized into four primary approaches, each offering distinct methodologies to handle variability and imprecision in decision-making processes. In Stochastic Programming Approach, the core idea is that some parameters are treated as random variables with known probability distributions. This approach is widely used in problems where uncertainty can be quantified statistically, such as financial planning, supply chain management, and energy systems. It optimizes the expected value of the objective function, balancing risk and reward.

In Fuzzy Programming Approach, the core idea is that certain variables are modeled as fuzzy numbers to capture imprecision or vagueness in data. This approach is particularly useful in scenarios where uncertainty is qualitative or subjective, such as human decision-making, environmental planning, and soft constraints in engineering. It seeks solutions that satisfy fuzzy constraints and objectives, often using membership functions to represent degrees of satisfaction.

In Stochastic Dynamic Programming Approach, the core idea is that it extends dynamic programming by incorporating random variables into multi-stage decision-making processes. This approach is essential in fields like inventory management, resource allocation, and control systems, where decisions are made sequentially over time. It optimizes decisions across multiple stages while accounting for probabilistic transitions between states.

In Robust Optimization Approach, the core idea is that it represents uncertainty by setting up different scenarios that reflect possible realizations of uncertain parameters. This approach is valuable in high-stakes environments, such as aerospace engineering, healthcare, and disaster management, where solutions must perform well under a wide range of conditions. It aims to find a robust solution that remains "close" to optimal across all specified scenarios, ensuring resilience to changing input data.

Each of these approaches provides unique tools for addressing uncertainty, catering to different types of problems and decision-making contexts. The choice of approach depends on the nature of the uncertainty, the structure of the problem, and the desired robustness or flexibility of the solution. By leveraging these methodologies, researchers and practitioners can develop more effective and resilient strategies for complex, uncertain environments.

In real-world decision-making, objectives, constraints, or parameters are often not precisely known, leading to decisions based on vague information or uncertain data. To address this, fuzzy set theory was introduced by Bellman and Zadeh (1970) as a framework for handling uncertainty and imprecision in decision-making problems. Fuzzy set theory allows the representation of inaccurate objectives and constraints (called fuzzy objectives and fuzzy constraints) using membership functions. These functions quantify the degree to which an element belongs to a fuzzy set, enabling the modeling of imprecise or subjective information. Tanaka et al. (1974) introduced the concept of fuzzy mathematical programming, extending traditional optimization methods to handle fuzzy objectives and constraints. Zimmermann (1978) formulated fuzzy linear programming for multi-objective problems, providing a structured approach to optimize under fuzzy conditions. Fuzzy System Integration investigated the integration of fuzzy theory with other methodologies, enhancing the flexibility and applicability of fuzzy decision-making models (Li, 2017). Combined fuzzy theory and fuzzy numbers with game theory models, enabling the handling of complex, multi-agent decision problems under uncertainty. Fuzzy theory is particularly effective in scenarios where data is qualitative or subjective. By integrating with other methods, fuzzy decision-making models can address a wider range of problems, from engineering design to financial planning. Fuzzy models provide solutions that are resilient to variations in input data, making them suitable for real-world applications.

Fuzzy set theory has revolutionized decision-making by providing tools to handle uncertainty and imprecision effectively. Its integration with mathematical programming, system integration, and game theory has expanded its applicability, making it a powerful approach for solving complex, real-world problems.

The integration of fuzzy multi-objective optimization, game theory, and dynamic programming with fuzzy parameters provides a powerful framework for modeling and solving complex real-world problems. These problems often involve two distinct types of uncertainty:

Uncertainty Due to Lack of Knowledge: This arises when concepts are well-defined in classical logic but insufficient information is available.
Uncertainty Due to Inherent Vagueness: This occurs when concepts themselves are imprecise or subjective.

Traditionally, these uncertainties have been addressed using probability theory and fuzzy set theory, respectively. However, in many real-world scenarios, both types of uncertainty coexist, necessitating a more integrated approach. Merging probability theory with fuzzy events allows for a more comprehensive model that captures both types of uncertainty. This is particularly useful in scenarios where the occurrence of fuzzy events is probabilistic, such as production systems where multiple conditions may arise with varying utilities. By assigning probabilities to fuzzy events, decision-makers can evaluate the likelihood of different outcomes and make integrated optimal decisions.

Dynamic Programming with Fuzzy Parameters incorporates fuzzy parameters into dynamic programming to handle sequential decision-making under uncertainty. Ideal for problems like inventory management, production planning, and control systems, where decisions are made over multiple stages. It optimizes decisions across time while accounting for the probabilistic and fuzzy nature of parameters.

In production systems, evaluating the probability of different conditions (e.g., machine failures, demand fluctuations) to optimize production schedules is vital. The integration of fuzzy multi-objective optimization, game theory, and dynamic programming with fuzzy parameters offers a robust framework for addressing complex real-world problems involving multiple types of uncertainty. By combining probability theory with fuzzy set theory, decision-makers can develop more effective and resilient strategies for handling imprecise, probabilistic, and multi-objective scenarios.

2.3 Proposed Method

The Utility Function Method, proposed by Keeney and Raiffa (1993), is a systematic approach for solving multi-objective decision-making problems. It focuses on constructing and optimizing a total utility function that represents the decision-maker's preferences across multiple objectives. The method aims to aggregate multiple objectives into a single total utility function, which reflects the decision-maker's overall satisfaction or preference. It assumes that the decision-maker's preferences can be modeled using utility functions, which quantify the value or desirability of different outcomes.

Steps in the Utility Function Method:

Step 1: Assign Conditional Utility Functions: For each objective, a conditional utility function is assigned. This function represents the utility of the objective while holding other objectives fixed at specific levels. In a production planning problem, one objective might be to maximize profit, while another is to minimize environmental impact. The conditional utility function for profit would measure the utility of profit levels, assuming environmental impact is fixed.
Step 2: Check Utility Independence: Utility independence is assessed to determine whether the utility of one objective is independent of the levels of other objectives. If an objective is utility independent, its conditional utility function can be constructed without considering the levels of other objectives. If the utility of profit is independent of environmental impact, the profit utility function can be defined separately.
Step 3: Combine Conditional Utility Functions: The conditional utility functions are combined to form the total utility function. This aggregation typically involves weighting the utilities of individual objectives based on their relative importance. The total utility function might be a weighted sum of the profit utility and environmental impact utility, reflecting the decision-maker's trade-offs between the two.
Step 4: Optimize the Total Utility Function: The final step is to optimize the total utility function, which involves finding the decision that maximizes the decision-maker's overall utility. In the production planning problem, this might involve selecting a production level that balances profit and environmental impact to maximize total utility.

The Utility Function Method is a powerful tool for solving multi-objective problems by systematically aggregating and optimizing the decision-maker's preferences. While it requires careful construction of utility functions and assessment of utility independence, it provides a transparent and flexible framework for making informed decisions in complex scenarios.

However, applying this method in real-world scenarios presents a significant challenge, as it fails to account for all potential conditions within the problem's environment. To illustrate this, consider a manufacturing company where maximizing profit is a top priority for managers. Typically, a single utility function can be applied to this objective, focusing on profit maximization during an 8-h primary shift. However, additional shifts often follow the main shift, and these overtime periods introduce unique complexities. Factors such as employee fatigue, environmental conditions, or company-specific policies often make the primary shift more desirable for workers compared to overtime. Consequently, the same utility function cannot be effectively used for overtime periods, even though the overarching goal remains profit maximization. Instead, a distinct utility function must be defined for overtime scenarios. Therefore, to achieve more accurate modeling and solutions that closely align with ideal outcomes, it is essential to develop separate utility functions for each objective, tailored to different situations and environments (Torkjazi & Fazlollahtabar, 2015a, 2015b).

In this chapter, we assume that each objective function has several utility functions, while researchers in the past used to consider just one utility function for each objective.

Planning for short or long time periods is a critical strategy managers use to control their companies and effectively compete in the market. Since these plans are designed for future periods, they cannot rely on precise data and must account for uncertainty. In corporate strategic planning, decision-makers often face multiple competing objectives under uncertain conditions. To address this complexity, a multi-stage utility optimization approach can be employed (Torkjazi & Fazlollahtabar, 2015a, 2015b). Each strategic objective (e.g., market expansion, cost reduction) is associated with multiple scenario-dependent utility representations. These account for varying environmental conditions the company might encounter. The inherent unpredictability of future conditions is captured through fuzzy set associations for each utility scenario and probability estimates for scenario occurrence probabilities. Partial Utility Analysis compute occurrence probabilities for each scenario-specific utility function. Objective-Level Synthesis aggregate partial utilities into unified conditional utility functions per objective. Global Optimization combine all conditional utilities into a total utility function and maximize through appropriate optimization techniques.

This methodology enables organizations to make robust strategic decisions that account for probabilistic future scenarios while maintaining mathematical rigor in preference aggregation. The approach particularly excels in situations requiring trade-off analysis between competing objectives under uncertainty.

This approach ensures that the planning process is robust, adaptable, and capable of addressing the uncertainties inherent in long-term decision-making.

The flowchart of the process is shown in Fig. 2.1.

Probability of a Fuzzy Event: The probability assessment of a given function requires consideration of all its constituent variables. Essentially, a function's realization depends on the simultaneous occurrence of every variable it encompasses. Therefore, determining the probability of function u necessitates computing the joint probability of its complete variable set M. This establishes that the probability of the function u is mathematically equivalent to the probability of its variable collection M.

For example, consider the function below:

$$u = ax_1 + bx_2 + cx_3 + h$$

and the probabilities of the variables are

$$P_{x_1} = p_1, \quad P_{x_2} = p_2, \quad P_{x_3} = p_3$$

$$M = \{x_1, x_2, x_3\}$$

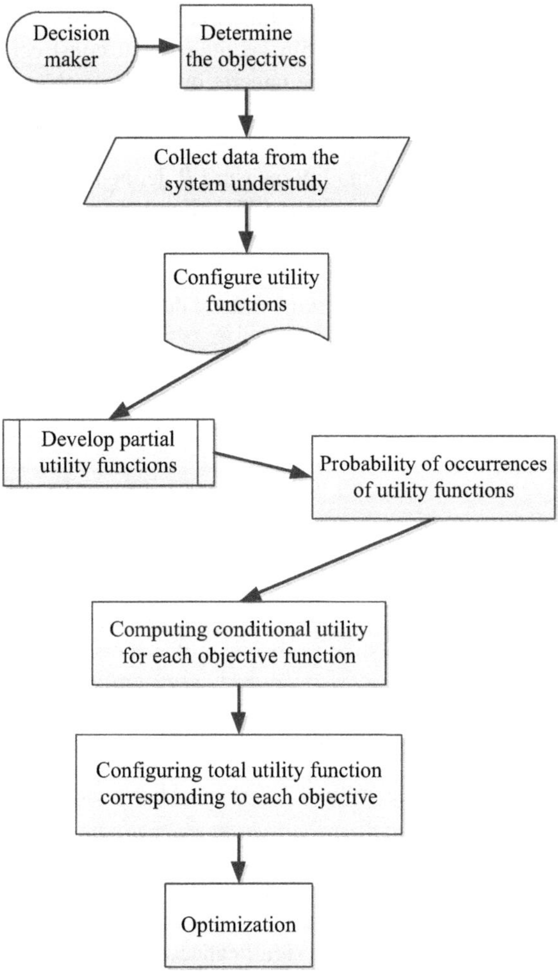

Fig. 2.1 Flowchart of the proposed multi-utility model

Because the variables in this function are crisp variables, the probability of the set M and the function u is

$$P(u) = P(M) = p_1 + p_2 + p_3$$

In this chapter, the utility functions operate within a fuzzy environment, where each function is linked to a fuzzy set that captures the uncertainty inherent in the problem. As a result, it is not possible to derive precise (crisp) results for their probabilities. To address this, Yager's (1984) method can be employed to calculate the probabilities of these fuzzy functions. The first step in this process involves defining a set M for each partial utility function. This is achieved by selecting variables from the crisp universe X, assigning each variable a degree of membership based on the conditions under which the function occurs. These variables, along with their respective degrees of membership, are then used to construct the set M for the function. This approach ensures that the uncertainty in the problem is systematically incorporated into the analysis, enabling a more accurate representation of the utility functions in a fuzzy environment.

Consider a probability measure P defined on the Borel set of all crisp subsets in universe X (Zimmermann, 1978). Let A be a fuzzy set representing a fuzzy event, where $\mu A(x)$ denotes the membership grade of element $x \in X$. The α-level sets (α-cuts) of A are denoted by $A\alpha$.

Yager (1984) proposed a natural definition for the probability of an α-level set:

$$P(A\alpha) = P(\{x \in X \,|\, \mu A(x) \geq \alpha\}) \tag{2.1}$$

Definition 1 For a fuzzy set A representing a fuzzy event, the probability of A is defined as

$$P(A) = \int_0^1 P(A\alpha)\, d\alpha \tag{2.2}$$

This represents the probability of achieving at least satisfaction level α for the condition A.

Yager (1984) introduced an alternative formulation through:

Definition 2 The truth of "$P(A) \geq w$" is represented by a fuzzy set with membership function:

$$\mu\{\Pi A\}(w) = \sup\{\alpha | P(A\alpha) \geq w\} \tag{2.3}$$

Here, w serves as the significance indicator instead of α. For the complement set A^c with α-levels $A^c\alpha$:

$$\mu\{\Pi\{A^c\}\}(w) = \sup\{\alpha | P(A^c\alpha) \geq w\}$$

representing the truth of "$P(A^c) \geq w$."

Definition 3 (Zimmermann, 1978) The possibility distribution for the exact probability w is

$$\mu\Delta(w) = \min(\mu\{\Pi A\}(w), \mu\{\Pi\{A^c\}\}(1 - w)) \tag{2.4}$$

This models the conjunction of "$P(A) \geq w$" and "$P(A) \leq w$" using the min-operator for possibility distributions.

$$\tilde{A} = \{(x_1, 1), (x_2, 0.7), (x_3, 0.6), (x_4, 0.2)\}$$

And the corresponding probabilities:

$$P_{x_1} = 0.1, \ P_{x_2} = 0.4, \ P_{x_3} = 0.3, \ P_{x_4} = 0.2$$

Then, for different α-levels, the exact probability is computed as shown in Table 2.1.

And the same computation for the complement of a fuzzy event is given in Table 2.2.

The probability $\overline{P}_y(\tilde{A})$ of the fuzzy event $\tilde{A}$ is now determined by the intersection of the fuzzy sets $P_y^*(\tilde{A})$ and $\overline{P}_y^*(\tilde{A})$ modeled by the min-operator as in Definition 3 (see Fig. 2.2):

$$\overline{P}_y(\tilde{A})(w) = \begin{Bmatrix} 0, & w = 0 \\ 0.2, & w \in [0, 0.2] \\ 0.6, & w \in [0.2, 0.8] \\ 0.2, & w \in [0.8, 1] \end{Bmatrix}$$

For example consider a problem with two objectives in which each objective corresponds to three partial utility functions.

Table 2.1 Computing exact probability of a fuzzy event

α	A_α	$P(A)_\alpha$	w	$P_y^*(\tilde{A})(w) = \sup \alpha$
[0, 0.2]	$[x_1, x_2, x_3, x_4]$	1	[0.8, 1]	0.2
[0.2, 0.6]	$[x_1, x_2, x_3]$	0.8	[0.5, 0.8]	0.6
[0.6, 0.7]	$[x_1, x_2]$	0.5	[0.1, 0.5]	0.7
[0.7, 1]	$[x_1]$	0.1	[0, 0.1]	1

Table 2.2 Computing exact probability for a complement fuzzy event

α	$(CA)_\alpha$	$P(CA)_\alpha$	w	$P_y^*(C\tilde{A})(w)$	$\overline{P}_y^*(\tilde{A})(w) = 1 - P_y(C\tilde{A})(w)$
0	$[x_1, x_2, x_3, x_4]$	1	[0.9, 1]	0	1
[0, 0.3]	$[x_2, x_3, x_4]$	0.9	[0.5, 0.9]	0.3	0.7
[0.3, 0.4]	$[x_3, x_4]$	0.5	[0.2, 0.5]	0.4	0.6
[0.4, 0.8]	$[x_4]$	0.2	[0, 0.2]	0.8	0.2
[0.8, 1]	0	0	0	1	0

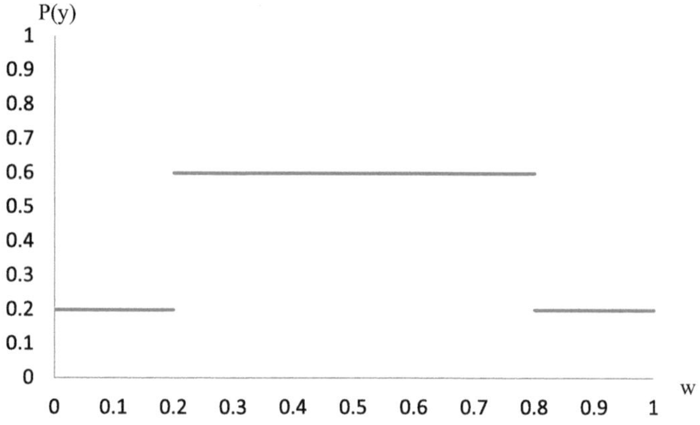

Fig. 2.2 Probability of a fuzzy event

$$f_1(x) \Rightarrow \begin{Bmatrix} \tilde{u}_{11} \\ \tilde{u}_{12} \\ \tilde{u}_{13} \end{Bmatrix} \begin{matrix} \Rightarrow \\ \Rightarrow \\ \Rightarrow \end{matrix} \begin{cases} \tilde{u}_{11} = a_1\tilde{x}_1 + b_1\tilde{x}_2 + c_1\tilde{x}_3 \\ \tilde{u}_{12} = a_2\tilde{x}_1 + b_2\tilde{x}_2 + c_2\tilde{x}_3 \\ \tilde{u}_{13} = a_3\tilde{x}_1 + b_3\tilde{x}_2 + c_3\tilde{x}_3 \end{cases}$$
$$\begin{matrix} \Rightarrow \\ \Rightarrow \\ \Rightarrow \end{matrix} \begin{cases} \tilde{M}_{u_{11}} = \{(x_1, \mu_1), (x_2, \mu_2), (x_3, \mu_3)\} \\ \tilde{M}_{u_{12}} = \{(x_1, \mu_1), (x_2, \mu_2), (x_3, \mu_3)\} \\ \tilde{M}_{u_{13}} = \{(x_1, \mu_1), (x_2, \mu_2), (x_3, \mu_3)\} \end{cases} \tag{2.5}$$

$$f_2(x) \Rightarrow \begin{Bmatrix} \tilde{u}_{21} \\ \tilde{u}_{22} \\ \tilde{u}_{23} \end{Bmatrix} \begin{matrix} \Rightarrow \\ \Rightarrow \\ \Rightarrow \end{matrix} \begin{cases} \tilde{u}_{21} = a_4\tilde{x}_1 + b_4\tilde{x}_2 + c_4\tilde{x}_3 \\ \tilde{u}_{22} = a_5\tilde{x}_1 + b_5\tilde{x}_2 + c_5\tilde{x}_3 \\ \tilde{u}_{23} = a_6\tilde{x}_1 + b_6\tilde{x}_2 + c_6\tilde{x}_3 \end{cases}$$
$$\begin{matrix} \Rightarrow \\ \Rightarrow \\ \Rightarrow \end{matrix} \begin{cases} \tilde{M}_{u_{21}} = \{(x_1, \mu_1), (x_2, \mu_2), (x_3, \mu_3)\} \\ \tilde{M}_{u_{22}} = \{(x_1, \mu_1), (x_2, \mu_2), (x_3, \mu_3)\} \\ \tilde{M}_{u_{23}} = \{(x_1, \mu_1), (x_2, \mu_2), (x_3, \mu_3)\} \end{cases} \tag{2.6}$$

Now, the probabilities of the partial utility functions can be computed by using Yager's method (1984).

Combining the Partial Utility Functions: To operationalize the decision model, we now combine the partial utility functions into unified conditional utility functions for each objective. This synthesis requires each partial utility function U_{ij} (representing scenario j for objective i) is assigned a weight c_{ij}. For objective i, its conditional utility U_i aggregates partial utilities.

Within this framework, partial utility functions are modeled using fuzzy probability sets, where each element reflects a temporal probability segment with an associated membership value. To operationalize this model, crisp coefficients must be derived for each utility function through strategic α-cut applications, guided by decision-maker priorities. Originally, precise probabilities were computed across undivided planning horizons. The current methodology introduces a segmented approach, partitioning the planning period into discrete intervals, each governed by distinct partial utility functions. This temporal segmentation requires targeted α-cut implementations on the fuzzy probability sets to maintain alignment with managerial goals.

Consider a manufacturing enterprise operating two daily shifts with profit maximization objectives. Analysis reveals the need for two complementary partial utility functions. Management imposes a critical threshold requiring $\geq 60\%$ profit attainment during the first shift to generate utility, consequently restricting the second shift's contribution to $\leq 40\%$ of total profitability. This constraint structure:

- Mandates minimum 60% temporal allocation for Shift 1's utility function.
- Caps Shift 2's utility function at 40% of the planning period.

The implementation employs α-cuts to transform these fuzzy temporal requirements into deterministic coefficients, ensuring adherence to the specified operational constraints while preserving the probabilistic nature of utility achievement.

To implement the manager's request, the probability sets are adjusted as follows. For the first partial utility function, only the members of its probability set within the 60–100% interval are selected. For the second partial utility function, only the members of its probability set within the 0–40% interval are selected. This approach ensures that the decision-maker's requirements are systematically incorporated into the planning process, enabling a more precise and tailored utility function for each partial period.

The calculation of weighting coefficients follows this computational sequence. For each partial utility function, compute the aggregate membership value "*s*" by summing the membership grades of all selected elements from their respective probability sets. For each partial utility function, compute the aggregate membership value "*s*" by summing the membership grades of all selected elements from their respective probability sets. Determine the normalized coefficient "*t*" for each partial utility function through the operation: $t = s/S$.

This process ensures proper weighting of each utility component, probabilistic consistency through membership aggregation, and dimensionless coefficients suitable for multi-objective comparison.

The resulting "*t*" coefficients represent the relative importance of each partial utility function within its parent objective's framework, enabling balanced integration into the total utility function.

For example, consider the following probabilities for two partial utility functions $\tilde{u}_1$ and $\tilde{u}_2$:

$$P_{u_1} = \{((0,\ 0.3),\ 0.2),\ ((0.3,\ 0.6),\ 0.3),\ ((0.6,\ 0.8),\ 0.1),\ ((0.8,\ 1),\ 0.3)\}$$
$$P_{u_2} = \{((0,\ 0.2),\ 0.3),\ ((0.2,\ 0.4),\ 0.4),\ ((0.4,\ 0.7),\ 0.1),\ ((0.7,\ 1),\ 0.2)\}$$

Consider that the manager requested that the function $\tilde{u}_1$ must occur at least 60% and the function $\tilde{u}_2$ must occur at most 40% of the whole period.

$$s_1 = \mu_{(0.6,0.8)} + \mu_{(0.8,1)} = 0.1 + 0.3 = 0.4$$
$$s_2 = \mu_{(0,0.2)} + \mu_{(0.2,0.4)} = 0.3 + 0.4 = 0.7$$
$$S = s_1 + s_2 = 0.4 + 0.7 = 1.1$$
$$t_1 = \frac{s_1}{S} = 0.36, \quad t_2 = \frac{s_2}{S} = 0.64$$

Following the established methodology, we can now compute the weighting coefficients for the specified utility functions. After performing all necessary calculations, the derived coefficients for functions (2.5) and (2.6) are obtained as follows:

$$f_1(x) \Rightarrow \begin{Bmatrix} \tilde{u}_{11} \\ \tilde{u}_{12} \\ \tilde{u}_{13} \end{Bmatrix} \begin{matrix} \Rightarrow \\ \Rightarrow \\ \Rightarrow \end{matrix} \begin{Bmatrix} t_{11} \\ t_{12} \\ t_{13} \end{Bmatrix} \qquad f_2(x) \Rightarrow \begin{Bmatrix} \tilde{u}_{21} \\ \tilde{u}_{22} \\ \tilde{u}_{23} \end{Bmatrix} \begin{matrix} \Rightarrow \\ \Rightarrow \end{matrix} \begin{Bmatrix} t_{21} \\ t_{22} \\ t_{23} \end{Bmatrix}$$

Since the partial utility functions are not utility independent from each other, for combining them and creating the conditional utility functions, one can simply add them together with their coefficients.

$$\tilde{U}_1 = t_{11}\tilde{u}_{11} + t_{12}\tilde{u}_{12} + t_{13}\tilde{u}_{13} \tag{2.7}$$

$$\tilde{U}_2 = t_{21}\tilde{u}_{21} + t_{22}\tilde{u}_{22} + t_{23}\tilde{u}_{23} \tag{2.8}$$

$\tilde{U}_1$ and $\tilde{U}_2$ are conditional utility functions for $f_1(x)$ and $f_2(x)$, respectively.

Combining the Conditional Utility Functions: A set of attributes exhibits mutual utility independence if every subset of attributes is utility independent of its complement. This fundamental property enables the multiplicative utility formulation (Keeney & Raiffa, 1993).

Definition 4 Attributes $X_1, X_2, \ldots, X_n$ are mutually utility independent if every subset of $\{X_1, X_2, \ldots, X_n\}$ is utility independent of its complement.

Theorem 1 *If attributes $X_1, X_2, \ldots, X_n$ are mutually utility independent, then*

$$\begin{aligned} u(x) = {} & \sum_{i=1}^{n} k_i u_i(x) + K \sum_{\substack{i=1 \\ j \succ i}}^{n} k_i \cdot k_j \cdot u_i(x_i) \cdot u_j(x_j) \\ & + K^2 \sum_{\substack{i=1 \\ j \succ i \\ \ell \succ j}}^{n} k_i \cdot k_j \cdot k_\ell \cdot u_i(x_i) \cdot u_j(x_j) \cdot u_\ell(x_\ell) \\ & + \cdots + K^{n-1} k_1 k_2 \cdots k_n \, u_1(x_1)\, u_2(x_2) \cdots u_n(x_n) \end{aligned} \tag{2.9}$$

where

1. *u is normalized by* $u\,(x_1^0, x_2^0, \ldots, x_n^0) = 0$ *and* $u\,(x_1^*, x_2^*, \ldots, x_n^*) = 1$.
2. $u_i(x_i)$ *is a conditional utility function on* X_i *normalized by* $u_i(x_i^0) = 0$ *and* $u_i(x_i^*) = 1$, $i = 1, 2, \ldots, n$.
3. $k_i = u\,(x_i^*, \overline{x}_i^0)$.
4. *K is a scaling constant that is a solution to*

$$1 + K = \prod_{i=1}^{n} (1 + K \cdot k_i)$$

2.4 Defuzzification Using Maximum Technique

In the problem we're considering, the conditional utility functions are fuzzy functions. When these functions are combined using a multiplicative formation, the resulting Total Utility Function becomes a nonlinear fuzzy function, which cannot be solved using existing methods. To address this, each conditional utility function must be defuzzified and converted into a crisp function. There are several methods for defuzzification, and one effective approach is the Maximum Technique. Since the variables in this problem are fuzzy numbers, this technique can be applied to defuzzify each variable by replacing them with their crisp equivalents. The Maximum Technique identifies the output with the highest membership function, providing a clear and precise value for further calculations.

Here's a step-by-step breakdown of the process:

Defuzzify Conditional Utility Functions: Use the Maximum Technique to convert each fuzzy conditional utility function into a crisp function.
Substitute Fuzzy Variables: Replace the fuzzy variables in the functions with their defuzzified crisp counterparts.
Combine Crisp Functions: Once all conditional utility functions are defuzzified, combine them to form the Total Utility Function.
Optimize the Total Utility Function: With the function now in a crisp form, apply optimization techniques to maximize the decision-maker's utility.

This approach ensures that the nonlinear fuzzy function is transformed into a solvable crisp function, enabling effective decision-making under uncertainty. This technique is given by algebraic expression as

$$\mu_A\left(x^*\right) \geq \mu_A(x) \qquad \text{for all } x \in X \tag{2.10}$$

where x^* is the defuzzified value. In other words, each one of the variables can be expressed as a fuzzy number as below:

$$\tilde{x}_1 = \{(x_{11}, \mu_{11}), (x_{12}, \mu_{12}), (x_{13}, \mu_{13})\}$$

Assume that

$$\mu_{12} \succ \mu_{11} \succ \mu_{13}.$$

So, the defuzzified result will be

$$x_1 = x_{12}.$$

x_1 is the defuzzified value of $\tilde{x}_1$ and from now on $\tilde{x}_1$ can be substituted with x_1 in the conditional utility function.

Calculating the Scaling Constants ($k_1, k_2, \ldots, K$): After the defuzzification process and obtaining the crisp conditional utility functions, the next step is to calculate the scaling constants required to combine these functions into the Total Utility Function. As shown in Eq. (2.9), each conditional utility function is associated with a scaling constant. These constants are essential for accurately reflecting the decision-maker's preferences and ensuring the combined utility function aligns with their objectives. For example, by asking the decision-maker, his indifference level is concluded as

$$(f_1, f_2) \approx (\beta, 0) \sim (0, \lambda).$$

Recall that $u(x_i)$ actually means $u(x_1^0, \ldots, x_{i-1}^0, x_i, x_{i+1}^0, \ldots, x_n^0)$. Hence, we can define

$$u(x_i) = k_i u_i(x_i). \tag{2.11}$$

So

$$U(f_1, 0) = k_1\, U_{f_1}(f_1, 0),$$
$$U(0, f_2) = k_2\, U_{f_2}(0, f_2).$$

So, by considering the indifference level of the decision-maker

$$\begin{aligned} U(\beta, 0) &= U(0, \lambda) \\ k_1\, U_{f_1}(\beta, 0) &= k_2\, U_{f_2}(0, \lambda). \end{aligned} \tag{2.12}$$

Within Eq. (2.12), we can directly substitute the computed values of the conditional utility functions. This substitution reduces the unknown parameters to only the scaling constants k_1 and k_2, establishing an explicit relationship between these two key variables:

$$1 + K = \prod_{i=1}^{n} (1 + K \cdot k_i)$$

Total Utility Function: The comprehensive utility function $U(F)$ is derived through systematic integration of all determined parameters and conditional utility components into Eq. (2.9). This synthesis yields a precisely defined, deterministic nonlinear function suitable for advanced optimization techniques.

$$U(F) = k_1 U_{(f_1)} + k_2 U_{(f_2)} + K \cdot k_1 \cdot k_2 \cdot U_{(f_1)} \cdot U_{(f_2)} \tag{2.13}$$

2.5 Unconstrained Model Implementation

To demonstrate the effectiveness of our approach in handling multi-objective optimization challenges, we examine a concrete example with two competing objectives:

$$\text{min:}\ \ f_1(x) = x_1^2 + 3x_1x_2 + 4x_2^2$$

$$\text{min:}\ \ f_2(x) = 2x_1^2 + x_1x_2 + x_2^2$$

To apply the utility function method, each objective in the problem should be solved individually to determine its optimum value. These optimum values represent the points where the utility functions equal 1, indicating that the decision-maker's utility level is maximized (Torkjazi et al., 2014).

Since the objectives in this problem are minimization goals, an upper bound must be established for each objective. In this case, the value 10 is chosen as the upper bound. This ensures that the utility functions are properly scaled and that the decision-maker's preferences are accurately reflected. Once the optimum values and upper bounds are determined, the utility functions can be constructed. The utility function for each objective will range from 0 (worst-case scenario, i.e., the upper bound) to 1 (best-case scenario, i.e., the optimum value). After constructing the utility functions for each objective, they can be combined into a Total Utility Function using appropriate scaling constants or aggregation methods. This combined function will reflect the decision-maker's overall utility level, balancing the trade-offs between the different objectives.

This structured approach ensures that the utility function method is applied systematically, enabling effective decision-making in multi-objective optimization problems. So, after solving each objective as a single-objective problem the following results will be achieved:

$$\text{min: } f_1(x) \quad \Rightarrow \quad \begin{cases} f_1^* = 0.00462 \\ x_1^* = 0.043 \\ x_2^* = -0.047 \end{cases} \quad \Rightarrow \quad \begin{cases} 0.00462 \leq f_1^* \leq 10 \\ P_{x_1} = 0.48 \\ P_{x_2} = 0.52, \end{cases}$$

$$\text{min: } f_2(x) \Rightarrow \begin{cases} f_2^* = 0.000063 \\ x_1^* = 0.0022 \\ x_2^* = 0.0063 \end{cases} \Rightarrow \begin{cases} 0.000063 \le f_2^* \le 10 \\ P_{x_1} = 0.26 \\ P_{x_2} = 0.74. \end{cases}$$

Through structured elicitation of decision-maker preferences regarding the problem's objectives, we establish the following utility framework:

$$f_1(x) \Rightarrow \begin{Bmatrix} \tilde{u}'_{11} \\ \tilde{u}'_{12} \end{Bmatrix} \begin{matrix} \Rightarrow \\ \Rightarrow \end{matrix} \begin{cases} \tilde{u}'_{11} = \tilde{x}_1 + 3\tilde{x}_2 & \Rightarrow \\ \tilde{u}'_{12} = 2\tilde{x}_1 + 2\tilde{x}_2 & \Rightarrow \end{cases} \begin{cases} \tilde{M}_{11} = \{(x_1, 0.4), (x_2, 0.5)\} \\ \tilde{M}_{12} = \{(x_1, 0.6), (x_2, 0.5)\}, \end{cases}$$

$$f_2(x) \Rightarrow \begin{Bmatrix} \tilde{u}'_{21} \\ \tilde{u}'_{22} \end{Bmatrix} \begin{matrix} \Rightarrow \\ \Rightarrow \end{matrix} \begin{cases} \tilde{u}'_{21} = 4\tilde{x}_1 + \tilde{x}_2 & \Rightarrow \\ \tilde{u}'_{22} = \tilde{x}_1 + \tilde{x}_2 & \Rightarrow \end{cases} \begin{cases} \tilde{M}_{21} = \{(x_1, 0.7), (x_2, 0.8)\} \\ \tilde{M}_{22} = \{(x_1, 0.3), (x_2, 0.2)\}. \end{cases}$$

Then the utility functions should be normalized:

$$\bar{\tilde{u}}_1 = \tilde{u}'_{11} + \tilde{u}'_{12} = 3\tilde{x}_1 + 5\tilde{x}_2,$$
$$\bar{\tilde{u}}_2 = \tilde{u}'_{21} + \tilde{u}'_{22} = 2\tilde{x}_1 + 3\tilde{x}_2.$$

$$\begin{cases} \tilde{u}_{11} = \dfrac{\tilde{u}'_{11}}{\bar{\tilde{u}}_1} = \dfrac{\tilde{x}_1 + 3\tilde{x}_2}{3\tilde{x}_1 + 5\tilde{x}_2} \\ \tilde{u}_{12} = \dfrac{\tilde{u}'_{12}}{\bar{\tilde{u}}_1} = \dfrac{2\tilde{x}_1 + 2\tilde{x}_2}{3\tilde{x}_1 + 5\tilde{x}_2} \end{cases}, \quad \begin{cases} \tilde{u}_{21} = \dfrac{\tilde{u}'_{21}}{\bar{\tilde{u}}_2} = \dfrac{4\tilde{x}_1 + \tilde{x}_2}{2\tilde{x}_1 + 3\tilde{x}_2} \\ \tilde{u}_{22} = \dfrac{\tilde{u}'_{22}}{\bar{\tilde{u}}_2} = \dfrac{\tilde{x}_1 + \tilde{x}_2}{2\tilde{x}_1 + 3\tilde{x}_2} \end{cases}.$$

The probabilities of the partial utility functions presented in Table 2.3 obtained by performing Yager's Method (1984).

After that the cuts over the probability sets should be applied based on the decision-maker's opinion to calculate the coefficients as given in Tables 2.4 and 2.5.

Table 2.3 Probabilities of the partial utility functions in unconstrained problem

$\tilde{u}$	$\overline{P}_y(\tilde{M})$
$\tilde{u}_{11}$	$\overline{P}_y(\tilde{M}_{11}) = \{(0, 0), ([0, 0.48], 0.4), ([0.48, 0.52], 0.5), ([0.52, 1], 0.4)\}$
$\tilde{u}_{12}$	$\overline{P}_y(\tilde{M}_{12}) = \{(0, 0), ([0, 1], 0.5)\}$
$\tilde{u}_{21}$	$\overline{P}_y(\tilde{M}_{21}) = \{(0, 0), ([0, 0.26], 0.7), ([0.26, 0.74], 0.8), ([0.74, 1], 0.7)\}$
$\tilde{u}_{22}$	$\overline{P}_y(\tilde{M}_{22}) = \{(0, 0), ([0, 1], 0.2)\}$

Table 2.4 Coefficiant of the partial utility functions for objective 1 in unconstrained problem

$\tilde{u}$	p	s	S_1	t
$\tilde{u}_{11}$	[0.5, 1]	$s_{11} = 0.9$	0.9	$t_{11} = 0.64$
$\tilde{u}_{12}$	[0, 0.5]	$s_{12} = 0.5$	1.4	$t_{12} = 0.36$

Table 2.5 Coefficiant of the partial utility functions for objective 2 in unconstrained problem

$\tilde{u}$	p	s	S_2	t
$\tilde{u}_{21}$	[0.6, 1]	$s_{21} = 1.5$	1.5	$t_{21} = 0.88$
$\tilde{u}_{22}$	[0, 0.6]	$s_{22} = 0.2$	1.7	$t_{22} = 0.12$

$$\tilde{U}_{(f_1)} = t_{11}\tilde{u}_{11} + t_{12}\tilde{u}_{12}$$

$$\tilde{U}_{(f_1)} = 0.64\,\tilde{u}_{11} + 0.36\,\tilde{u}_{12}$$

$$\tilde{U}_{(f_1)} = \frac{1.36\tilde{x}_1 + 2.64\tilde{x}_2}{3\tilde{x}_1 + 5\tilde{x}_2}$$

$$\tilde{U}_{(f_2)} = t_{21}\tilde{u}_{21} + t_{22}\tilde{u}_{22}$$

$$\tilde{U}_{(f_2)} = 0.88\,\tilde{u}_{21} + 0.12\,\tilde{u}_{22}$$

$$\tilde{U}_{(f_2)} = \frac{3.64\tilde{x}_1 + \tilde{x}_2}{2\tilde{x}_1 + 3\tilde{x}_2}$$

The subsequent phase in our analytical procedure involves the transformation of fuzzy utility functions into precise, crisp values through defuzzification. The maximum technique, which processes each variable independently, yields the following defuzzified conditional utility functions:

$$\tilde{U}_{(f_1)} = \frac{1.36\tilde{x}_1 + 2.64\tilde{x}_2}{3\tilde{x}_1 + 5\tilde{x}_2} \quad \Rightarrow \quad U_{(f_1)} = \frac{1.36x_1 + 2.64x_2}{3x_1 + 5x_2}$$

$$\tilde{U}_{(f_2)} = \frac{3.64\tilde{x}_1 + \tilde{x}_2}{2\tilde{x}_1 + 3\tilde{x}_2} \quad \Rightarrow \quad U_{(f_2)} = \frac{3.64x_1 + x_2}{2x_1 + 3x_2}$$

After defuzzification of the utility functions, the scaling constants should be calculated:

$$\begin{cases} U_{(F)}\,(10,\ 10) = 0 \\ U_{(F)}\,(0.00462,\ 0.000063) = 1 \end{cases} \quad \Rightarrow \quad \begin{cases} U_{(F)}\,(0.00462,\ 0) = k_1 \\ U_{(F)}\,(0,\ 0.000063) = k_2 \end{cases}$$

$$\begin{cases} U_{(F)}\,(f_1,\ 0) = k_1 U_{(f_1)}(f_1,\ 0) \\ U_{(F)}\,(0,\ f_2) = k_2 U_{(f_2)}(0,\ f_2). \end{cases}$$

The indifference level of the decision-maker has been described as

$$(f_1, f_2) \approx (1.5,\ 0) \sim (0,\ 0.000063)$$

$$U_{(F)}(1.5,\ 0) = U_{(F)}(0,\ 0.000063).$$

So, by using Eq. (2.12)

$$k_1 \cdot U_{(f_1)}(1.5,\ 0) = k_2 \cdot U_{(f_2)}(0,\ 0.000063)$$
$$k_1 \cdot U_{(f_1)}(1.5,\ 0) = k_2.$$

On the other hand, we know that

$$\begin{cases} U_{(f_1)}\ (10,\ 10) = 0 \\ U_{(f_1)}\ (0.00462,\ 0) = 1 \end{cases}$$

$$\begin{cases} U_{(f_2)}\ (10,\ 10) = 0 \\ U_{(f_2)}\ (0.000063,\ 0) = 1 \end{cases}$$

and

$$U_{(f_1)}(1.5,\ 0) = 0.85$$
$$k_2 = 0.85k_1.$$

So

$$1 + K = \prod_{i=1}^{3} (1 + K \cdot k_i)$$
$$1 + K = (1 + K \cdot k_1) \cdot (1 + K \cdot k_2)$$
$$K = \frac{1 - k_1 - k_2}{k_1 k_2} = \frac{1 - 1.85k_1}{0.85k_1^2}$$

In order to calculate k_1, the following procedure can be used. By asking from the decision-maker, we conclude that

$$\{(0.00462,\ 0.000063) : (10,\ 10)\} \sim (3,\ 5);$$

$$(3,\ 5) \begin{array}{ccc} \Rightarrow & \frac{1}{2} & \Rightarrow\ (0.00462,\ 0.000063) \\ \Rightarrow & \frac{1}{2} & \Rightarrow\ (10,\ 10) \end{array}$$

$$U_{(F)}\ (3,\ 5) = \frac{1}{2}\ U_{(F)}\ (0.00462,\ 0.000063) + \frac{1}{2}\ U_{(F)}\ (10,\ 10) = 0.5$$

So, by considering this value in Eq. (2.13)

$$U(F) = 0.5 = k_1 U_{(f_1=3)} + k_2 U_{(f_2=5)} + K \cdot k_1 \cdot k_2 \cdot U_{(f_1=3)} \cdot U_{(f_2=5)}. \tag{2.14}$$

On the other hand, we know that

$$U_{(f_1)}(3,\ 0) = 0.7$$
$$U_{(f_2)}(0,\ 5) = 0.5.$$

So, by solving Eq. (2.14)

$$k_1 = 0.32, \quad k_2 = 0.27, \quad K = 4.69.$$

At last, the Total Utility Function will be expressed as

$$U(F) = U(f_1, f_2) = 0.32\big(U_{(f_1)}\big) + 0.27\big(U_{(f_2)}\big) + 0.4\big(U_{(f_1)} \cdot U_{(f_2)}\big)$$
$$U(f_1, f_2) = \left(\frac{0.43x_1 + 0.84x_2}{3x_1 + 5x_2}\right) + \left(\frac{0.98x_1 + 0.27x_2}{2x_1 + 3x_2}\right) + \left(\frac{1.98x_1^2 + 4.39x_1x_2 + 1.01x_2^2}{6x_1^2 + 19x_1x_2 + 15x_2^2}\right)$$

and the problem is changed in to the following form:

$$\max \quad U(f_1, f_2). \tag{2.15}$$

Solving the nonlinear problem (2.15) results in the following optimum values for the objectives and the utility level of the decision-maker:

$$U^*(f_1, f_2) = 0.96$$
$$x^* = (2.4691,\ 0)$$
$$F^* = (6.096,\ 12.193)$$

The numerical example discussed is based on a production system, with data derived from a normal production shift. The utility functions are constructed in alignment with management policies, ensuring they reflect the decision-maker's priorities and constraints. This example is highly adaptable and can be extended or modified to address other utilities or objectives, whether they involve minimization or maximization. The example can be redesigned to incorporate different utility functions or objectives, depending on the specific context or problem. The procedure remains consistent, allowing for repetition and application to various scenarios. The problem can be replaced with a more complex constrained multi-objective problem, involving numerous factors or constraints. The complexity of the problem does not hinder the solution capability of the algorithm, as the underlying methodology remains robust and scalable. The utility functions are tailored to reflect management policies, ensuring that the solutions align with organizational goals and priorities. This approach ensures that the decision-making process is both practical and policy-driven.

The procedure can be generalized and applied to other systems or industries, making it a versatile tool for decision-making under uncertainty. Whether the objectives are simple or complex, the methodology provides a structured and systematic approach to finding optimal solutions. Suppose the production system is replaced with a supply chain optimization problem involving multiple objectives such as minimizing costs, maximizing delivery efficiency, and reducing environmental impact. The same procedure can be applied:

- Define utility functions for each objective based on management policies.
- Solve each objective individually to determine its optimum value.
- Combine the utility functions using appropriate scaling constants.
- Apply the algorithm to find the optimal solution, even with added complexity.

This demonstrates the flexibility and robustness of the methodology, making it suitable for a wide range of applications.

2.6 Constrained Model Implementation

To demonstrate the practical efficacy of the proposed methodology, a comprehensive numerical case study is presented, illustrating the framework's capacity to resolve complex multi-objective optimization challenges. This implementation utilizes empirical data from a manufacturing environment, where multiple competing objectives—including production cost minimization, quality maximization, and delivery time reduction—must be simultaneously optimized under operational constraints and uncertainty.

The case study incorporates the core methodological innovations: multiple utility functions per objective, fuzzy set representations for uncertain parameters, and probabilistic weighting through Yager's approach. Computational experiments examine both independent and interdependent variable scenarios, with results validating the method's robustness across different problem structures. Performance metrics confirm that the Multi-Choice Goal Programming (MCGP) variant consistently outperforms standard Goal Programming (GP) in solution quality, particularly in environments characterized by high uncertainty and conflicting objectives.

The numerical analysis further demonstrates the approach's scalability to problems of varying complexity, maintaining solution fidelity while accommodating additional objectives, constraints, and utility representations. This empirical validation provides compelling evidence of the methodology's practical utility for decision-makers operating in complex, uncertain environments across engineering, logistics, and resource management domains.

Consider the following problem with two objectives:

$$\max\ f_1(x) = 2x_1 + 2.4x_2 + 3x_3$$

$$\max\ f_2(x) = 1.8x_1 + 1.5x_2$$

$$\text{s.t.}\ \left.\begin{aligned} 3x_1 + x_2 + 1.5x_3 &\le 400 \\ x_1 + x_2 + 2x_3 &\le 250 \\ 1.2x_2 + 0.4x_3 &\le 150 \\ x_1, x_2, x_3 &\ge 0 \end{aligned}\right\} \rightarrow Z$$

To apply the utility function approach, each objective must first be solved individually as a single-objective optimization problem to determine its ideal solution. These optimal values define the points at which the corresponding utility functions reach their maximum value of 1, representing full decision-maker satisfaction. In other words, when an objective achieves its independently optimized value, the associated utility level is considered maximized ($U = 1$), establishing a normalized scale where performance can be meaningfully compared and aggregated. This preliminary step ensures consistent benchmarking across all objectives before integrating them into a multi-criteria decision framework.

After solving each of the objectives as a single-objective problem, the following results are achieved:

$$\begin{cases} \max\ f_1(x) \\ \text{s.t.}\ x \in Z \end{cases} \Rightarrow \begin{cases} f_1^* = 520.435 \\ x_1^* = 81.52 \\ x_2^* = 116.3 \\ x_3^* = 26.09 \end{cases} \Rightarrow \begin{cases} 0 \le f_1^* \le 520.435 \\ P_{x_1} = 0.36 \\ P_{x_2} = 0.52 \\ P_{x_3} = 0.12 \end{cases}$$

$$\begin{cases} \max\ f_2(x) \\ \text{s.t.}\ x \in Z \end{cases} \Rightarrow \begin{cases} f_2^* = 352.5 \\ x_1^* = 91.67 \\ x_2^* = 125 \\ x_3^* = 0 \end{cases} \Rightarrow \begin{cases} 0 \le f_2^* \le 352.5 \\ P_{x_1} = 0.42 \\ P_{x_2} = 0.58 \\ P_{x_3} = 0 \end{cases}$$

Through structured elicitation of decision-maker preferences and requirements, this research establishes a triple-utility framework for each objective, significantly advancing beyond conventional single-utility approaches. This tripartite representation enables more nuanced modeling of decision-maker priorities across varying operational contexts and uncertainty conditions.

Implementation Framework:

Each objective now incorporates three distinct utility functions that collectively capture:

Baseline Performance Requirements—Minimum acceptable achievement levels.
Target Performance Goals—Realistically attainable performance targets.
Aspirational Performance Levels—Ideal achievement scenarios under optimal conditions.

Methodological Advantages:

Enhanced flexibility in capturing decision-maker risk tolerance and preference structures.
Improved accommodation of contextual factors influencing utility perceptions.
More robust uncertainty representation through multiple fuzzy set characterizations.
Better alignment with real-world decision processes involving compromise and trade-offs.

Based on the questions asked from the decision-maker pertaining to the objectives of the problem, it has been decided to define three utility functions for each objective.

$$f_1(x) \Rightarrow \begin{Bmatrix} \tilde{u}'_{11} \\ \tilde{u}'_{12} \\ \tilde{u}'_{13} \end{Bmatrix} \begin{matrix} \Rightarrow \\ \Rightarrow \\ \Rightarrow \end{matrix} \begin{cases} \tilde{u}'_{11} = 2\tilde{x}_1 + 3\tilde{x}_2 + 2\tilde{x}_3 \\ \tilde{u}'_{12} = \tilde{x}_1 + 2\tilde{x}_2 + \tilde{x}_3 \\ \tilde{u}'_{13} = \tilde{x}_1 + \tilde{x}_2 + 3\tilde{x}_3 \end{cases}$$

$$\begin{matrix} \Rightarrow \\ \Rightarrow \\ \Rightarrow \end{matrix} \begin{cases} \tilde{M}_{11} = \{(x_1, 0.5), (x_2, 0.3), (x_3, 0.4)\} \\ \tilde{M}_{12} = \{(x_1, 0.3), (x_2, 0.4), (x_3, 0.2)\} \\ \tilde{M}_{13} = \{(x_1, 0.2), (x_2, 0.3), (x_3, 0.4)\} \end{cases}$$

$$f_2(x) \Rightarrow \begin{Bmatrix} \tilde{u}'_{21} \\ \tilde{u}'_{22} \\ \tilde{u}'_{23} \end{Bmatrix} \begin{matrix} \Rightarrow \\ \Rightarrow \\ \Rightarrow \end{matrix} \begin{cases} \tilde{u}'_{21} = \tilde{x}_1 + 2\tilde{x}_2 \\ u'_{22} = \tilde{x}_1 + \tilde{x}_2 \\ \tilde{u}'_{23} = 2\tilde{x}_1 + \tilde{x}_2 \end{cases} \begin{matrix} \Rightarrow \\ \Rightarrow \\ \Rightarrow \end{matrix} \begin{cases} \tilde{M}_{21} = \{(x_1, 0.6), (x_2, 0.3)\} \\ \tilde{M}_{22} = \{(x_1, 0.2), (x_2, 0.4)\} \\ \tilde{M}_{23} = \{(x_1, 0.2), (x_2, 0.3)\} \end{cases}$$

Then, these utility functions should be normalized:

$$\bar{\tilde{u}}_1 = \tilde{u}'_{11} + \tilde{u}'_{12} + \tilde{u}'_{13} = 4\tilde{x}_1 + 6\tilde{x}_2 + 6\tilde{x}_3,$$
$$\bar{\tilde{u}}_2 = \tilde{u}'_{21} + \tilde{u}'_{22} + \tilde{u}'_{23} = 4\tilde{x}_1 + 4\tilde{x}_2.$$

$$\begin{cases} \tilde{u}_{11} = \dfrac{\tilde{u}'_{11}}{\bar{\tilde{u}}_1} = \dfrac{2\tilde{x}_1 + 3\tilde{x}_2 + 2\tilde{x}_3}{4\tilde{x}_1 + 6\tilde{x}_2 + 6\tilde{x}_3} \\ \tilde{u}_{12} = \dfrac{\tilde{u}'_{12}}{\bar{\tilde{u}}_1} = \dfrac{\tilde{x}_1 + 2\tilde{x}_2 + \tilde{x}_3}{4\tilde{x}_1 + 6\tilde{x}_2 + 6\tilde{x}_3} \\ \tilde{u}_{13} = \dfrac{\tilde{u}'_{13}}{\bar{\tilde{u}}_1} = \dfrac{\tilde{x}_1 + \tilde{x}_2 + 3\tilde{x}_3}{4\tilde{x}_1 + 6\tilde{x}_2 + 6\tilde{x}_3} \end{cases} \qquad \begin{cases} \tilde{u}_{21} = \dfrac{\tilde{u}'_{21}}{\bar{\tilde{u}}_2} = \dfrac{\tilde{x}_1 + 2\tilde{x}_2}{4\tilde{x}_1 + 4\tilde{x}_2} \\ \tilde{u}_{22} = \dfrac{\tilde{u}'_{22}}{\bar{\tilde{u}}_2} = \dfrac{\tilde{x}_1 + \tilde{x}_2}{4\tilde{x}_1 + 4\tilde{x}_2} \\ \tilde{u}_{23} = \dfrac{\tilde{u}'_{23}}{\bar{\tilde{u}}_2} = \dfrac{2\tilde{x}_1 + \tilde{x}_2}{4\tilde{x}_1 + 4\tilde{x}_2} \end{cases}$$

The quantitative likelihoods associated with each partial utility function, as systematically presented in Table 2.6, were derived through the rigorous application of Yager's fuzzy probability measure. This methodological approach enables the transformation of qualitative, uncertainty-laden utility representations into mathematically tractable probabilistic weights, forming the foundation for subsequent multi-objective optimization (Yager, 1984).

Following the probabilistic assessment of partial utility functions, the methodology advances to implementing strategic α-cuts across the probability sets, directly incorporating decision-maker preferences to derive the final weighting coefficients documented in Tables 2.7 and 2.8. This process represents a critical translation of managerial judgment into mathematical parameters that govern the multi-objective optimization.

$$\tilde{U}_{(f_1)} = t_{11}\tilde{u}_{11} + t_{12}\tilde{u}_{12} + t_{13}\tilde{u}_{13}$$

$$\tilde{U}_{(f_1)} = 0.2\ \tilde{u}_{11} + 0.67\ \tilde{u}_{12} + 0.13\ \tilde{u}_{13}$$

$$\tilde{U}_{(f_1)} = \frac{1.2\tilde{x}_1 + 2.07\tilde{x}_2 + 1.46\tilde{x}_3}{4\tilde{x}_1 + 6\tilde{x}_2 + 6\tilde{x}_3}$$

$$\tilde{U}_{(f_2)} = t_{21}\tilde{u}_{21} + t_{22}\tilde{u}_{22} + t_{23}\tilde{u}_{23}$$

$$\tilde{U}_{(f_2)} = 0.3\ \tilde{u}_{21} + 0.2\ \tilde{u}_{22} + 0.5\ \tilde{u}_{23}$$

$$\tilde{U}_{(f_2)} = \frac{1.5\tilde{x}_1 + 1.3\tilde{x}_2}{4\tilde{x}_1 + 4\tilde{x}_2}$$

The methodological progression now advances to the defuzzification stage, where we transform the fuzzy utility functions into precise, crisp formulations. Utilizing the maximum technique—which processes each variable independently—we obtain the following defuzzified conditional utility functions:

$$\tilde{U}_{(f_1)} = \frac{1.2\tilde{x}_1 + 2.07\tilde{x}_2 + 1.46\tilde{x}_3}{4\tilde{x}_1 + 6\tilde{x}_2 + 6\tilde{x}_3} \Rightarrow U_{(f_1)} = \frac{1.2x_1 + 2.07x_2 + 1.46x_3}{4x_1 + 6x_2 + 6x_3}$$

$$\tilde{U}_{(f_2)} = \frac{1.5\tilde{x}_1 + 1.3\tilde{x}_2}{4\tilde{x}_1 + 4\tilde{x}_2} \Rightarrow U_{(f_2)} = \frac{1.5x_1 + 1.3x_2}{4x_1 + 4x_2}$$

After defuzzification of the utility functions, the scaling constants should be calculated:

$$\begin{cases} U_{(F)}\,(0,\ 0) = 0 \\ U_{(F)}\,(520.435,\ 352.5) = 1 \end{cases} \Rightarrow \begin{cases} U_{(F)}\,(520.435,\ 0) = k_1 \\ U_{(F)}\,(0,\ 352.5) = k_2 \end{cases}$$

Table 2.6 Probabilities of the partial utility functions in constrained problem

$\tilde{u}$	$\overline{P}_y(\tilde{M})$
$\tilde{u}_{11}$	$\overline{P}_y(\tilde{M}_{11}) = \{(0, 0), ([0, 1], 0.3)\}$
$\tilde{u}_{12}$	$\overline{P}_y(\tilde{M}_{12}) = \{(0, 0), ([0, 0.12], 0.2), ([0.12, 0.48], 0.3), ([0.48, 0.52], 0.4), ([0.52, 0.88], 0.3), ([0.88, 1], 0.2)\}$
$\tilde{u}_{13}$	$\overline{P}_y(\tilde{M}_{13}) = \{(0, 0), ([0, 0.36], 0.2), ([0.36, 0.64], 0.3), ([0.64, 1], 0.2)\}$
$\tilde{u}_{21}$	$\overline{P}_y(\tilde{M}_{21}) = \{(0, 0), ([0, 1], 0.3)\}$
$\tilde{u}_{22}$	$\overline{P}_y(\tilde{M}_{22}) = \{(0, 0), ([0, 1], 0.2)\}$
$\tilde{u}_{23}$	$\overline{P}_y(\tilde{M}_{13}) = \{(0, 0), ([0, 0.42], 0.2), ([0.42, 0.58], 0.3), ([0.58, 1], 0.2)\}$

Table 2.7 Coefficiant of the partial utility functions for objective 1 in constrained problem

$\tilde{u}$	p	s	S_1	t
$\tilde{u}_{11}$	[0.6, 1]	$s_{11} = 0.3$	0.3	$t_{11} = 0.2$
$\tilde{u}_{12}$	[0.3, 0.6]	$s_{12} = 1$	1.3	$t_{12} = 0.67$
$\tilde{u}_{13}$	[0, 0.3]	$s_{13} = 0.2$	1.5	$t_{13} = 0.13$

Table 2.8 Coefficiant of the partial utility functions for objective 2 in constrained problem

$\tilde{u}$	p	s	S_2	t
$\tilde{u}_{21}$	[0.7, 1]	$s_{21} = 0.3$	0.3	$t_{21} = 0.3$
$\tilde{u}_{22}$	[0.5, 0.7]	$s_{22} = 0.2$	0.5	$t_{22} = 0.2$
$\tilde{u}_{23}$	[0, 0.5]	$s_{22} = 0.5$	1	$t_{22} = 0.5$

and

$$\begin{cases} U_{(F)}\,(f_1,\, 0) = k_1 U_{(f_1)}(f_1,\, 0) \\ U_{(F)}\,(0,\, f_2) = k_2 U_{(f_2)}(0,\, f_2). \end{cases}$$

The indifference level of the decision-maker has been described as

$$(f_1,\, f_2) \approx (250,\ 0) \sim (0,\ 352.5)$$

$$U_{(F)}(250,\ 0) = U_{(F)}(0,\ 352.5)$$

$$\begin{aligned} k_1 \cdot U_{(f_1)}(250,\ 0) &= k_2 \cdot U_{(f_2)}(0,\ 352.5) \\ k_1 \cdot U_{(f_1)}(250,\ 0) &= k_2. \end{aligned}$$

On the other hand, as known

$$\begin{cases} U_{(f_1)}\,(0,\ 0) = 0 \\ U_{(f_1)}\,(520.435,\ 0) = 1 \end{cases}$$

$$\begin{cases} U_{(f_2)}\,(0,\ 0) = 0 \\ U_{(f_2)}\,(0,\ 352.5) = 1 \end{cases}$$

and

$$\frac{1}{520.435} = \frac{U_{(f_1)}(250,\ 0)}{250}$$

$$U_{(f_1)}(250,\ 0) = 0.48$$
$$k_2 = 0.48k_1$$

So,

$$1 + K = \prod_{i=1}^{2} (1 + K \cdot k_i)$$
$$1 + K = (1 + K \cdot k_1) \cdot (1 + K \cdot k_2)$$
$$K = \frac{1 - 1.48k_1}{0.48k_1^2}$$

In order to calculate k_1, the following procedure can be used. By asking the decision-maker, the following can be concluded:

$$\{(520.435,\ 352.5) : (0,\ 0)\} \sim (320,\ 160);$$

$$(320,\ 160) \begin{array}{l} \Rightarrow \frac{1}{2} \Rightarrow \quad (520.435,\ 352.5) \\ \Rightarrow \frac{1}{2} \Rightarrow \qquad\qquad (0,\ 0) \end{array}$$

$$U_{(F)}\,(320,\ 160) = \frac{1}{2}\,U_{(F)}\,(520.435,\ 352.5) + \frac{1}{2}\,U_{(F)}\,(0,\ 0) = 0.5$$

So by considering this value in Eq. (2.12)

$$U(F) = 0.5 = k_1 U_{(f_1=320)} + k_2 U_{(f_2=160)} + K \cdot k_1 \cdot k_2 \cdot U_{(f_1=320)} \cdot U_{(f_2=160)}. \tag{2.16}$$

On the other hand, it is known that

$$U_{(f_1)}(320,\ 0) = 0.61$$
$$U_{(f_2)}(0,\ 160) = 0.45$$

So by solving Eq. (2.16)

$$k_1 = 0.54, \quad k_2 = 0.26, \quad K = 1.43$$

At last, the Total Utility Function will be expressed as

$$U(F) = U(f_1, f_2) = 0.54\big(U_{(f_1)}\big) + 0.26\big(U_{(f_2)}\big) + 0.2\big(U_{(f_1)} \cdot U_{(f_2)}\big)$$

$$U(f_1, f_2) = \left(\frac{0.65x_1 + 1.12x_2 + 0.79x_3}{4x_1 + 6x_2 + 6x_3}\right) + \left(\frac{0.39x_1 + 0.34x_2}{4x_1 + 4x_2}\right) + \left(\frac{0.36x_1^2 + 0.93x_1x_2 + 0.54x_2^2 + 0.38x_2x_3 + 0.44x_1x_3}{16x_1^2 + 40x_1x_2 + 24x_2^2 + 24x_1x_3 + 24x_2x_3}\right)$$

and the problem is changed to the following form:

$$\begin{aligned}
&\max \quad U(f_1, f_2) \\
\text{s.t.}\quad & 3x_1 + x_2 + 1.5x_3 \leq 400 \\
& x_1 + x_2 + 2x_3 \leq 250 \\
& 1.2x_2 + 0.4x_3 \leq 150 \\
& x_1, x_2, x_3 \geq 0.
\end{aligned} \tag{2.17}$$

Solving the nonlinear problem (2.17) results in the following optimum values for the objectives and the utility level of the decision-maker:

$$\begin{aligned}
& U^*(f_1, f_2) = 0.294 \\
& x^* = (0,\ 125,\ 0) \\
& F^* = (300,\ 187.5)
\end{aligned}$$

The numerical example discussed is based on a production system, with data derived from a normal production shift. The utility functions are constructed in alignment with management policies, ensuring they reflect the decision-maker's priorities and constraints. This example is highly adaptable and can be extended or modified to address other utilities or objectives, whether they involve minimization or maximization. The example can be redesigned to incorporate different utility functions or objectives, depending on the specific context or problem. The procedure remains consistent, allowing for repetition and application to various scenarios. The problem can be replaced with a more complex constrained multi-objective problem, involving numerous factors or constraints. The complexity of the problem does not hinder the solution capability of the algorithm, as the underlying methodology remains robust and scalable. The utility functions are tailored to reflect management policies, ensuring that the solutions align with organizational goals and priorities. This approach ensures that the decision-making process is both practical and policy-driven. The procedure can be generalized and applied to other systems or industries, making it a versatile tool for decision-making under uncertainty. Whether the objectives are simple or complex, the methodology provides a structured and systematic approach to finding optimal solutions. This demonstrates the flexibility and robustness of the methodology, making it suitable for a wide range of applications.

2.7 Conclusions

This chapter presents an innovative methodology that fundamentally advances multi-objective optimization by employing multiple utility functions for each objective, representing a significant departure from classical utility theory that typically relies on single-function representations. This multi-faceted approach enables a more nuanced and realistic modeling of decision-maker preferences, particularly in environments characterized by uncertainty and variability.

The proposed framework systematically addresses uncertainty through fuzzy set integration, allowing for the direct incorporation of imprecise or incomplete information that commonly arises in practical decision contexts. By adapting Yager's fuzzy probabilistic methodology, the approach provides a mathematically rigorous foundation for handling the inherent vagueness in both environmental parameters and preference structures. The conditional utility functions are combined using Keeney and Raiffa's (1993) multiplicative formulation, preserving the theoretical integrity of utility theory while extending its applicability to uncertain domains.

A critical innovation lies in the application of the Maximum Technique for defuzzification, which transforms the fuzzy conditional utility representations into crisp values suitable for conventional optimization methods. This process maintains the essential characteristics of the original fuzzy system while enabling efficient solution identification. The resulting total utility function undergoes optimization to identify solutions that maximize decision-maker satisfaction while respecting the complex interplay between multiple objectives and their associated uncertainties.

The methodology demonstrates particular strength in scenarios where:

- Decision-maker preferences cannot be adequately captured by single utility representations.
- Environmental uncertainty significantly impacts objective achievement.
- Multiple stakeholder perspectives must be reconciled within a unified framework.
- Traditional deterministic approaches yield oversimplified or impractical solutions.

By bridging the gap between classical utility theory and fuzzy decision analysis, this approach offers a comprehensive framework for addressing complex multi-objective problems in fields ranging from engineering design to resource allocation and strategic planning. The integration of multiple utility representations with fuzzy uncertainty management provides decision-makers with a more expressive and practically relevant analytical tool than previously available.

The problem is framed as an unconstrained multi-objective optimization problem, with each objective represented by multiple utility functions. Utility functions are defined in a fuzzy environment to account for uncertainty and variability. The multiplicative combination of conditional utility functions is used to aggregate individual utilities into a total utility function. The fuzzy utility functions are defuzzified using the Maximum Technique, and the total utility function is optimized to find the best solutions. A numerical example from an industrial unit is used to demonstrate the method's effectiveness. The problem is formulated as an unconstrained multi-objective optimization problem, and the proposed methodology is applied to optimize the objectives under uncertainty. Experimental results confirm that the method effectively supports decision-making in such scenarios. While the current work focuses on linear utility functions, future research could investigate exponential utility functions for more complex decision-making scenarios. Incorporating linguistic variables based on decision-makers' preferences could enhance the method's applicability in subjective and qualitative contexts. Investigating the fuzzy transition matrix among multiple utility functions could provide deeper insights into the relationships and interactions between objectives. The proposed method offers a robust and flexible framework for solving unconstrained multi-objective problems under uncertainty, leveraging multiple utility functions and fuzzy logic. Its adaptability and effectiveness are demonstrated through a numerical example, and the outlined future research directions highlight its potential for further development and application in diverse fields.

References and Further Reading

Bellman, R. E., & Zadeh, L. A. (1970). Decision-making in a fuzzy environment. *Management Science, 17*, 141–164.

Fazlollahtabar, H., Akbari, F., & Mahdavi, I. (2013). A fuzzy goal programming for optimizing service industry market using virtual intelligent agent. *Journal of Industrial and Production Engineering, 30*(1), 20–29.

Garg, H., Rani, M., Sharma, S. P., & Vishwakarma, Y. (2014a). Bi-objective optimization of the reliability-redundancy allocation problem for series-parallel system. *Journal of Manufacturing Systems, 33*(2), 353–367.

Garg, H., Rani, M., Sharma, S. P., & Vishwakarma, Y. (2014b). Intuitionistic fuzzy optimization technique for solving multi-objective reliability optimization problems in interval environment. *Expert Systems with Applications, 41*(7), 3157–3167.

Jozefowiez, N., Semet, F., & Talbi, E. (2008). Multi-objective vehicle routing problems. *European Journal of Operational Research, 189*, 293–309.

Keeney, & Raiffa. (1993). *Decisions with multiple objectives* (pp. 289–305). Cambridge University Press.

Li, D.-F. (2017). *Theoretical and practical advancements for fuzzy system integration*. IGI Global.

Rani, D., Gulati, T. R., & Garg, H. (2016). Multi-objective non-linear programming problem in intuitionistic fuzzy environment: Optimistic and pessimistic view point. *Expert Systems with Applications, 64*(1), 228–238.

Tanaka, H., Okuda, T., & Asai, K. (1974). On fuzzy mathematical programming. *Journal of Cybernetics, 3*, 37–46.

Torkjazi, M., & Fazlollahtabar, H. (2015a). Applying a utility based fuzzy probabilistic α-cut method to optimize a constrained multi objective model. *Creative Mathematics and Informatics, 24*(2), 221–230.

Torkjazi, M., & Fazlollahtabar, H. (2015b). A fuzzy probabilistic maximum technique to optimize an unconstrained utility based multi objective model. *Industrial Engineering and Management, 4*(147). ISSN: 2169-0316.

Torkjazi, M., Fazlollahtabar, H., & Mahdavi, I. (2014). Optimizing an unconstrained multi objective model using a utility based fuzzy probabilistic α-cut method. *Journal of Intelligent & Fuzzy Systems, 26*(6), 2927–2936.

Yager, R. R. (1984). A representation of the probability of fuzzy subsets. *Fuzzy Sets and Systems, 13*, 273–283.

Zimmermann, H. J. (1978). Fuzzy programming and linear programming with several objective functions. *Fuzzy Sets and Systems, 1*, 45–55.

Chapter 3
Integrated Fuzzy PROMETHEE and Fuzzy Linear Multi-objective Program

Abstract This chapter introduces an integrated decision-making framework for supplier evaluation in a fuzzy environment. The methodology begins by constructing a fuzzy decision matrix, followed by pairwise comparisons using fuzzy PROMETHEE. Attribute weights are then derived via fuzzy linear programming (FLP) and the fuzzy outputs are defuzzified using the fuzzy analytic hierarchy process (FAHP). A case study demonstrates the model's applicability and validity, offering insights for businesses to enhance competitive advantage and profitability through optimized supplier selection.

Keywords Supplier selection · Fuzzy PROMETHEE · Fully fuzzy linear programming

3.1 Introduction

Effective supply chain management hinges on selecting suppliers that align with customer demands while minimizing costs and maximizing efficiency (Ghodsypour & O'Brien, 2001). Traditional supplier selection often prioritized pricing, but reliance on a single criterion can undermine organizational performance. Studies have identified up to 23 criteria for evaluation, emphasizing the need for a holistic approach (Fazlollahtabar et al., 2010). Supplier selection involves identifying vendors capable of delivering quality products/services at optimal prices, quantities, and timelines.

Research in this domain falls into two categories: (1) assessing supplier selection criteria from the buyer's perspective and (2) developing methodologies to evaluate potential suppliers (Lin, 2009). Modern problem-solving techniques aid in defining criteria, qualifying suppliers, and making informed decisions (Mojaveri & Fazlollahtabar, 2012). Due to the presence of conflicting criteria, multi-criteria decision-making (MCDM) methods are essential for effective supplier evaluation.

Supplier selection frameworks vary in performance metrics, necessitating integrated qualitative–quantitative approaches. Companies often prioritize certain criteria over others, underscoring the utility of MCDM tools like AHP and ANP for

H. Fazlollahtabar, *Uncertain Multi-objective Decision Making: Methods and Industrial Models*, https://doi.org/10.1007/978-981-95-6035-6_3

weight assignment. Addressing these complexities can significantly advance supplier selection practices.

As a critical procurement activity, supplier selection directly influences product quality and supply chain performance (Thrulogachantar & Zailani, 2011). It also facilitates assessing suppliers' potential for collaboration. The process narrows down a pool of candidates to the most viable options (Fazlollahtabar et al., 2011), relying on both quantitative and qualitative evaluations. Uncertainty arises from subjective judgments, multiple decision-makers, and limited historical data (Amindoust et al., 2012).

Fuzzy set theory combined with MCDM methods—such as Fuzzy AHP (Fazlollahtabar & Tehranchian, 2015) and Fuzzy TOPSIS—effectively addresses such uncertainties by handling imprecise criteria. Agarwal et al. (2011) analyzed 68 studies (2000–2011) on supplier selection, revealing the prevalence of Data Envelopment Analysis (DEA: 30%), mathematical programming (17%), and AHP (15%), with fuzzy methods accounting for 10%.

Fuzzy PROMETHEE (F-PROMETHEE) is particularly suited for supplier selection due to its ability to manage fuzzy data. Developed by Brans et al. (1984), PROMETHEE ranks alternatives using preference functions and criteria weights. Its advantages include

- User-friendly outranking.
- Proven applicability in real-world scenarios.
- Flexibility in partial (PROMETHEE I) or total (PROMETHEE II) rankings.

The method requires two inputs: criterion weights and decision-makers' preference functions. Few studies have explored F-PROMETHEE; Goumas and Lygerou (2000) addressed fuzzy number comparisons using a weighted index based on membership functions.

This chapter presents a novel fuzzy-based methodology for supplier selection and ranking under uncertainty.

3.2 Statement of the Problem and Modeling

Due to the strategic significance of supplier selection, this area has been the focus of extensive research, with growing interest in recent years. Fazlollahtabar et al. (2011) highlight that supplier selection is a critical component of product and service management within supply chains. In manufacturing, for instance, raw materials and components can account for up to 70% of total production costs, making procurement a key driver of cost efficiency. Supplier evaluation thus plays a pivotal role in purchasing management.

The supplier selection process is inherently a multi-criteria decision-making (MCDM) problem, involving both quantitative and qualitative factors, some of which may conflict. Two primary sourcing strategies exist:

- Single Sourcing: The buyer selects the best supplier from a pool where all candidates meet requirements (demand, quality, delivery, etc.). The decision is straightforward—choosing the optimal supplier.
- Multiple Sourcing: Constraints such as limited supplier capacity or variable quality necessitate procuring from multiple vendors to fulfill demand (Fazlollahtabar, 2016). Here, buyers must decide (a) which suppliers to engage and (b) order allocation across them.

To address supplier evaluation under uncertainty, we employ the following steps:

Fuzzy Decision Matrix: Constructed using fuzzy PROMETHEE for pairwise supplier comparisons based on predefined attributes.
Weight Assignment: Fuzzy linear programming (FLP) determines criterion weights.
Ranking: Suppliers are ranked using the derived weights (Fazlollahtabar, 2016). The research process is illustrated in Fig. 3.1.

We introduce a novel approach for supplier evaluation under uncertainty. Consider two fuzzy numbers:

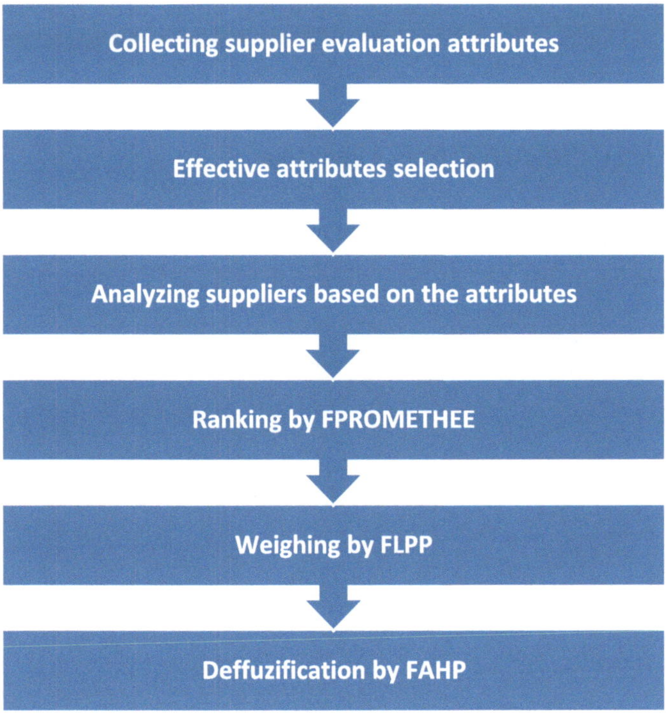

Fig. 3.1 Research process

$$\tilde{A} = \left(a_1 \; a_2 \; a_3 \right)$$

$$\tilde{B} = \left(b_1 \; b_2 \; b_3 \right)$$

$$\widetilde{P_{AB}} = (\mathrm{PL}_{AB}, \mathrm{PM}_{AB}, \mathrm{PU}_{AB})$$

Then P is the weight for comparing two fuzzy numbers. Like a triangular membership function, we have

$$\mathrm{PL}_{AB} = \begin{cases} 0 & \frac{a_1+a_2+a_3}{3} \le \frac{b_1+b_2+b_3}{3} \\ a_1 - b_3 & \text{otherwise} \end{cases},$$

$$\mathrm{PM}_{AB} = \begin{cases} 0 & \frac{a_1+a_2+a_3}{3} \le \frac{b_1+b_2+b_3}{3} \\ a_2 - b_2 & \text{otherwise} \end{cases},$$

$$\mathrm{PU}_{AB} = \begin{cases} 0 & \frac{a_1+a_2+a_3}{3} \le \frac{b_1+b_2+b_3}{3} \\ a_3 - b_1 & \text{otherwise} \end{cases}.$$

The subsequent steps of the algorithm would follow here, retaining original equations and logic.

3.2.1 *Fuzzy PROMETHEE Algorithm*

For each attribute, a pairwise comparison matrix is in hand.

Step 1: Decision matrix

DM	C_1	C_2	…	C_n
A_1	$\widetilde{x_{11}}$	$\widetilde{x_{12}}$	…	$\widetilde{x_{1n}}$
A_2	$\widetilde{x_{21}}$	$\widetilde{x_{22}}$	…	$\widetilde{x_{2n}}$
⋮	⋮	⋮	…	⋮
A_X	$\widetilde{x_{m1}}$	$\widetilde{x_{m2}}$	…	$\widetilde{x_{mn}}$
w_j	$\widetilde{w_1}$	$\widetilde{w_2}$	…	$\widetilde{w_n}$

$$\widetilde{x_{ij}} = \left(L_{ij}, M_{ij}, U_{ij} \right).$$

Step 2: Pairwise matrix

For any criterion, the costs are compared.

C_1	A_1	A_2	…	A_m
A_1	$\widetilde{p_{11}^{c_1}}$	$\widetilde{p_{12}^{c_1}}$	…	$\widetilde{p_{1m}^{c_1}}$
A_2				
⋮				
p_m	$\widetilde{p_{m_1}^{c_1}}$	$\widetilde{p_{m_2}^{c_1}}$	…	$\widetilde{p_{m_n}^{c_1}}$

$$\widetilde{p_{ij}^{c_k}} = \left(\widetilde{\mathrm{PL}_{ij}^{c_k}}, \widetilde{\mathrm{PM}_{ij}^{c_k}}, \widetilde{\mathrm{PU}_{ij}^{c_k}}\right)$$

$$m\colon 1, \ldots, j$$

$$n\colon 1, \ldots, k.$$

Step 3: Preference matrix

Π	A_1	A_2	A_m	$\check{\varphi}^{+}$	$\tilde{\varphi}$
A_1	$\sum_{ij}^{N} \widetilde{p_{11}^{c_1}}$		$\sum_{k=1}^{m} \widetilde{p_{1m}^{c1}} w_k$	$\varphi_1^{+} = \sum_{j=1}^{m} \sum_{k=1}^{n} \widetilde{p^{c1}} w_k$	$\tilde{\varphi}_1 = \tilde{\varphi}_1^{+} - \tilde{\varphi}_1$
A_2					
⋮					
A_m	$\sum_{ij}^{N} \widetilde{p_{m1}^{cm}}$		$\sum_{k=1}^{m} \widetilde{p_{mm}^{ck}} w_k$		
$\tilde{\varphi}^{-}$	$\varphi_1^{-} = \sum_{j=1}^{m} \sum_{k=1}^{n} \widetilde{p^{c1}} w_k$				

3.2.2 *Fuzzy Linear Program*

To complement the fuzzy PROMETHEE approach, we employ a fully fuzzy linear programming (FFLP) formulation for supplier selection. This methodology utilizes a specialized ranking function that operates on triangular fuzzy numbers to effectively model decision-making uncertainty. The general form of our FFLP problem is expressed as follows:

$$\begin{aligned} &\text{Max/Min } \tilde{z} = \tilde{c}^t \cdot \tilde{x} \\ &\text{s.t.} \\ &\tilde{A}\tilde{x} \le \tilde{b} \\ &\tilde{x} \ge 0 \end{aligned}$$

where $\tilde{z}, \tilde{c}^t = (\tilde{c}_1, \ldots, \tilde{c}_n)$, $\tilde{x} = (\tilde{x}_1, \ldots, \tilde{x}_n)$, $\tilde{A} = \left[\tilde{a}_{ij}\right]_{m \times n}$, $\tilde{b} = (\tilde{b}_1, \ldots, \tilde{b}_m)^t$ denote the triangular fuzzy numbers for objective function, fuzzy objective function coefficients, fuzzy technical coefficients, and fuzzy resource constraints of the linear programming.

The solution yields fuzzy rankings that require defuzzification for practical implementation. For this transformation, we adapt principles from Fuzzy Analytic Hierarchy Process (FAHP). Notably, both the priority degree and dignity degree share the same measurement scale, enabling the application of FAHP logic. However, we address a critical limitation in conventional FAHP implementations where premature defuzzification occurs—our methodology maintains the fuzzy nature of calculations throughout the entire solution process until final results are obtained (Fazlollahtabar, 2016).

3.3 Numerical Example

To demonstrate the practical application of our proposed methodology, we examine a supplier selection problem involving six potential vendors:

- LG.
- Samsung.
- Snowa.
- Techno.
- Bosch.
- Daewoo.

The evaluation is conducted based on five critical attributes:

1. Price competitiveness.
2. Product quality standards.
3. On-time delivery performance.
4. Service quality metrics.
5. Brand reputation.

A panel of 150 industry experts, comprising sales specialists and customer relationship management professionals contributed to completing the decision matrix. The implementation follows our established methodological framework:

1. Initial assessment phase:
   - Construction of the fuzzy decision matrix.
   - Execution of pairwise comparisons using fuzzy PROMETHEE.
2. Weight determination phase:
   - Application of Fully Fuzzy Linear Programming (FFLP) to derive attribute weights.
3. Final evaluation phase:
   - Transformation of fuzzy outputs to crisp values through fuzzy AHP.
   - Comprehensive ranking of supplier alternatives.

The complete decision matrix, representing the experts' evaluations, is presented in Table 3.1. This matrix serves as the foundation for all subsequent analytical procedures in our supplier selection process.

The attributes pairwise comparison is shown in Tables 3.2, 3.3, 3.4, 3.5 and 3.6.

The resulted supplier/attribute matrix is shown in Table 3.7.

The fuzzy linear program to obtain the weights is given below:

$$\text{Min}\, Z = (1-\lambda)^2 M \sum_j \sum_i y_{ij}^2 + z_{ij}^2$$

s.t.

$$x_i - x_j - \lambda\left(\ln \frac{m_{ij}}{l_{ij}}\right) + y_{ij} \geq 0,$$

$$-x_i + x_j - \lambda\left(\ln \frac{u_{ij}}{m_{ij}}\right) + z_{ij} \geq 0,$$

$$w_i = \frac{\text{Exp}(x_i)}{\sum_i x_i},$$

$$x_i, y_{ij}, z_{ij}, \lambda \geq 0,$$

where M is a large number and the aim of the linear model is to find the weights.

Solving the model in lingo optimization software, we have

```
Local optimal solution found.
Objective value:                    0.1232595E-27
Infeasibilities:                    0.3178013E-14
Total solver iterations:                       41
```

Table 3.1 The decision matrix

	Price			Product quality			On-time delivery			Service quality			Brand		
LG	0.55	0.72	0.90	0.60	0.44	0.60	0.76	0.47	0.63	0.37	0.59	0.74	0.54	0.70	0.83
Samsung	0.27	0.40	0.57	0.42	0.60	0.69	0.07	0.22	0.38	0.32	0.59	0.71	0.44	0.62	0.79
Snowa	0.71	0.83	1.00	0.93	0.98	0.99	0.66	0.81	1.00	0.48	0.65	0.80	0.63	0.76	0.86
Techno	0.70	0.82	0.90	0.88	0.91	0.94	0.85	0.94	0.99	0.52	0.70	0.83	0.30	0.45	0.62
Bosch	0.50	0.65	0.80	0.64	0.82	0.90	0.87	0.91	0.97	0.53	0.71	0.79	0.29	0.41	0.54
Daewoo	0.63	0.81	0.91	0.24	0.39	0.59	0.66	0.79	0.97	0.37	0.63	0.76	0.43	0.54	0.64
Weight	0.21	0.39	0.54	0.56	0.67	0.77	0.74	0.86	0.93	0.47	0.67	0.80	0.77	0.84	0.90

Table 3.2 Pairwise comparison for price attribute

Price	LG			Samsung			Snowa			Techno			Bosch			Daewoo		
LG	0.00	0.00	0.00	− 0.02	0.32	0.64	0.00	0.00	0.00	0.00	0.00	0.00	− 0.25	0.07	0.40	− 0.25	0.00	0.00
Samsung	0.00	0.00	0.00	0.00	0.00	0.00	0.00	0.00	0.00	0.00	0.00	0.00	0.00	0.00	0.00	0.00	0.00	0.00
Snowa	− 0.19	0.12	0.45	0.14	0.44	0.73	0.00	0.00	0.00	− 0.19	0.02	0.30	− 0.09	0.18	0.50	− 0.09	0.02	0.37
Techno	− 0.20	0.10	0.35	0.13	0.42	0.64	0.00	0.00	0.00	0.00	0.00	0.00	− 0.10	0.17	0.40	− 0.10	0.01	0.28
Bosch	0.00	0.00	0.00	− 0.07	0.25	0.53	0.00	0.00	0.00	0.00	0.00	0.00	0.00	0.00	0.00	0.00	0.00	0.00
Daewoo	− 0.28	0.09	0.35	0.06	0.41	0.64	0.00	0.00	0.00	0.00	0.00	0.00	− 0.17	0.16	0.41	0.00	0.00	0.00

Table 3.3 Pairwise comparison for product quality

Product quality	LG			Samsung			Snowa			Techno			Bosch			Daewoo		
LG	0.00	0.00	0.00	0.00	0.00	0.00	0.00	0.00	0.00	0.00	0.00	0.00	0.00	0.00	0.00	0.00	0.06	0.37
Samsung	− 0.18	0.15	0.09	0.00	0.00	0.00	0.00	0.00	0.00	0.00	0.00	0.00	0.00	0.00	0.00	0.00	0.21	0.45
Snowa	0.33	0.53	0.38	0.24	0.38	0.56	0.00	0.00	0.00	− 0.01	0.00	0.11	0.03	0.16	0.35	0.03	0.59	0.75
Techno	0.28	0.47	0.34	0.19	0.32	0.52	0.00	0.00	0.00	0.00	0.00	0.00	− 0.02	0.09	0.30	− 0.02	0.53	0.70
Bosch	0.04	0.38	0.29	− 0.05	0.22	0.47	0.00	0.00	0.00	0.00	0.00	0.00	0.00	0.00	0.00	0.00	0.43	0.66
Daewoo	0.00	0.00	0.00	0.00	0.00	0.00	0.00	0.00	0.00	0.00	0.00	0.00	0.00	0.00	0.00	0.00	0.00	0.00

Table 3.4 Pairwise comparison for on-time delivery

On-time delivery	LG			Samsung			Snowa			Techno			Bosch			Daewoo		
LG	0.0	0.0	0.0	0.4	0.2	0.6	0.0	0.0	0.0	0.0	0.0	0.0	0.0	0.0	0.0	0.0	0.0	0.0
Samsung	0.0	0.0	0.0	0.0	0.0	0.0	0.0	0.0	0.0	0.0	0.0	0.0	0.0	0.0	0.0	0.0	0.0	0.0
Snowa	0.0	0.3	0.2	0.3	0.6	0.9	0.0	0.0	0.0	0.0	0.0	0.0	0.0	0.0	0.0	0.0	0.0	0.3
Techno	0.2	0.5	0.2	0.5	0.7	0.9	− 0.2	0.1	0.3	0.0	0.0	0.0	− 0.1	0.0	0.1	− 0.1	0.1	0.3
Bosch	0.2	0.4	0.2	0.5	0.7	0.9	− 0.1	0.1	0.3	0.0	0.0	0.0	0.0	0.0	0.0	0.0	0.1	0.3
Daewoo	0.0	0.3	0.2	0.3	0.6	0.9	0.0	0.0	0.0	0.0	0.0	0.0	0.0	0.0	0.0	0.0	0.0	0.0

Table 3.5 Pairwise comparison for service quality

Service quality	LG			Samsung			Snowa			Techno			Bosch			Daewoo		
LG	0.00	0.00	0.00	− 0.34	0.00	0.41	0.00	0.00	0.00	0.00	0.00	0.00	0.00	0.00	0.00	0.00	0.00	0.00
Samsung	0.00	0.00	0.00	0.00	0.00	0.00	0.00	0.00	0.00	0.00	0.00	0.00	0.00	0.00	0.00	0.00	0.00	0.00
Snowa	− 0.26	0.06	0.43	− 0.23	0.06	0.48	0.00	0.00	0.00	0.00	0.00	0.00	0.00	0.00	0.00	0.00	0.02	0.43
Techno	− 0.22	0.11	0.47	− 0.19	0.11	0.51	− 0.28	0.05	0.35	0.00	0.00	0.00	− 0.27	− 0.01	0.30	− 0.27	0.07	0.46
Bosch	− 0.21	0.12	0.43	− 0.18	0.12	0.47	− 0.27	0.06	0.31	0.00	0.00	0.00	0.00	0.00	0.00	0.00	0.08	0.42
Daewoo	− 0.37	0.04	0.39	− 0.34	0.04	0.43	0.00	0.00	0.00	0.00	0.00	0.00	0.00	0.00	0.00	0.00	0.00	0.00

Table 3.6 Pairwise comparison for brand

Brand	LG			Samsung			Snowa			Techno			Bosch			Daewoo		
LG	0.00	0.00	0.00	− 0.25	0.07	0.38	0.00	0.00	0.00	− 0.08	0.24	0.53	0.00	0.29	0.53	0.00	0.16	0.40
Samsung	0.00	0.00	0.00	0.00	0.00	0.00	0.00	0.00	0.00	− 0.17	0.17	0.49	− 0.09	0.22	0.50	− 0.09	0.09	0.36
Snowa	− 0.20	0.06	0.32	− 0.17	0.13	0.42	0.00	0.00	0.00	0.01	0.31	0.56	0.09	0.35	0.57	0.09	0.22	0.43
Techno	0.00	0.00	0.00	0.00	0.00	0.00	0.00	0.00	0.00	0.00	0.00	0.00	− 0.24	0.05	0.32	− 0.24	0.00	0.00
Bosch	0.00	0.00	0.00	0.00	0.00	0.00	0.00	0.00	0.00	0.00	0.00	0.00	0.00	0.00	0.00	0.00	0.00	0.00
Daewoo	0.00	0.00	0.00	0.00	0.00	0.00	0.00	0.00	0.00	− 0.19	0.08	0.34	− 0.11	0.13	0.34	0.00	0.00	0.00

Table 3.7 Supplier/attribute matrix

	LG			Samsung			Snowa			Techno			Bosch			Daewoo		
LG	0.00	0.00	0.00	− 0.08	0.40	1.54	0.00	0.00	0.00	− 0.06	0.21	0.48	− 0.05	0.27	0.70	− 0.05	0.18	0.64
Samsung	− 0.10	0.10	0.07	0.00	0.00	0.00	0.00	0.00	0.00	− 0.13	0.14	0.45	− 0.07	0.18	0.45	− 0.07	0.22	0.68
Snowa	− 0.10	0.79	1.40	0.13	1.09	2.45	0.00	0.00	0.00	− 0.04	0.26	0.75	0.07	0.47	1.05	0.07	0.63	1.82
Techno	0.18	0.83	1.03	0.39	1.07	2.00	− 0.24	0.14	0.58	0.00	0.00	0.00	− 0.43	0.18	1.10	− 0.43	0.53	1.36
Bosch	0.10	0.71	0.76	0.23	0.92	1.86	− 0.22	0.12	0.53	0.00	0.00	0.00	0.00	0.00	0.00	0.00	0.45	1.13
Daewoo	− 0.20	0.34	0.70	0.06	0.68	1.53	0.00	0.00	0.00	− 0.14	0.07	0.31	− 0.12	0.17	0.53	0.00	0.00	0.00

The defuzzification process yields the following crisp weight values for each supplier:

LG: 0.21 (Rank 1).
Daewoo: 0.18 (Rank 2).
Bosch: 0.175 and Snowa: 0.175 (Rank 3).
Samsung: 0.14 (Rank 5).
Techno: 0.12 (Rank 6).

These results demonstrate clear differentiation among suppliers, with LG emerging as the top performer followed closely by Daewoo.

3.4 Conclusions

This study presents an innovative hybrid methodology that combines fuzzy PROMETHEE with fully fuzzy linear programming (FFLP) for supplier selection under uncertainty. The proposed framework implements a systematic three-phase approach:

1. Fuzzy Evaluation Phase:
   - Construction of the fuzzy decision matrix.
   - Implementation of pairwise comparisons through fuzzy PROMETHEE.
2. Weight Optimization Phase:
   - Determination of optimal attribute weights using FFLP.
   - Resolution of conflicting criteria through mathematical programming.
3. Decision Finalization Phase:
   - Transformation of fuzzy outputs to actionable crisp values via fuzzy AHP.
   - Comprehensive ranking of supplier alternatives.

The case study validation confirms the method's practical applicability and robustness in real-world decision-making scenarios. Key advantages of our integrated approach include

- Effective handling of uncertainty through fuzzy set theory.
- Mathematical rigor in weight determination.
- Balanced consideration of both qualitative and quantitative factors.
- Transparent conversion of fuzzy evaluations to concrete rankings.

This methodology provides supply chain managers with a scientifically grounded yet practical tool for making optimal supplier selection decisions in uncertain environments.

References and Further Reading

Agarwal, P., Sahai, M., Mishra, V., Bag, M., & Singh, V. (2011). A review of multi-criteria decision making techniques for supplier evaluation and selection. *International Journal of Industrial Engineering Computations, 2*, 801–810.

Amindoust, A., Ahmed, S., Saghafinia, A., & Bahreininejad, A. (2012). Sustainable supplier selection: A ranking model based on fuzzy inference system. *Applied Soft Computing, 12*, 1668–1677.

Brans, J. P., Mareschal, B., & Vincke, P. (1984). PROMETHEE: A new family of outranking methods in MCDM. In J. P. Brans (Ed.), *Operational Research IFORS 84* (pp. 477–490). North-Holland.

Fazlollahtabar, H. (2016). An integration between fuzzy PROMETHEE and fuzzy linear program for supplier selection problem: Case study. *Journal of Applied Mathematical Modelling and Computing, 1*(1).

Fazlollahtabar, H., Eslami, H., & Salmani, H. (2010). Designing a fuzzy expert system to evaluate alternatives in fuzzy analytic hierarchy process. *Journal of Software Engineering and Applications, 3*(04), 409.

Fazlollahtabar, H., Mahdavi, I., Ashoori, M. T., Kaviani, S., & Mahdavi-Amiri, N. (2011). A multi-objective decision-making process of supplier selection and order allocation for multi-period scheduling in an electronic market. *The International Journal of Advanced Manufacturing Technology, 52*, 1039–1052.

Fazlollahtabar, H., & Tehranchian, A. M. (2015). Integration between mathematical programming and fuzzy logic to optimize consumers behavior. In *Research methods: Concepts, methodologies, tools, and applications* (pp. 1600–1615). IGI Global.

Ghodsypour, S. H., & O'Brien, C. (2001). The total cost of logistics in supplier selection, under conditions of multiple sourcing, multiple criteria and capacity constraint. *International Journal of Production Economics, 73*(1), 15–27.

Goumas, M., & Lygerou, V. (2000). An extension of the PROMETHEE method for decision making in fuzzy environment: Ranking of alternative energy exploitation projects. *European Journal of Operational Research, 123*, 606–613.

Lin, R.-H. (2009). An integrated FANP–MOLP for supplier evaluation and order allocation. *Applied Mathematical Modelling, 33*(6), 2730–2736.

Mojaveri, H. S., & Fazlollahtabar, H. (2012). Designing an integrated AHP based fuzzy expert system and SWOT analysis to prioritize development strategies of Iran agriculture. *Revista de Management Comparat International, 13*(1), 117.

Thrulogachantar, P., & Zailani, S. (2011). The influence of purchasing strategies on manufacturing performance. *Journal of Manufacturing Technology Management, 22*, 641–663.

Chapter 4
Fuzzy Multi-objective AHP-TOPSIS Method

Abstract In large-scale enterprises, supplier selection is a critical process due to the involvement of numerous uncontrollable and unavoidable factors. To address this challenge, this research proposes a resilient supply chain framework that enhances flexibility and mitigates disruptions. The study identifies five main criteria for supplier selection, tailored to vulnerable areas and production requirements, and employs fuzzy multi-objective linear programming (FMOLP) and fuzzy analytic hierarchy process (FAHP) to handle the vagueness of information and prioritize criteria. Supplier selection involves both qualitative (e.g., reliability, reputation) and quantitative (e.g., cost, delivery time) attributes, making it a multi-faceted decision-making problem. Information related to supplier performance and criteria importance is often imprecise or ambiguous, requiring robust methods to handle uncertainty. A resilient supply chain is essential to overcome disruptions, ensuring continuity and flexibility in material procurement. The criteria are selected based on vulnerable areas and production requirements to enhance flexibility. While the specific criteria are not detailed here, they likely include factors such as cost, quality, delivery reliability, sustainability, and innovation.

Keywords Resilient supply chain · Supplier selection · Disruption · Fuzzy multi-objective · FAHP

4.1 Introduction

Supply Chain Management (SCM) is a complex process that encompasses activities such as material procurement, conversion of raw materials into finished goods, and distribution of final products to customers. Among these, supplier selection is a critical first step, as it directly impacts the efficiency, cost, and quality of the entire supply chain. Supplier selection is the first step in SCM, influencing all subsequent activities, including production, logistics, and customer satisfaction. Selecting the right suppliers can reduce operational costs, improve product quality, and enhance market competitiveness. Effective supplier selection ensures timely delivery and high-quality

H. Fazlollahtabar, *Uncertain Multi-objective Decision Making: Methods and Industrial Models*, https://doi.org/10.1007/978-981-95-6035-6_4

products, enabling businesses to meet customer demands effectively. Supplier selection involves evaluating multiple criteria, such as cost, quality, delivery reliability, sustainability, and innovation, which are often conflicting in nature. The process is complicated by unpredictable factors such as market fluctuations, geopolitical risks, and natural disasters.

The need to balance qualitative (e.g., supplier reputation) and quantitative (e.g., cost) criteria makes supplier selection a complex MCDM problem. Define and prioritize the key criteria for supplier selection based on business objectives and supply chain requirements. Assess potential suppliers against the identified criteria using ranking and evaluation methods. Use MCDM techniques (e.g., Analytic Hierarchy Process (AHP), TOPSIS, or VIKOR) to rank and weigh the criteria, ensuring a balanced evaluation of conflicting factors. Based on the evaluation and ranking, select the supplier that best aligns with the business's strategic goals and supply chain needs. MCDM methods enable a comprehensive assessment of suppliers by considering multiple criteria simultaneously. By systematically ranking and weighting criteria, MCDM reduces bias and ensures transparent decision-making. MCDM techniques can incorporate fuzzy logic or probabilistic models to handle uncertainty and variability in supplier performance.

Supplier selection is a pivotal process in SCM, with far-reaching implications for operational efficiency, product quality, and customer satisfaction. By framing it as an MCDM problem, businesses can systematically evaluate and rank suppliers based on multiple criteria, ensuring the selection of the best possible partners. This approach not only enhances supply chain resilience but also supports long-term strategic goals.

In the complex and dynamic environment of supply chains, disruptions can occur at any stage, posing significant challenges to operations and continuity. To address this, resilience has emerged as a critical concept in supply chain management. According to Scholten et al. (2014), resilience in the context of supply chains is defined as the capacity of supply chains to react effectively to unavoidable and unexpected disruptions. The ability to prepare for, adapt to, and recover from disruptions, ensuring continuity and minimizing impact. Resilience in supply chains is built on several critical components, as outlined by Scholten et al. (2014):

Horizontal collaboration: Coordination among entities at the same level of the supply chain (e.g., suppliers collaborating with other suppliers).

Vertical collaboration: Coordination across different levels of the supply chain (e.g., suppliers, manufacturers, distributors).

These collaborations enhance information sharing, resource pooling, and joint problem-solving during disruptions. The process of designing or redesigning supply chain structures to improve flexibility, efficiency, and responsiveness. This includes strategies such as dual sourcing, inventory buffering, and modular production systems. The ability to adapt quickly to changes in demand, supply, or environmental conditions. Flexibility can be achieved through multi-sourcing, agile manufacturing, and dynamic routing. Proactively identifying, assessing, and monitoring potential risks in the supply chain.

This involves risk mapping, scenario planning, and continuous monitoring to anticipate and mitigate disruptions. The systematic collection, sharing, and utilization of knowledge to improve decision-making and response strategies. This includes lessons learned from past disruptions, best practices, and training programs for supply chain stakeholders.

A resilient supply chain ensures that suppliers can maintain operations and fulfill commitments even during disruptions. Resilience minimizes the impact of disruptions on production, delivery, and customer satisfaction. Resilient supply chains are better equipped to handle uncertainties, providing a strategic edge in volatile markets.

Resilience is a critical capability for supply chains operating in complex and dynamic environments. By incorporating horizontal and vertical collaboration, supply chain (re)engineering, flexibility, risk awareness, and knowledge management, businesses can effectively mitigate disruptions and ensure continuity. Supplier selection plays a pivotal role in building resilience, as it directly impacts the supply chain's ability to adapt and recover from unforeseen challenges.

In today's complex and unpredictable business environment, a lack of preparation for inevitable events can severely disrupt material, information, and cash flows, leading to lost sales, increased costs, or both. Building a resilient supply chain is essential for organizations to effectively navigate threats and maintain operational continuity. Since suppliers are a critical component of the supply chain, ensuring their resilience is paramount.

In the context of supplier selection, especially in industries like the steel supply chain, quantifying the benefits of various options is a critical challenge for managers. The presence of uncertainty, vagueness, and incomplete information further complicates decision-making (Mahdavi et al., 2008a, 2008b). To address these issues, fuzzy set theory has emerged as a powerful tool for handling imprecise data and making informed decisions. Decision-makers often lack accurate and complete information about criteria and constraints, making it difficult to evaluate suppliers objectively. Information about supplier capabilities, performance, and risks is often imprecise or subjective. The steel industry is particularly susceptible to disruptions such as material damage and production breakdowns due to non-conformity of raw materials with production requirements. Fuzzy set theory provides a mathematical framework to model and analyze vague and imprecise information, making it suitable for supplier selection problems. It allows decision-makers to incorporate qualitative judgments and linguistic variables (e.g., "high," "medium," "low") into the evaluation process.

By quantifying uncertainty, fuzzy set theory helps in making more informed and realistic decisions. Suppliers must provide raw materials that conform to production requirements to avoid disruptions like material damage or production breakdowns. Suppliers should demonstrate a track record of delivering materials on-time and in the required quantities. Suppliers must be capable of adapting to the company's changing needs, such as adjustments in material specifications or delivery schedules. Suppliers should have robust risk management strategies to mitigate potential disruptions, such as diversified sourcing or contingency plans. While ensuring quality and reliability, suppliers should offer competitive pricing to maintain cost efficiency in the supply chain.

Techniques like fuzzy AHP (analytic hierarchy process) or Fuzzy TOPSIS (technique for order of preference by similarity to ideal solution) can be used to evaluate suppliers based on multiple criteria. Decision-makers can use linguistic terms (e.g., "very good," "good," "fair") to rate suppliers, which are then converted into fuzzy numbers for analysis. Fuzzy aggregation methods combine the evaluations of different criteria to rank suppliers, providing a clear decision framework. By incorporating uncertainty and vagueness, fuzzy set theory leads to more accurate and realistic supplier evaluations. It allows decision-makers to use qualitative and quantitative data simultaneously, making it adaptable to various scenarios. Selecting suppliers based on fuzzy evaluations ensures that the supply chain is better prepared to handle disruptions.

In the steel supply chain, where disruptions like material damage and production breakdowns are common, selecting the right suppliers is crucial for maintaining operational continuity. Fuzzy set theory provides an effective framework for handling the uncertainty and vagueness inherent in supplier selection processes. By focusing on criteria such as quality, reliability, adaptability, risk management, and cost efficiency, and leveraging fuzzy MCDM techniques, decision-makers can make informed and resilient supplier choices.

Fuzzy Multi-Objective Linear Programming (FMOLP) serves as a tool to optimize the distribution of orders across suppliers, tackling the inherent ambiguity in data and harmonizing various goals such as reducing costs and improving quality. Meanwhile, Fuzzy Analytic Hierarchy Process (FAHP) is utilized to determine the relative significance of each evaluation criterion, integrating the subjective preferences of decision-makers within a fuzzy framework. The practical value of this approach is illustrated through a case study conducted in the steel industry. This study assesses and chooses suppliers based on predefined criteria, employs FMOLP to optimize order distribution, and uses FAHP to rank the importance of these criteria. The findings underscore the method's capability to strengthen supply chain resilience and improve the precision of decision-making in uncertain conditions.

This approach emphasizes building a resilient supply chain that ensures flexibility and uninterrupted material procurement. By incorporating fuzzy logic into FMOLP and FAHP, the method effectively handles uncertainty and imprecision in data, enhancing the robustness of decision-making processes. A case study conducted in the steel industry confirms the method's practicality and efficacy in real-world applications. Integrating FMOLP and FAHP with advanced techniques like machine learning or simulation could further improve the method's scalability and accuracy. Additionally, testing the approach in other industries, such as automotive or electronics, could showcase its broader applicability. The research presents a comprehensive framework for supplier selection, designed to tackle the uncertainty and complexity faced by large-scale enterprises. Through the combined use of FMOLP and FAHP, the method ensures optimal order allocation and prioritizes critical criteria, as evidenced by the steel industry case study.

4.2 Literature Review

Supplier selection is a critical area of research, recognized for its complexity due to the involvement of multiple, often conflicting criteria. These include procurement cost, material quality, delivery lead time, and supplier dependability, among others. According to Sonmez (2006), supplier selection is the process of identifying suppliers capable of providing high-quality products and/or services at a reasonable price, in the right quantities, and at the right time. The criteria for supplier selection can vary across organizations, influenced by factors such as the size of the buyer organization. Researchers have proposed a range of qualitative and quantitative parameters to address this challenge, with studies identifying 23 distinct criteria for supplier selection. Among these, quality, delivery, and performance have been highlighted as the most critical factors. Specifically, quality is the most widely used criterion, followed by delivery and price/cost.

In this study, five key criteria—quality, flexibility, time, cost, and conformity of raw material with production requirements—have been selected to evaluate and choose suppliers within a resilient supply chain (Fazlollahtabar, 2008). These criteria align with the need for adaptability and continuity in supply chain operations, ensuring that suppliers can meet the dynamic demands of production environments.

The process involves balancing multiple, often competing factors, making it a multi-faceted decision-making challenge. Quality, delivery, and cost are consistently identified as the most important factors in supplier evaluation. The selected criteria (quality, flexibility, time, cost, and material conformity) emphasize the importance of building a resilient supply chain capable of withstanding disruptions. The choice of criteria can vary based on the specific needs and scale of the buyer organization. This approach provides a structured framework for supplier selection, addressing both operational efficiency and resilience in supply chain management.

The supplier selection problem has been extensively studied in literature, with a variety of models and approaches proposed to address its complexities. These approaches are broadly categorized into individual approaches and hybrid approaches, each offering unique perspectives and methodologies. Individual approaches focus on standalone techniques to evaluate and select suppliers. The most widely used methods include: Developed by Saaty (1980), AHP is a systematic method for decision-making involving multiple criteria and alternatives. Its ability to handle both qualitative and quantitative criteria makes it particularly suitable for supplier selection problems. Data Envelopment Analysis (DEA) helps organizations classify suppliers into efficient and inefficient categories based on performance metrics (Mahdavi et al., 2008a, 2008b). Mathematical Programming includes techniques such as Linear Programming (LP), Integer Programming (IP), Mixed Integer Nonlinear Programming (MINLP), Multi-objective Programming, and Goal Programming (GP). These methods are effective in optimizing supplier selection based on specific objectives and constraints.

Artificial Intelligence (AI)-based techniques like Neural Networks (NN), Genetic Algorithms (GA), Fuzzy Set Theory (FST), Expert Systems (ES), and Case-Based

Reasoning (CBR) have also been widely applied to supplier selection problems, offering advanced capabilities for handling complex and uncertain data.

Recognizing that no single technique can provide a comprehensive solution to supplier selection problems, researchers have developed hybrid approaches. These methods combine multiple techniques to leverage their strengths and address their limitations. For example, integrating AHP with mathematical programming or fuzzy set theory can enhance decision-making by incorporating both subjective judgments and objective optimization. Hybrid approaches are particularly effective in handling the multi-criteria and uncertainty inherent in supplier selection (Kazemirash & Fazlollahtabar, 2022).

The supplier selection problem is addressed through a wide range of methods, each with its own strengths and applications. While individual approaches are effective for specific aspects of supplier selection, hybrid approaches provide a more holistic solution by combining multiple techniques. The choice of approach depends on the specific needs of the organization, the complexity of the problem, and the availability of data.

Hybrid approaches were used by past researchers to handle a more effective supplier selection model making use of multiple advantages of any model (Table 4.1).

In large-scale enterprises, disruptions often arise from the complexity and uncertainty of supply chain environments. A supply network disruption refers to an unexpected and unplanned event that interrupts the regular movement of goods and materials, posing significant operational and financial risks to businesses. Resilience has emerged as a unique source of sustainable competitive advantage for suppliers. At its core, supplier resilience is the ability to manage risks effectively, enabling them to outperform competitors in handling disruptions. To address disruptions and adapt to fluctuating demands, flexibility is a crucial capability for suppliers, which can be enhanced through strategies like high redundancy, improved conformity, and process standardization (Fazlollahtabar, 2022).

To mitigate the impact of supply chain disruptions and build a resilient supply chain, this research focuses on strategic supplier selection and material procurement based on identified vulnerable areas. One critical vulnerability in material procurement is the non-conformity of materials with production requirements. To address this, the study incorporates material conformity—specifically, the alignment of raw materials with the physical and chemical properties of the manufacturer—as a key factor in analyzing flexible production parameters. Ignoring this factor can lead to significant damage to devices and products.

Additionally, the environmental conditions of manufacturers, such as light, temperature, and warehouse settings, are treated as fixed factors. Therefore, suppliers must be selected based on their ability to provide materials that align with these conditions. In essence, input materials should be compatible with the devices, equipment, and environmental conditions of the manufacturer. To solve the supplier selection problem, the research employs the Fuzzy Analytic Hierarchy Process (FAHP) to determine the weight of criteria and Fuzzy Multi-Objective Linear Programming (FMOLP) to optimize the selection process. Furthermore, Fuzzy TOPSIS is utilized to rank suppliers and determine their exact priorities (Katiraee et al., 2017).

Table 4.1 A review of criteria used in literature for supplier selection and evaluation (Singh, 2014)

No.	Criteria	No.	Criteria
1	Financial capability	21	Service innovation
2	Price (cost)	22	Age and position in market
3	Purchasing cost	23	Political and economical stability
4	Delivery	24	Compatibility of organizational cultures
5	Flexibility	25	Competitive pressure
6	Reliability	26	Evaluation and certification system
7	Consistency	27	Organization and management
8	Geographical location	28	Supplier strategic objective
9	Reserve capacity	29	Training and education
10	Service level	30	Sustainability
11	Product quality	31	Environmental effects and preventive actions
12	Commitment to improvement and cost reduction	32	Reputation for integrity
13	Warranty and after sales support	33	Work safety and labor health
14	Process/manufacturing capability	34	Packaging capability
15	Technological capability	35	Maintainability
16	Trust and information sharing	36	Demand
17	Ease of communication	37	Impression
18	Long-term relationship	38	Business attempt
19	Top management support	39	Costumer focus
20	Reciprocal arrangement	40	Operational control

4.3 Proposed Approach

To mitigate vulnerability in supply chains, this research emphasizes the importance of building a resilient supply chain. A key component of this approach is the selection of suppliers through a specialized framework, referred to as "resilient suppliers." In this study, a resilient supplier is defined as a supplier capable of delivering high-quality materials that align with the manufacturer's specifications and contribute to long-lasting production. Furthermore, these suppliers must exhibit flexibility in responding to disruptions without causing significant increases in operational costs or lead times.

The study begins by pinpointing vulnerabilities in the material supply chain to understand potential risks and challenges. Based on production requirements, five key criteria are identified to achieve a resilient supply chain in material procurement. These criteria are designed to ensure suppliers meet essential standards for quality, conformity, and adaptability. Fuzzy Analytic Hierarchy Process (FAHP) is used to determine the weights of the selected criteria, enabling a balanced evaluation of supplier capabilities. Fuzzy multi-objective model is developed to address the supplier selection and order allocation problem, optimizing for resilience and efficiency. Fuzzy TOPSIS is proposed to rank suppliers and determine their exact priorities, facilitating informed decision-making.

Suppliers must align with the manufacturer's requirements and demonstrate the ability to adapt to disruptions. A critical attribute for suppliers to maintain operational efficiency during disruptions. The combination of FAHP, fuzzy multi-objective modeling, and fuzzy TOPSIS provides a robust framework for selecting and prioritizing resilient suppliers. This approach not only enhances supply chain resilience but also ensures that suppliers are selected based on their ability to meet specific operational and quality standards (Katiraee et al., 2017). The main structure of the proposed approach is summarized in Fig. 4.1.

4.3.1 FAHP Method

In this study, the importance of criteria is calculated by FAHP as it is a popular approach for multiple criteria decision-making. Triangular fuzzy numbers were used for pairwise comparison scale of FAHP in order to develop a new approach for handling FAHP. In this study, the extent fuzzy AHP is utilized. The steps of the FAHP procedure are as follows:

Step 1: Drawing hierarchy process diagram for the problem.
Step 2: Definition of the fuzzy numbers in order to establish pairwise comparison matrix (see Table 4.2).
Step 3: Establish pairwise comparison matrix:

$$\tilde{A} \cong \begin{bmatrix} 1 & \tilde{a}_{12} & \cdots & \tilde{a}_{1n} \\ & \vdots & \ddots & \vdots \\ \tilde{a}_{n1} & \tilde{a}_{n2} & \cdots & 1 \end{bmatrix}.$$

Let $(\tilde{A})$ represent the $n * n$-judgment matrix containing triangle fuzzy number (TFN) $\tilde{a}_{ij}$ for all $i, j \in \{1, 2, ..., n\}$ which is shown as follows:

$$\tilde{a}_{ij} = \begin{cases} 1 & i = j, \\ \tilde{1}, \tilde{3}, \tilde{5}, \tilde{7}, \tilde{9}\ or \tilde{1}^{-1}, \tilde{3}^{-1}, \tilde{5}^{-1}, \tilde{7}^{-1}, \tilde{9}^{-1}, & i \neq j, \end{cases}$$

where $\tilde{a}_{ij} = (l_{ij}, m_{ij}, u_{ij})$ with l_{ij} is the lower and u_{ij} is the upper limit and m_{ij} is the most likely value.

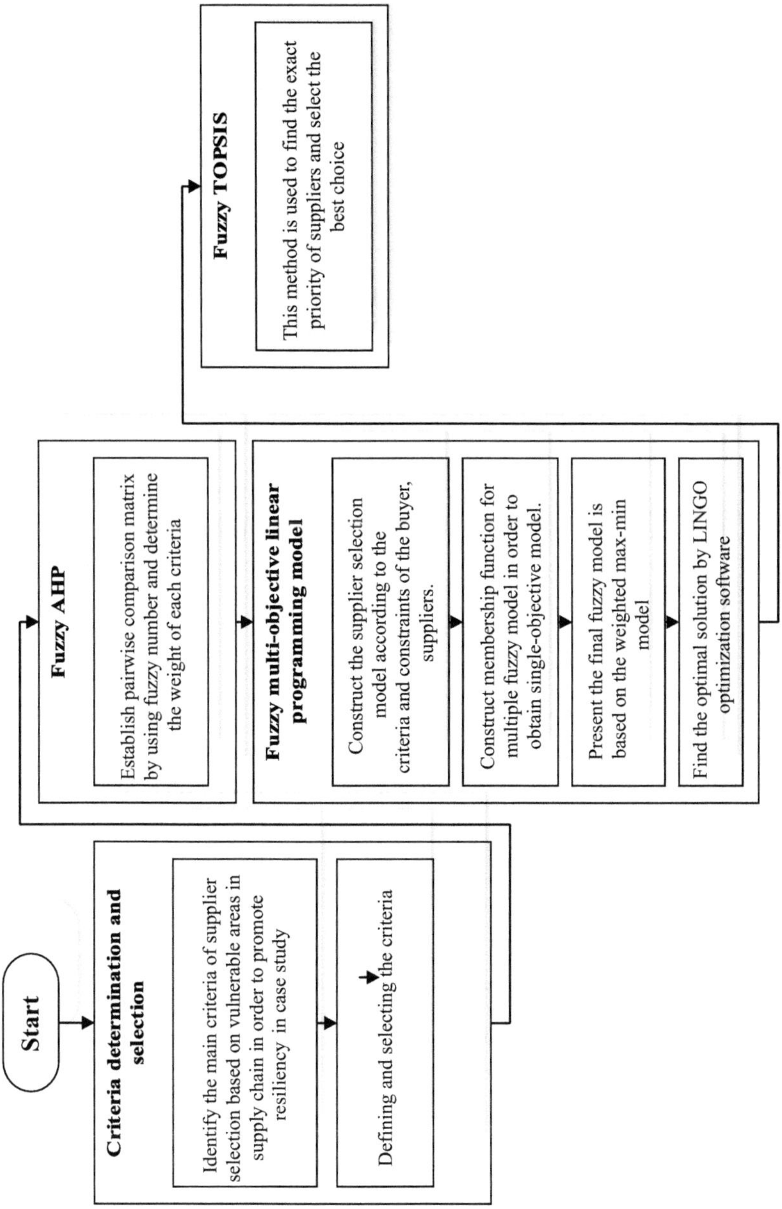

Fig. 4.1 Research framework

Table 4.2 Fuzzy pairwise comparison matrix

Fuzzy number	Definition	Fuzzy triangular scale
$\tilde{9}$	Extremely more importance	(7, 9, 9)
$\tilde{7}$	Very strong importance	(5, 7, 9)
$\tilde{5}$	Strong importance	(3, 5, 7)
$\tilde{3}$	Moderate importance	(1, 3, 5)
$\tilde{1}$	Equal importance	(1, 1, 3)
1	Exactly equal	(1, 1, 1)

Step 4: Calculation of fuzzy synthetic extents:

The value of fuzzy synthetic extent with respect to each object is defined as

$$S_i = \sum_{j=1}^{m} M_{gi}^{j} \otimes \left[\sum_{i=1}^{n} \sum_{j=1}^{m} M_{gi}^{j} \right]^{-1},$$

where M_{gj}^{j} $(j = 1, 2, 3, \ldots, m)$ are all triangular fuzzy numbers and $\otimes$ denotes the extended multiplication for two fuzzy members. In order to obtain $\sum_{j=1}^{m} M_{gi}^{j}$, we perform the addition of m extent analysis values for a particular matrix such that

$$\sum_{j=1}^{m} M_{gi}^{j} = \left(\sum_{j=1}^{m} l_j, \sum_{j=1}^{m} m_j, \sum_{j=1}^{m} u_j \right),$$

and to obtain $\left[\sum_{i=1}^{n} \sum_{j=1}^{m} M_{gi}^{j} \right]^{-1}$, we perform the fuzzy addition operation of $\tilde{M}_{gi}^{j}$ $(j = 1, 2, \ldots, m)$ values such that

$$\sum_{i=1}^{n} \sum_{j=1}^{m} M_{gi}^{j} = \left(\sum_{i=1}^{n} l_i, \sum_{i=1}^{n} m_i, \sum_{i=1}^{n} u_i \right).$$

Then, the inverse of the vector is computed as

$$\sum_{i=1}^{n} \sum_{j=1}^{m} M_{gi}^{j}]^{-1} = \left(\frac{1}{\sum_{i=1}^{n} u_i}, \frac{1}{\sum_{i=1}^{n} m_i}, \frac{1}{\sum_{i=1}^{n} l_i} \right).$$

Step 5: For two fuzzy triangular numbers, M_1 and M_2, the degree of possibility that $M_2 = (l_2, m_2, u_2) \geq M_1 = (l_1, m_1, u_1)$ is defined as

$$V(M_2 \geq M_1) = hgt(M_1 \cap M_2) = \mu_{m_2}(d) = \begin{cases} 1 & \text{if } m_2 \geq m_1 \\ 0 & \text{if } l_1 \geq u_2 \\ \dfrac{l_1 - u_2}{(m_2 - u_2) - (m_1 - l_1)} & \end{cases}.$$

Step 6: The degree of possibility for a convex fuzzy number M to be greater than k fuzzy numbers is defined as follows:

$$V(M \geq M_1, M_2, ..., M_K) = V[(M \geq M_1) \text{ and } (M \geq M_2) \text{ and} \ldots \text{and}(M \geq M_K)]$$
$$= \text{Min}\, V(M \geq M_i) \quad i = 1, 2, \ldots, k.$$

Step 7: Calculate the weight of criteria in pairwise comparison matrix:

Assume that $d'(A_i) = \text{Min}\, V(S_1 \geq S_K)$ for $K = 1, 2, ..., n, k \neq i$.

Then $W' = (d'(A_1), d'(A_2), \ldots, d'(A_n))^T \; A_i \; (i = 1, 2, \ldots, n)$ is the weight vector for $k = 1, 2, ..., n$.

Step 8: Normalization of weight vector. The normalized weight vector is given by

$$W_i = \frac{W_i}{\sum W_i}.$$

4.4 The Multi-objective Supplier Selection Model

In order to solve supplier selection problem in resilient supply chain, the multiple objective models is developed. In this model, it is assumed the buyer would like to choose the best suppliers among the n vendors whose capacities are limited (Katiraee et al., 2017). To formulate the model, the following notations, objective function, and constraints are defined.

Notations

X_i = The order quantity from supplier i.
S_{ij} = Performance of supplier i in terms of criterion j.
D = Demand over fixed period.
n = Number of suppliers.
f_i = Percentage of flexibility by the supplier i.
f_a = Buyers' minimum acceptable flexibility percentage.
q_i = Percentage of quality by the supplier i.
a_i = Percentage of material adaptation with production requirements.

p_i = Per unit net purchase cost and transportation cost from supplier i.
t_i = On-time deliveries by the supplier i.
c_i = Capacity of each supplier.

Objective Function

$$\text{Max/Min}\, Z_j = \sum_{i=1}^{n} S_{ij}X_i \quad \forall j = 1, 2, \ldots, m.$$

Note that, depending on the concept of criterion included in the model the function is either maximization or minimization.

Constraints

Demand constraint: The amounts of materials ordered should be equal to the demand received to produce, i.e., lost sale or inventory holding cost is not allowed.

$$\sum_{i=1}^{n} X_i = D.$$

Capacity constraint: This constraint means that order quantity of each supplier should be equal to or less than its capacity.

$$X_i \le C_i, \qquad \forall i = 1, 2, \ldots, n.$$

Flexibility constraint: In this model, flexibility constraint is used in order to achieve resilient suppliers. As f_a is the buyers' minimum acceptable perfect rate and f_i is the perfect rate of the supplier i, the flexibility constraint can be shown as

$$\sum_{i=1}^{n} f_i X_i \ge f_a\ .$$

Five objective functions are formulated to minimize cost and delivery time and maximize quality, flexibility, and conformity of raw materials with production requirements. Also the model consists of three set of constraints, based on buyer's demand, supplier's capacity, and supplier's flexibility.

The final typical multi-objective model for purchasing and selecting suppliers among several vendors with conflicting criteria is given below:

$$\text{Max}\, Z_1 = \sum_{i=1}^{n} q_i X_i,$$

$$\text{Max}\, Z_2 = \sum_{i=1}^{n} f_i X_i,$$

$$\text{Min}\, Z_3 = \sum_{i=1}^{n} p_i X_i,$$

$$\text{Min}\, Z_4 = \sum_{i=1}^{n} t_i X_i,$$

$$\text{Max}\, Z_5 = \sum_{i=1}^{n} a_i X_i,$$

s.t.

$$\sum_{i=1}^{n} f_i X_i \geq f_a \ ,$$

$$\sum_{i=1}^{n} X_i = D \, ,$$

$$X_i \leq C_i \, , \quad i = 1, 2, \ldots, n,$$

$$X_i \geq 0, \quad i = 1, 2, \ldots, n \, .$$

4.5 Fuzzy Model for Supplier Selection

To address the limitations of a standalone multi-objective model in solving the supplier selection problem, this research proposes an integrated approach combining Fuzzy Analytic Hierarchy Process (FAHP) and Fuzzy Multi-Objective Linear Programming (FMOLP). This integration aims to handle the uncertainty and vagueness inherent in the information, enabling the determination of the optimal order quantity for each selected supplier.

Phase 1: The study adopts Zimmermann's approach to construct the membership function for the fuzzy multi-objective model. Zimmermann (1978) extended the principles of fuzzy linear programming to address fuzzy multi-objective linear programming problems, providing a robust framework for handling imprecise data. This approach ensures a more effective and nuanced solution to the supplier selection problem, leveraging the strengths of both FAHP and FMOLP.

In this approach, the membership function of objectives is formulated by separating every objective function into its maximum and minimum values. The linear membership functions for minimization goals (Z_k) and maximization goals (Z_1) are given as follows:

$$\mu_{zk}(x) = \begin{cases} 1 & \text{for } z_k \leq z_k^- \\ f_{\mu z_k} = \left(z_k^+ - z_k(x)\right) / \left(z_k^+ - z_k^-\right), & \text{for } z_k^- \leq z_k(x) \leq z_k^+ \ (k = 1, 2, \ldots, p), \\ 0 & \text{for } z_k \geq z_k^+ \end{cases}$$

$$\mu_{zl}(x) = \begin{cases} 1 & \text{for } z_l \geq z_l^+ \\ f_{\mu z_l} = \left(z_l(x) - z_l^-\right) / \left(z_l^+ - z_l^-\right), & \text{for } z_l^- \leq z_l(x) \leq z_l^+ \\ 0 & \text{for } z_l \leq z_l^- \end{cases}$$

$$(l = p + 1, p + 2, \ldots, q),$$

Phase 2: After describing membership function and determining the weight of criteria by FAHP, the final fuzzy model is presented based on weighted max–min model which was used by Zimmermann (1978) for fuzzy multi-objective problems. This model is used for selecting the best suppliers for achieving optimal order allocation.

$$\text{Max } \lambda$$

$$\text{Subject to :}$$

$$w_k\lambda \leq \left(z_k^+ - z_k(x)\right) / \left(z_k^+ - z_k^-\right) \quad \text{(for all min objective functions)}$$

$$w_l\lambda \leq \left(z_l(x) - z_l^-\right) / \left(z_l^+ - z_l^-\right) \quad \text{(for all max objective functions)}$$

$$\sum_{i=1}^{n} f_i x_i \geq f_a$$

$$X_i \leq C_i \quad i = 1, 2, \ldots, n$$

$$\sum_{i=1}^{n} X_i = D$$

$$\lambda \in [0, 1]$$

$$\sum_{j=1}^{q} w_j = 1\,, \quad w_i \geq 0 \quad \forall j$$

$$x_i \geq 0 \quad i = 1, \ldots, n.$$

To obtain the optimal solution of this single-objective model the optimization is performed using LINGO software. Then the fuzzy TOPSIS is applied to find the exact priority of suppliers.

4.5.1 *Fuzzy TOPSIS*

The main steps of fuzzy TOPSIS are as follows:

Step 1: Construct the fuzzy decision-making matrix.
Step 2: Calculate the normalized fuzzy decision matrix.
Step 3: Construct the weighted normalized fuzzy decision matrix.
Step 4: Determine the fuzzy positive-ideal solution (FPIS, A^+) and fuzzy negative-ideal solution (FNIS, A^-).
Step 5: Calculate the distance of each alternative from FPIS and FNIS.
Step 6: Obtain the closeness coefficient (CC) and rank the order of alternatives.

4.6 Numerical Example

To apply the proposed methodology, a case study was conducted at Maziyar Manufacturing Industry. This organization initially started with limited facilities and small-scale operations. Over time, it has evolved significantly, now equipped with state-of-the-art machinery and a diverse range of high-quality products. The company is actively working to establish and expand new units to meet growing customer demands. The company specializes in the production of the following items:

- Concrete beams.
- Lighting foundations.
- Electrical cabinets.
- Foundations.
- Masts.
- Branching boxes.

In this context, suppliers must demonstrate resilience by being adaptable to various risks and disruptions. They should possess the capability to minimize delays, control costs, and enhance product quality and flexibility. Here, resilience is defined as the ability to effectively respond to disruptions and maintain operational stability. Five key criteria were identified for raw material procurement and supplier selection, as outlined in Table 4.3. These criteria were carefully chosen to ensure the selection of suppliers who can contribute to a resilient and efficient supply chain. This approach ensures that suppliers are evaluated based on their ability to meet the organization's operational and quality standards while maintaining adaptability in the face of disruptions.

To calculate the weight of each criteria the FAHP method is employed, and the results are summarized in Table 4.4. Also, Criteria's importance is shown in Fig. 4.2 according to their weights.

To identify the effect of these factors on procurement unit, four suppliers' performances are considered. The quantity information of suppliers is shown in Table 6. After determining the weight of criteria, the fuzzy multi-objective linear formulation

Table 4.3 Supplier selection criteria definition

Criteria	Definition
Quality	Containing innovation features (technology development and R & D units), standards and certificates, control and quality tests
Flexibility	Flexible contracts with suppliers in terms of number, time, and order; suppliers' ability in supplying alternative material to not to stop production line. Also flexibility refers to a company's ability to respond to disruption without significantly increasing its operational costs
Cost	Including raw material price and transportation cost
Time	Delivery time
Conformity	Conformity of raw materials (physical and chemical features of material) with manufacturer is assumed as a factor in analysis of the parameters of supplier selection

Table 4.4 Fuzzy pairwise comparison matrix and fuzzy weights

Criteria	Conformity with production requirement	Flexibility	Delivery time	Cost	Quality	Weights
Conformity with production requirement	(1, 1, 1)	(2/5, 1/2, 2/3)	(1/2, 2/3, 1)	(1/2, 1, 3/2)	(1/2, 2/3, 1)	0.062
Flexibility	(3/2, 2, 5/2)	(1, 1, 1)	(3/2, 2, 5/2)	(2, 5/2, 3)	(3/2, 2, 5/2)	0.44
Delivery time	(1, 3/2, 2)	(2/5, 1/2, 2/3)	(1, 1, 1)	(1, 3/2, 2)	(1, 3/2, 2)	0.239
Cost	(2/3, 1, 2)	(1/3, 2/5, 1/2)	(1/2, 2/3, 1)	(1, 1, 1)	(1/2, 2/3, 1)	0.084
Quality	(1, 3/2, 2)	(2/5, 1/2, 2/3)	(1/2, 2/3, 1)	(1, 3/2, 2)	(1, 1, 1)	0.175

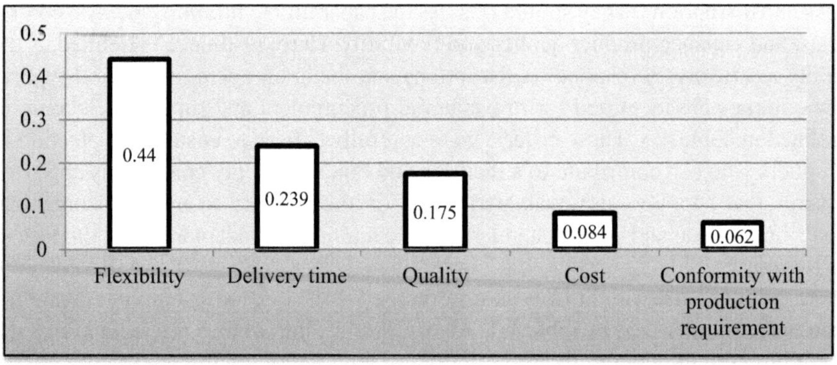

Fig. 4.2 Priority of each criterion

of numerical example is presented below:

$$
\begin{aligned}
&\text{Max } \tilde{Z}_1 = 0.937X_1 + 0.875X_2 + 0.961X_3 + 0.902X_4 \tilde{\lesssim} Z_1^0,\\
&\text{Max } \tilde{Z}_2 = 0.931X_1 + 0.87X_2 + 0.915X_3 + 0.957X_4 \tilde{\lesssim} Z_2^0,\\
&\text{Min } \tilde{Z}_3 = 17.625X_1 + 16.087X_2 + 19.2X_3 + 17.1X_4 \tilde{\lesssim} Z_3^0,\\
&\text{Min } \tilde{Z}_4 = 21.25X_1 + 17.875X_2 + 15.125X_3 + 12.5X_4 \tilde{\lesssim} Z4,\\
&\text{Max } \tilde{Z}_5 = 0.89X_1 + 0.909X_2 + 0.925X_3 + 0.921X_4 \tilde{\lesssim} Z_5^0,\\
&\text{s.t. :}\\
&0.931X_1 + 0.87X_2 + 0.915X_3 + 0.957X_4 \geq 0.90,\\
&X_1 \leq 0.55,\\
&X_2 \leq 0.5,\\
&X_3 \leq 0.45,\\
&X_4 \leq 0.6,\\
&X_1 + X_2 + X_3 + X_4 = 1,\\
&X_i \geq 0 \qquad i = 1, \ldots, 4.
\end{aligned}
$$

Five objective functions Z_1, Z_2, Z_3, Z_4, and Z_5 are quality, flexibility, cost, delivery time, and conformity of raw material with production requirements, respectively. Based on the weighted max–min models and according to the dataset for membership functions, which are calculated in Table 4.5, the single-objective formulation for this problem is obtained as follows:

$$
\begin{aligned}
&\text{Max } \lambda\\
&\text{Subject to :}\\
&0.175\lambda \leq ((0.937X_1 + 0.875X_2 + 0.961X_3 + 0.902X_4) - 0.888)/0.059,\\
&0.44\lambda \leq ((0.931X_1 + 0.87X_2 + 0.915X_3 + 0.957X_4) - 0.9)/0.046,\\
&0.084\lambda \leq (18.333 - (17.625X_1 + 16.087X_2 + 19.2X_3 + 17.1X_4))/1.74,\\
&0.239\lambda \leq (19.731 - (21.25X_1 + 17.875X_2 + 15.125X_3 + 12.5X_4))/6.181,\\
&0.062\lambda \leq ((0.89X_1 + 0.909X_2 + 0.925X_3 + 0.921X_4) - 0.898)/0.024,\\
&0.931X_1 + 0.87X_2 + 0.915X_3 + 0.957X_4 \geq 0.90,\\
&X_1 \leq 0.55,\\
&X_2 \leq 0.5,\\
&X_3 \leq 0.45,\\
&X_4 \leq 0.6,\\
&X_1 + X_2 + X_3 + X_4 = 1,\\
&X_i \geq 0, \qquad i = 1, \ldots, 4.
\end{aligned}
$$

Table 4.5 Dataset for membership functions

	$\mu = 0$	$\mu = 1$	$\mu = 0$
(Quality)Z_1	0.888	0.947	–
(Flexibility)Z_2	0.9	0.946	–
(Cost)Z_3	–	16.593	18.333
(Delivery time)Z_4	–	13.55	19.731
(Conformity)Z_5	0.898	0.922	–

The single-objective model is solved by LINGO 14.0 software. The optimal solution for the above formulation is as follows:

$$\begin{cases} \lambda = 1 \\ X_1 = 0.55\,, \quad X_2 = X_3 = 0\,, \quad X_4 = 0.45. \\ Z_1 = 0.921\,, \quad Z_2 = 0.942\,, \quad Z_3 = 17.38, \quad Z_4 = 17.31, \quad Z_5 = 0.9\,. \end{cases}$$

To find the exact priority of suppliers, the problem is solved by fuzzy TOPSIS. According to the closeness coefficient, ranking the preference order of these alternatives is shown in Table 4.7. In addition, all fuzzy TOPSIS steps are shown in Tables 4.8, 4.9, 4.10, 4.11 and 4.12.

Table 4.8 Fuzzy decision-making matrix

Alternatives	Quality	Flexibility	Cost	Delivery time	Conformity
Supplier 1	0.7, 0.9, 1)	(0.7, 0.9, 1)	(0.33, 0.42, 0.6)	(0.42, 0.6, 1)	(0.3, 0.5, 0.7)
Supplier 2	(0.5, 0.7, 0.9)	(0.3, 0.5, 0.7)	(0.3, 0.33, 0.42)	(0.33, 0.42, 0.6)	(0.3, 0.5, 0.7)
Supplier 3	(0.9, 1, 1)	(0.7, 0.9, 1)	(0.42, 0.6, 1)	(0.3, 0.33, 0.42)	(0.5, 0.7, 0.9)
Supplier 4	(0.7, 0.9, 1)	(0.9, 1, 1)	(0.33, 0.42, 0.6)	(0.3, 0.3, 0.33)	(0.7, 0.9, 1)

Table 4.9 Normalized fuzzy decision matrix

	Quality	Flexibility	Cost	Delivery time	Conformity
Supplier1	(0.1225, 0.1575, 0.175)	(0.308, 0.396, 0.44)	(0.0277, 0.0352, 0.0504)	(0.1003, 0.1434, 0.239)	(0.0186, 0.031, 0.0434)
Supplier 2	(0.0875, 0.1225, 0.1575)	(0.132, 0.22, 0.308)	(0.0252, 0.0277, 0.0352)	(0.0788, 0.1003, 0.1443)	(0.0186, 0.031, 0.0434)
Supplier 3	(0.1575, 0.175, 0.175)	(0.308, 0.396, 0.44)	(0.0352, 0.0504, 0.084)	(0.0717, 0.0788, 0.1003)	(0.031, 0.0434, 0.0558)
Supplier 4	(0.1225, 0.1575, 0.175)	(0.396, 0.44, 0.44)	(0.0277, 0.0352, 0.0504)	(0.0717, 0.0717, 0.0788)	(0.0434, 0.0558, 0.062)

Table 4.10 The fuzzy positive-ideal solution (FPIS, A^+) and negative-ideal solution (FNIS, A^-)

	Quality	Flexibility	Cost	Delivery time	Conformity
A^+	(0.1575, 0.175, 0.175)	(0.396, 0.44, 0.44)	(0.0252, 0.0277, 0.0352)	(0.0717, 0.0717, 0.0788)	(0.0434, 0.0558, 0.062)
A^-	(0.0875, 0.1225, 0.1575)	(0.132, 0.22, 0.308)	(0.0352, 0.0504, 0.084)	(0.1003, 0.1434, 0.1443)	(0.0186, 0.031, 0.0434)

Table 4.11 The distance of each alternative from positive-ideal solution (FPIS)

Alternatives	Quality	Flexibility	Cost	Delivery time	Conformity
Supplier 1	0.0225	0.0568	0.0098	0.102	0.0229
Supplier 2	0.0515	0.0212	0	0.0414	0.0229
Supplier 3	0	0.0568	0.0316	0.0130	0.0107
Supplier 4	0	0	0.0098	0	0

Table 4.12 The distance of each alternative from negative-ideal solution (FNIS)

Alternatives	Quality	Flexibility	Cost	Delivery time	Conformity
Supplier 1	0.0303	0.162	0.0217	0.0546	0
Supplier 2	0	0	0.00577	0.0278	0
Supplier 3	0.0515	0.162	0	0.0480	0.0124
Supplier 4	0.0303	0.212	0.0217	0.0584	0.0229

After analyzing the results, the fuzzy TOPSIS approach identifies Supplier 4 as the top choice, followed by Supplier 3, Supplier 1, and Supplier 2 in the second, third, and fourth positions, respectively.

The primary goal of this research is to enhance the capability of suppliers to adapt to turbulence and disruptions in the supply chain. To achieve this, the main criteria for supplier selection were established based on areas prone to disruptions. The importance of each criterion was calculated using the proposed model, creating a framework for selecting resilient suppliers in a resilient supply chain.

The final model for material supply and procurement was developed after validating the model and incorporating critical criteria that address vulnerabilities. This ensures the model's effectiveness in real-world applications. The proposed model is illustrated in Fig. 4.3, providing a clear and actionable framework for decision-making.

Fuzzy TOPSIS provides a systematic ranking of suppliers based on their resilience and adaptability. The model emphasizes selecting suppliers capable of handling disruptions and maintaining supply chain stability. The validated model offers a reliable tool for organizations to optimize their supplier selection process. This approach ensures that suppliers are evaluated not only on traditional metrics but also on their ability to navigate challenges and contribute to a resilient supply chain.

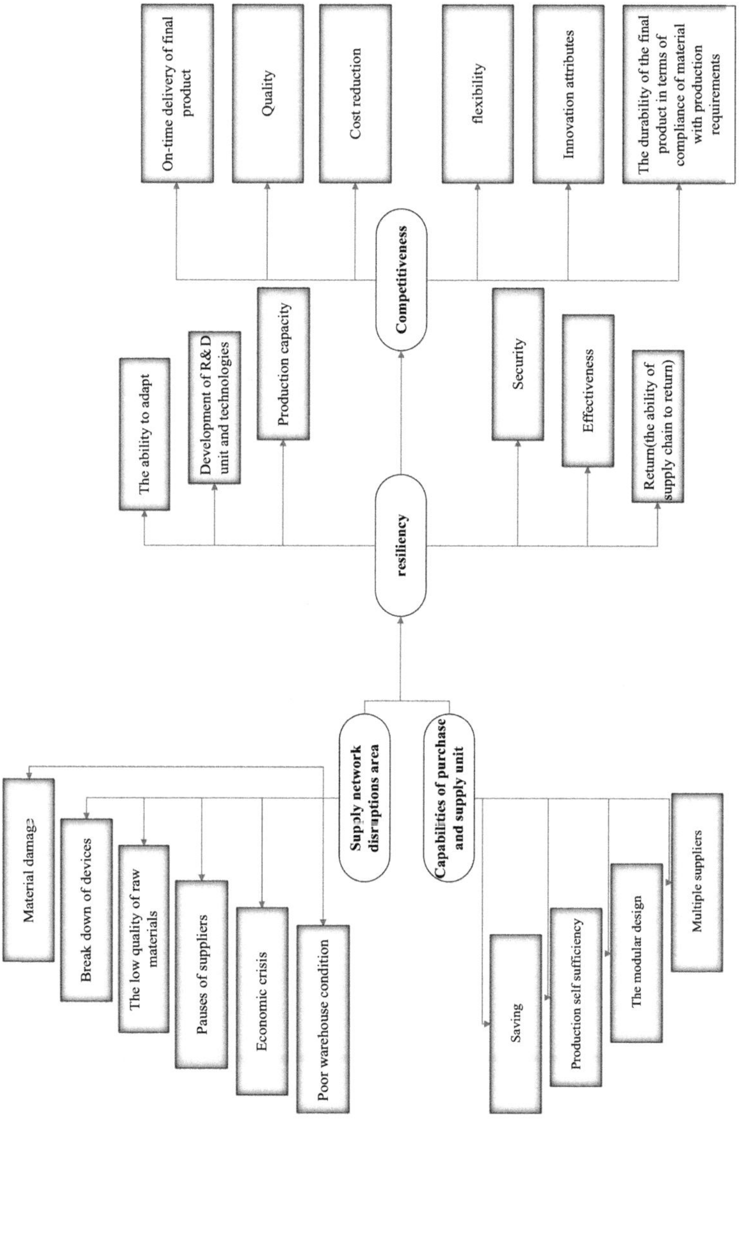

Fig. 4.3 The final model of resilient supply chain with regard to the disruption areas, for the competitiveness of Maziyar Manufacturing Industry

4.7 Conclusions

To foster long-term relationships with suppliers and address disruptions, supplier selection has emerged as a critical component of supply chain management. This study focuses on the resilient supply chain framework, using the steel supply chain as a case study. Four alternative suppliers were evaluated using Fuzzy Multi-Objective Linear Programming (FMOLP) for supplier selection, and their priorities were determined through the fuzzy TOPSIS method. The study identifies vulnerable areas in the supply network and selects suppliers based on five primary criteria: quality, flexibility, delivery time, cost, conformity of raw materials with production requirements. These criteria are chosen to ensure suppliers can adapt to requirements and disruptions. Ignoring the physical and chemical properties of materials can lead to production breakdowns, material wastage, and increased total costs due to factors like warehouse conditions and other stable contributors in manufacturing.

Given the nature of corporate products and materials (e.g., iron, steel), the proposed order preference and criteria classification can serve as a suitable model for other manufacturing plants, particularly in the steel industry. This is especially relevant as few studies have explored the conformity of material properties with production requirements.

To address the multi-criteria supplier selection problem, it is essential to monitor disruptions and their impacts. This study employs the FAHP-FMOLP model as a tool to solve the problem, followed by the fuzzy TOPSIS method to determine supplier priorities. Given the rapid changes, developments, and complex relationships within the supply chain, it is crucial to periodically revise and update the supply chain framework. Manufacturing units are advised to establish a dedicated workgroup composed of experts and professionals. By enhancing resiliency potential, enterprises can better react and adapt to changes in the business environment, ensuring sustained success and operational efficiency.

References and Further Reading

Fazlollahtabar, H. (2008). Applying multiple-criteria decision making methods for developing information technology industry. *International Journal of Information and Decision Sciences, 1*(1), 115–131.

Fazlollahtabar, H. (2022). Intelligent marketing decision model based on customer behavior using integrated possibility theory and K-means algorithm. *Journal of Intelligent Management Decision, 1*(2), 88–96.

Katiraee, N., Shirazi, B., & Fazlollahtabar, H. (2017). Fuzzy multi-objective supplier selection considering production requirements in resilient supply chain: Case study in steel industry. *International Journal of Information Systems and Supply Chain Management (IJISSCM), 10*(3), 65–83.

Kazemirash, N., & Fazlollahtabar, H. (2022). A multidimensional decision making for supplier selection in the presence of information systems. *Croatian Operational Research Review, 13*(1), 31–47.

Mahdavi, I., Fazlollahtabar, H., Heidarzade, A., Mahdavi-Am, N., & Rooshan, Y. I. (2008a). A heuristic methodology for multi-criteria evaluation of web-based e-learning systems based on user satisfaction. *Journal of Applied Sciences, 8*(24), 4603–4609.

Mahdavi, I., Fazlollahtabar, H., Mozaffari, E., Heidari, M., & Mahdavi-Amiri, N. (2008b). Data envelopment analysis based comparison of two hybrid multi-criteria decision-making approaches for mobile phone selection: A case study in Iranian telecommunication environment. *International Journal of Information and Decision Sciences, 1*(2), 194–220.

Saaty, T. L. (1980). *The analytic hierarchy process*. McGraw-Hill.

Scholten, K., Scott, P. S., & Fynes, B. (2014). Mitigation processes—Antecedents for building supply chain resilience. *Supply Chain Management, 19*(2), 211–228.

Singh, A. (2014). Supplier evaluation and demand allocation among suppliers in a supply chain. *Journal of Purchasing and Supply Management, 20*(3), 167–176.

Sonmez, M. (2006). A review and critique of supplier selection process and practices. Loughborough University Business School occasional papers series, No. 2006:1.

Zimmermann, H. J. (1978). Fuzzy programming and linear programming with several objective functions. *Fuzzy Sets and Systems, 1*, 45–55.

Chapter 5
Fuzzy Multi-objective Mathematical Programming Using Ranking Method

Abstract Supply chain design has garnered significant attention in recent years, as integrated management of the supply chain can mitigate the spread of disruptions across the network and substantially influence the profitability of its stakeholders. Uncertainties, particularly in demand and associated costs, are often inherent in such systems. Traditional approaches typically model these uncertainties as random variables, relying on probability theory for analysis. In this study, we introduce a fuzzy mathematical programming framework tailored for a supply chain that incorporates multiple depots, vehicles, products, customers, and time periods. Unlike conventional models, this approach treats not only demand and costs but also decision variables as fuzzy entities. Two ranking functions are employed to solve the model, which aims to optimize depot selection from potential candidates, allocate orders to depots and vehicles, and assign returning vehicles to depots, all with the objective of minimizing total costs. The model's validity is demonstrated through numerical experiments and sensitivity analysis. Additionally, a regression model is utilized to evaluate the effectiveness of the applied fuzzy ranking methods.

Keywords Supply chain network · Fuzzy mathematical programming · Ranking functions

5.1 Introduction

Over the past decade, supply chain management (SCM) has become a focal point of research for operations research practitioners. In the context of an increasingly globalized and competitive market, collaboration among enterprises is essential to achieve shared objectives, such as reducing delivery delays, minimizing holding costs, and optimizing transportation expenses. A supply chain can be described as an interconnected network involving suppliers, manufacturers, wholesalers, distributors, retailers, and customers, through which materials and products are sourced,

H. Fazlollahtabar, *Uncertain Multi-objective Decision Making: Methods and Industrial Models*, https://doi.org/10.1007/978-981-95-6035-6_5

processed, and delivered to end consumers. As a result, a growing number of companies are adopting and refining SCM strategies to enhance overall operational efficiency. Effective SCM necessitates a shift from managing isolated functions to integrating activities into cohesive supply chain processes (Fazlollahtabar et al., 2011).

Given the high complexity and inherent uncertainties in industrial supply chains, traditional centralized decision-making systems often struggle to efficiently manage information flows and operational actions. Delays in decision-making can significantly extend process timelines, ultimately diminishing a company's competitive edge. To mitigate such delays, supply chain participants must respond swiftly and effectively. Consequently, a supply chain can be viewed as a logistics network composed of partially independent decision-makers. The essence of supply chain management lies in the coordination of decisions across this network (Fazlollahtabar et al., 2012).

In supply chain management, ordering and inventory decisions are among the most critical challenges faced by managers. Typically, orders are placed based on forecasted customer demand, often overlooking the uncertainties inherent in industrial environments. To address these complexities, mathematical programming models have emerged as powerful analytical tools for optimizing intricate decision-making processes, particularly in supply chain planning. Over time, a wide range of deterministic mathematical programming models have been developed to tackle the design of supply chain networks. Notable examples include the works of Goetschalckx et al. (2002) and Yan et al. (2003), which have significantly contributed to this field.

In most cases, critical supply chain design parameters—such as customer demands, prices, and resource capacities—are inherently uncertain. This uncertainty has made the design of supply chains a prominent area of research (Esmaeilpour et al., 2013). Over the years, numerous stochastic programming models have been developed to address strategic and tactical planning challenges. Notable examples include the works of Guillén et al. (2005), Min and Zhou (2002), and Landeghen and Vanmaele (2002). However, the assumption of precise probability distribution parameters has been increasingly scrutinized. These parameters are often statistically estimated based on historical demand data, yet demand is rarely static. When external factors, such as technological innovations or shifts in consumer preferences, undergo significant changes, the mean and variance of demand distributions may also shift. Additionally, accurately determining true parameter values is often unfeasible, particularly in scenarios with limited data, such as the demand for new products. To address these challenges, fuzzy variables—derived from expert experience—are increasingly used to model such uncertainties. In this context, random variables with imprecise parameters are referred to as random fuzzy variables (Akbari et al., 2013).

Random fuzzy programming refers to the optimization theory applied in environments where both randomness and fuzziness coexist. This approach enhances the flexibility of supply chain design plans, particularly when the parameters of coefficient distributions are uncertain. In practice, fuzzy parameter values are often derived from expert experience, allowing for a more adaptable and realistic representation of uncertainties. The number of fuzzy variables used can reflect varying conditions that influence the parameters of probability distributions. It's important to note that the concept of random fuzzy variables differs from the definition proposed by Nattier (2001). Despite its potential, research on programming and solving supply chain design problems within a random fuzzy environment remains limited, highlighting a significant gap in the field.

Here, we propose a comprehensive supply chain model that incorporates multiple depots, vehicles, products, customers, and time periods. The supplier receives orders and forwards them to depots handling multiple products. Given the inherent fluctuations in demand, uncertainty is a critical factor to address. To model this, we represent demands as fuzzy numbers. Additionally, the variability in consideration costs—due to diverse environmental and geographical conditions—is also treated as fuzzy numbers. These uncertainties extend to the output decision variables, which are similarly modeled as fuzzy numbers.

The model involves selecting a set of depots from a pool of candidates. Each depot evaluates its capacity level and decides whether to accept or reject the order. Based on the location of customers, depots determine the most suitable vehicles for delivery. Each vehicle has an associated traveling cost. Furthermore, after delivering orders to customers, the model includes an allocation mechanism for returning vehicles to depots, ensuring efficient resource utilization.

5.2 Problem Description

The problem addressed in this chapter focuses on servicing multiple customers through a single supplier. The supplier offers a variety of products, which are stored across different depots. The initial challenge lies in selecting the most suitable depots from a pool of candidate locations. Each depot utilizes distinct types of vehicles to fulfill customer orders, with all depots already stationed at their respective locations. The proposed model adopts a multi-echelon supply chain network comprising one supplier, multiple depots, and customers, while also incorporating multi-commodity flows with fuzzy demand to account for uncertainties. Each depot is equipped with a set of vehicles stationed at its location. Depots can store multiple products, enabling them to respond to diverse customer orders. Customer orders can be fulfilled by one or multiple depots simultaneously, depending on product availability and demand. Each selected vehicle is limited to transporting only one product at a time, ensuring efficient and focused delivery. In scenarios where a depot lacks specific vehicles during a particular period but needs to respond to an order, returning vehicles are reallocated to the depot. This mechanism enhances resource utilization and operational flexibility (Fazlollahtabar et al., 2013). The configuration of this supply chain network is visually represented in Fig. 5.1, providing a clear illustration of the relationships and flows between the supplier, depots, vehicles, and customers.

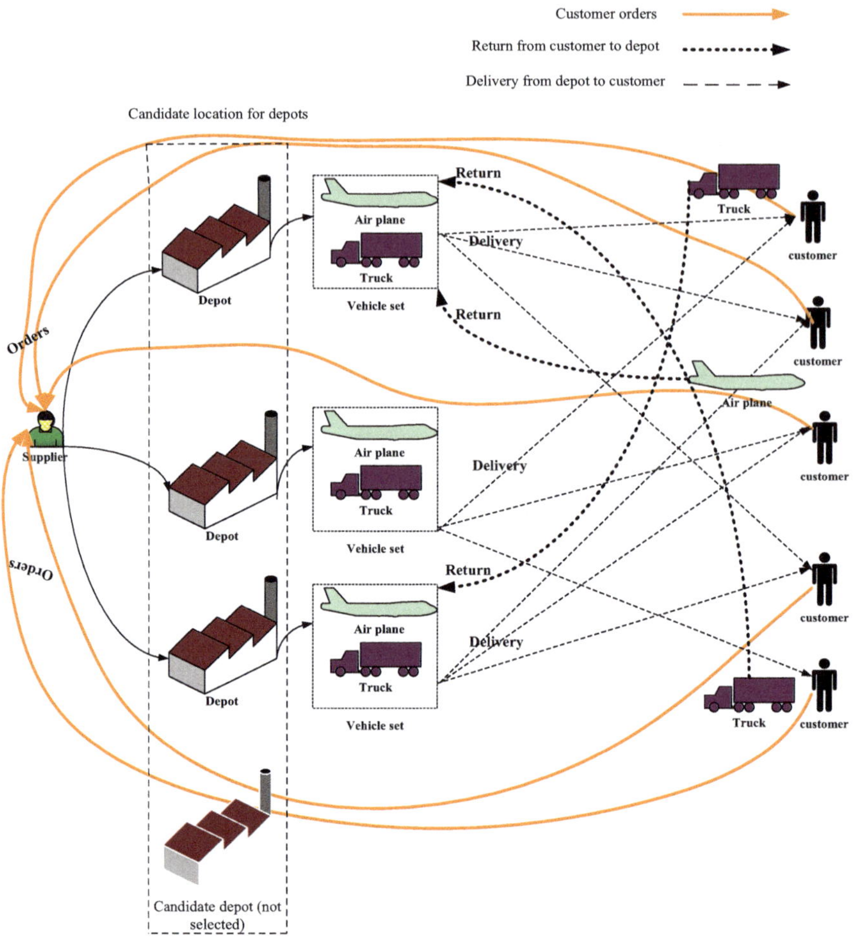

Fig. 5.1 A configuration of the proposed supply chain network

5.3 Fuzzy Mathematical Model

Here, we present a primal framework of the fuzzy variable linear programming (FVLP) to obtain the solution of the linear programming problem with fuzzy cost coefficients, Fazlollahtabar et al., (2013). Consider the FVLP problem:

$$\begin{aligned} \max\ & \tilde{z} \underset{R}{=} c\tilde{x} \\ \text{s.t.}\ & A\tilde{x} \underset{R}{\le} \tilde{b} \\ & \tilde{x} \underset{R}{\ge} \tilde{0}. \end{aligned}$$

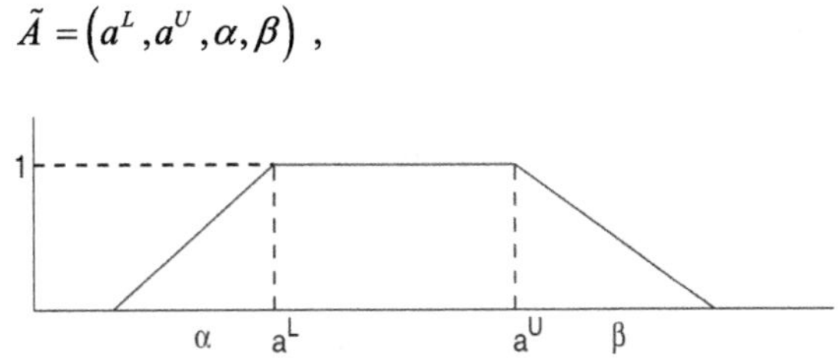

Fig. 5.2 Trapezoidal fuzzy number

The proposed fuzzy parameters are supposed as trapezoidal fuzzy numbers. A mathematical notation and a configuration of a trapezoidal fuzzy number are shown in Fig. 5.2.

$$\tilde{A} = \left(a^L, a^U, \alpha, \beta\right).$$

5.3.1 Ranking Function

In the context of solving fuzzy linear programming problems, various methods have been explored, including those reported by Maleki et al. (2000). One prominent approach involves the use of ranking functions to order and defuzzify fuzzy numbers, thereby transforming the problem into a more tractable form. In this study, we employ two ranking functions to defuzzify fuzzy numbers, as proposed by Fazlollahtabar et al. (2013). These ranking functions play a critical role in converting fuzzy parameters into crisp values, enabling the application of traditional linear programming techniques.

1. Bisector of area method (BOA): The proposed method in this category is reported by Yager (1981) which reduces to

$$R(\tilde{a}) = \frac{a^L + a^U}{2} + \frac{(\beta - \alpha)}{4}.$$

2. Mean of maximum method (MOM) which reduces to

$$R(\tilde{a}) = \frac{a^L + a^U}{2}.$$

Fuzzy Mathematical Programming

The fuzzy linear mathematical program is given below. In our proposed problem, demand is a fuzzy parameter, and also costs are assumed to be fuzzy value. The model is as follows:

Notations

P: Set of products.
I: Set of depots stationed.
J: Set of customers.
T: Set of time units.
V: Set of all vehicles.
$\tilde{D}_{jpt}$: The fuzzy demand of product p for customer j at time t.
ND_{ivt}: The number of existing vehicles v in depot i at time t.
NC_{jvt}: The number of existing vehicles v in customer j at the end of time t.
VL_{vp}: Capacity of vehicle v for product p.
$\tilde{C}T_{ijv}$: The fuzzy traveling cost per mile from depot i to customer j using vehicle v.
d_{ij}: Distance between depot i and customer j.
M: A large number.
$\tilde{C}h_{ip}$: The fuzzy holding cost for product p in depot i.
$\tilde{C}s_{ipt}$: The fuzzy supplying cost for product p in depot i at time t.
CO_i: The opening cost of depot i.
$\mathrm{cap\_}p_{ip}$: Maximum capacity of product p in depot i.

Decision Variables

x_{ijpvt}: {1, If depot i deliver product p to customer j using vehicle v at time t; 0; o.w}.
y_{jivt}: {1, if customer j deliver vehicle v to depot i at time t; 0; o.w}.
z_{ipt}: {1, if depot i receive product p at time t;0;o.w}.
w_{it}: {1, if depot i is active at time t; 0; o.w}.
TH_{ipt}: Amount of received product p in depot i at time t.
S_{ipt}: Amount of stored product p in depot i at the end of time t.
f_{ijpvt}: Frequency of traveling between depot i and customer j by vehicle v at time t.
$\tilde{Q}P_{ijpt}$: The fuzzy Quantity of product p can be satisfied by depot i to customer j at time t.
NR_{jivt}: Number of vehicle v delivered to depot i by customer j at time t.
DV_{ivt}: Demand of vehicle v for depot i at time t.

Objective Functions

$$\mathrm{Minimize}(F) = \mathrm{Min}(f_1 + f_2 + f_3 + f_4 + f_5)$$

$$f_1 = \sum_{i\in I}\sum_{j\in J}\sum_{p\in P}\sum_{v\in V}\sum_{t\in T} x_{ijpvt} \cdot f_{ijpvt} \cdot d_{ij} \cdot \tilde{C}T_{ijv}. \tag{5.1}$$

$$f_2 = \sum_{j\in J}\sum_{i\in I}\sum_{v\in V}\sum_{t\in T} y_{jivt} \cdot \mathrm{NR}_{jivt} \cdot d_{ij} \cdot \tilde{C}T_{ijv}. \tag{5.2}$$

$$f_3 = \sum_{i\in I}\sum_{p\in P}\sum_{t\in T} S_{ipt}\cdot \tilde{C}h_{ip}. \tag{5.3}$$

$$f_4 = \sum_{i\in I}\sum_{p\in P}\sum_{t\in T} \mathrm{TH}_{ipt}\cdot \tilde{C}s_{ipt}. \tag{5.4}$$

$$f_5 = \sum_{i\in I}\sum_{t\in T} (w_{it+1} - w_{it})\cdot \mathrm{CO}_i. \tag{5.5}$$

Constraints

$$\sum_{p\in P} z_{ipt} \le M\cdot w_{it}, \qquad \forall i\in I,\ \forall t\in T, \tag{5.6}$$

$$\sum_{p\in P} z_{ipt} \ge w_{it}, \qquad \forall i\in I,\ \forall t\in T, \tag{5.7}$$

$$\mathrm{TH}_{ipt} \le M\cdot z_{ipt}, \qquad \forall i\in I,\ \forall p\in P,\ \forall t\in T, \tag{5.8}$$

$$\mathrm{TH}_{ipt} \ge z_{ipt}, \qquad \forall i\in I,\ \forall p\in P,\ \forall t\in T, \tag{5.9}$$

$$w_{it+1} \ge w_{it}, \qquad \forall i\in I,\ \forall t\in T, \tag{5.10}$$

$$\sum_{i\in I} \mathrm{QP}_{ijpt} = D_{jpt}, \qquad \forall j\in J,\ \forall p\in P,\ \forall t\in T, \tag{5.11}$$

$$\sum_{j\in J} \mathrm{QP}_{ijpt} \le \mathrm{TH}_{ipt} + S_{ipt-1}, \qquad \forall i\in I,\ \forall p\in P,\ \forall t\in T, \tag{5.12}$$

$$S_{ipt-1} + \mathrm{TH}_{ipt} - \sum_{j\in J} \mathrm{QP}_{ijpt} = S_{ipt}, \qquad \forall i\in I,\ \forall p\in P,\ \forall t\in T, \tag{5.13}$$

$$\mathrm{TH}_{ipt} + S_{ipt-1} \le \mathrm{cap\_}p_{ip}, \qquad \forall i\in I,\ \forall p\in P,\ \forall t\in T, \tag{5.14}$$

$$\mathrm{QP}_{ijpt}\cdot\Big(1 - \sum_{v\in V} x_{ijpvt}\Big) = 0, \qquad \forall i\in I,\ \forall j\in J,\ \forall p\in P,\ \forall t\in T, \tag{5.15}$$

$$\mathrm{QP}_{ijpt} \ge \sum_{v\in V} x_{ijpvt}, \qquad \forall i\in I,\ \forall j\in J,\ \forall p\in P,\ \forall t\in T, \tag{5.16}$$

$$\left\lceil ((\mathrm{QP}_{ijpt} \div VL_{vp})\cdot x_{ijpvt}) + 0.999 \right\rceil = f_{ijpvt}, \qquad \forall i\in I,\ \forall j\in J,\ \forall p\in P,\ \forall v\in V,\ \forall t\in T, \tag{5.17}$$

$$\mathrm{ND}_{ivt-1} - \sum_{j \in J} \sum_{p \in P} f_{ijpvt} + \sum_{j \in J} \mathrm{NR}_{jivt} = \mathrm{ND}_{i,v,t}, \quad \forall i \in I, \quad \forall v \in V, \quad \forall t \in T, \tag{5.18}$$

$$\sum_{j \in J} \sum_{p \in P} f_{ijpvt} \leq \mathrm{ND}_{ivt-1}, \quad \forall i \in I, \quad \forall v \in V, \quad \forall t \in T, \tag{5.19}$$

$$DV_{ivt} \leq M \cdot w_{it}, \quad \forall i \in I, \quad \forall v \in V, \quad \forall t \in T, \tag{5.20}$$

$$\sum_{j \in J} \mathrm{NR}_{jivt} = \mathrm{DV}_{ivt}, \quad \forall i \in I, \quad \forall v \in V, \quad \forall t \in T, \tag{5.21}$$

$$\mathrm{NC}_{jvt-1} = \sum_{i \in I} \mathrm{NR}_{jivt}, \quad \forall j \in J, \quad \forall v \in V, \quad \forall t \in T, \tag{5.22}$$

$$\mathrm{NR}_{jivt} \leq M \cdot y_{jivt}, \quad \forall j \in J, \quad \forall i \in I, \quad \forall v \in V, \quad \forall t \in T, \tag{5.23}$$

$$\mathrm{NR}_{jivt} \geq y_{jivt}, \quad \forall j \in J, \quad \forall i \in I, \quad \forall v \in V, \quad \forall t \in T, \tag{5.24}$$

$$\mathrm{NC}_{jvt-1} + \sum_{i \in I} \sum_{p \in P} f_{ijpvt} - \sum_{i \in I} \mathrm{NR}_{jivt} = \mathrm{NC}_{jvt}, \quad \forall j \in J, \ \forall v \in V, \ \forall t \in T. \tag{5.25}$$

MOM constraints for fuzzy consideration:

$$\left\lceil \frac{\left(\tilde{D}_{jpta^l} + \tilde{D}_{jpta^u}\right)}{2} \right\rceil = D_{jpt}, \quad \forall j \in J, \quad \forall p \in P, \quad \forall t \in T, \tag{5.26}$$

$$\left\lceil \frac{\left(\tilde{C}T_{ijva^l} + \tilde{C}T_{ijva^u}\right)}{2} \right\rceil = \mathrm{CT}_{ijv}, \quad \forall i \in I, \quad \forall j \in J, \quad \forall v \in V, \tag{5.27}$$

$$\left\lceil \frac{\left(\tilde{C}s_{ipta^l} + \tilde{C}s_{ipta^u}\right)}{2} \right\rceil = \mathrm{Cs}_{ipt}, \quad \forall i \in I, \quad \forall p \in P, \quad \forall t \in T, \tag{5.28}$$

$$\left\lceil \frac{\left(\tilde{C}h_{ipa^l} + \tilde{C}h_{ipa^u}\right)}{2} \right\rceil = \mathrm{Ch}_{ip}, \quad \forall i \in I, \quad \forall p \in P, \tag{5.29}$$

$$\left\lceil \frac{\left(\tilde{Q}P_{ijpta^l} + \tilde{Q}P_{ijpta^u}\right)}{2} \right\rceil = \mathrm{QP}_{ijpt}, \quad \forall i \in I, \quad \forall j \in J, \quad \forall p \in P, \quad \forall t \in T, \tag{5.30}$$

$$\tilde{Q}P_{ijpta^l} \leq \tilde{Q}P_{ijpta^u}, \quad \forall i \in I, \quad \forall j \in J, \quad \forall p \in P, \quad \forall t \in T. \tag{5.31}$$

Yager constraints for fuzzy consideration:

$$\left\lceil \frac{\left(\tilde{D}_{jpta^l} + \tilde{D}_{jpta^u}\right)}{2} + \frac{\left(\tilde{D}_{jpt\beta} - \tilde{D}_{jpt\alpha}\right)}{4} \right\rceil = D_{jpt}, \quad \forall j \in J, \quad \forall p \in P, \quad \forall t \in T, \tag{5.32}$$

$$\left\lceil \frac{\left(\tilde{C}T_{ijva^l} + \tilde{C}T_{ijva^u}\right)}{2} + \frac{\left(\tilde{C}T_{ijv\beta} - \tilde{C}T_{ijv\alpha}\right)}{4} \right\rceil = \mathrm{CT}_{ijv}, \quad \forall i \in I, \quad \forall j \in J, \quad \forall v \in V, \tag{5.33}$$

$$\left\lceil \frac{\left(\tilde{C}s_{ipta^l} + \tilde{C}s_{ipta^u}\right)}{2} + \frac{\left(\tilde{C}s_{ipt\beta} - \tilde{C}s_{ipt\alpha}\right)}{4} \right\rceil = \mathrm{Cs}_{ipt}, \quad \forall i \in I, \quad \forall p \in P, \quad \forall t \in T, \tag{5.34}$$

$$\left\lceil \frac{\left(\tilde{C}h_{ipa^l} + \tilde{C}h_{ipa^u}\right)}{2} + \frac{\left(\tilde{C}h_{ip\beta} - \tilde{C}h_{ip\alpha}\right)}{4} \right\rceil = \mathrm{Ch}_{ip}, \quad \forall i \in I, \quad \forall p \in P, \tag{5.35}$$

$$\left\lceil \frac{\left(\tilde{Q}P_{ijpta^l} + \tilde{Q}P_{ijpta^u}\right)}{2} + \frac{\left(\tilde{Q}P_{ijpt\beta} - \tilde{Q}P_{ijpt\alpha}\right)}{4} \right\rceil = \mathrm{QP}_{ijpt}, \quad \forall i \in I, \quad \forall j \in J, \quad \forall p \in P, \quad \forall t \in T, \tag{5.36}$$

$$\tilde{Q}P_{ijpta^l} \le \tilde{Q}P_{ijpta^u}, \quad \forall i \in I, \quad \forall j \in J, \quad \forall p \in P, \quad \forall t \in T. \tag{5.37}$$

Integrity and non-negativity constraints:

$$x_{ijpvt} \in \{0, 1\}, \quad \forall i \in I, \ \forall j \in J, \ \forall p \in P, \ \forall v \in V, \ \forall t \in T, \tag{5.38}$$

$$y_{jivt} \in \{0, 1\}, \quad \forall j \in J, \ \forall i \in I, \ \forall v \in V, \ \forall t \in T, \tag{5.39}$$

$$z_{ipt} \in \{0, 1\}, \quad \forall i \in I, \ \forall p \in P, \ \forall t \in T, \tag{5.40}$$

$$w_{it} \in \{0, 1\}, \quad \forall i \in I, \ \forall t \in T, \tag{5.41}$$

$$f_{ijpvt} \ge 0, \quad \forall i \in I, \ \forall j \in J, \ \forall p \in P, \ \forall v \in V, \ \forall t \in T, \tag{5.42}$$

$$\tilde{Q}P_{ijpt}, \quad \text{Integer}, \quad \forall i \in I, \ \forall j \in J, \ \forall p \in P, \ \forall t \in T, \tag{5.43}$$

$$\mathrm{NR}_{jivt}, \quad \text{Integer}, \quad \forall j \in J, \ \forall i \in I, \ \forall v \in V, \ \forall t \in T, \tag{5.44}$$

$$\mathrm{TH}_{ipt} \geq 0, \quad i \in I, \quad \forall p \in P, \quad \forall t \in T, \tag{5.45}$$

$$\mathrm{DV}_{ivt} \geq 0, \quad \forall i \in I, \quad \forall v \in V, \quad \forall t \in T. \tag{5.46}$$

The problem described involves a multi-objective optimization model for a supply chain network, focusing on minimizing various costs while ensuring operational constraints are satisfied. Objective functions are total cost minimization including forward distance cost minimized in Eq. (5.1) and backward distance cost minimized in Eq. (5.2). Storage cost is minimized in Eq. (5.3). Supply cost is minimized in Eq. (5.4). Depot opening cost is minimized in Eq. (5.5).

Constraints (5.6) and (5.7) ensure depots can only be supplied if activated. Constraints (5.8) and (5.9) ensure non-negative product allocation to depots. Constraint (5.10) prevents depots from changing their status more than once. Constraint (5.11) guarantees all customer demands are met for all products and periods.

Constraint (5.12) shows flow conservation at depots. Constraint (5.13) is amount of stored product at the end of a period. Constraint (5.14) is capacity restriction for depots. Constraints (5.15) and (5.16) ensure delivery is accomplished by only one vehicle. Constraint (5.17) shows frequency of traveling between depots and customers. Constraint (5.18) is number of remaining vehicles at the end of a period. Constraint (5.19) limits the frequency of traveled vehicles to the number of stationed vehicles. Constraint (5.20) allows activated depots to order vehicles. Constraint (5.21) ensures all depot vehicle demands are met. Constraint (5.22) flow balance of stationed vehicles at the end of a period. Constraints (5.23) and (5.24) ensure vehicle delivery from customer to depot is accomplished if the corresponding path is selected.

Constraint (5.25): Number of remaining vehicles stationed at the customer at the end of a period. Equations (5.26)–(5.30) defuzzify demand, traveling cost, supplying cost, holding cost, and product quantity using the Mean of Maximum (MOM) ranking function. Equations (5.32)–(5.36) defuzzify the same variables using the Bisector of Area (BOA) ranking function proposed by Yager. Constraints (5.31) and (5.37) ensure the first fuzzy number of the decision variable Q_P (quantity of product) in trapezoidal form is always less than or equal to its second fuzzy number. Constraints (5.38)–(5.41) restrict Q_P to binary values. Constraints (5.42)–(5.46) ensure all other variables take non-negative values.

The model integrates fuzzy logic to handle uncertainties in demand, costs, and product quantities, using MOM and BOA ranking functions for defuzzification. It ensures operational efficiency by minimizing costs while adhering to constraints related to depot activation, product allocation, vehicle allocation, and customer demand fulfillment. The use of binary and non-negative variables ensures the model's feasibility and alignment with real-world supply chain scenarios.

5.4 Numerical Experiments

To demonstrate the effectiveness of the proposed fuzzy mathematical model, we present a numerical example with the following setup:

Customers: 3, Products: 3, Candidate Depots: 7, Vehicles: 2, Main Periods: 3 (with demand).
Additional Period: 1 (Period 4, with no demand, to account for vehicle returns).

Fuzzy orders and principal orders for different customers across time periods ($\tilde{D}_{jpt}$) are provided in Table 5.1.

Ranking functions are used to defuzzify fuzzy numbers. Results are rounded to the nearest integer to align with real-world scenarios. The proposed fuzzy mathematical model incorporates uncertainties in customer orders and operational parameters. Uses ranking functions to convert fuzzy numbers into crisp values for practical application. Period 4 is introduced to handle vehicle returns, ensuring the model accounts for all operational phases. Decimal results from defuzzification are rounded to integers, reflecting real-life constraints and practices. The steps of implementation follow here. Extract fuzzy and principal orders from Table 5.1. Define parameters for depots, vehicles, and products. Apply ranking functions to defuzzify fuzzy numbers. Round the results to integers for practical use. Solve the fuzzy mathematical model using the prepared data. Analyze the results to evaluate the model's effectiveness.

Expected outcomes are minimized total costs for forward and backward distances, storage, supply, and depot opening. The operational efficiency of the model ensures satisfied customer demands across all periods and efficient allocation of vehicles and products to depots.

Table 5.1 The fuzzy orders in the first time period for different customers

	First period			
	Principle	Order	Yager	MOM
Customer 1				
Product 1	40	(38, 42, 2, 5)	41	40
Product 2	45	(42, 45, 3, 3)	44	44
Product 3	60	(50, 58, 4, 5)	55	54
Customer 2				
Product 1	70	(65, 68, 1, 3)	67	67
Product 2	30	(28, 29, 2, 2)	29	29
Product 3	50	(45, 56, 5, 3)	50	51
Customer 3				
Product 1	0	(0, 0, 0, 0)	0	0
Product 2	20	(18, 22, 4, 2)	20	20
Product 3	30	(27, 31, 2, 4)	30	29

Table 5.2 The distance between customers and depots

Distance	Customer 1	Customer 2	Customer 3
Depot 1	20	25	10
Depot 2	10	15	17
Depot 3	14	12	13
Depot 4	10	15	12
Depot 5	16	22	24
Depot 6	13	16	20
Depot 7	14	15	16

This numerical example highlights the effectiveness of the proposed fuzzy mathematical model in addressing real-world supply chain challenges. By incorporating fuzzy logic and ranking functions, the model provides a robust framework for optimizing operations under uncertainty.

The distance between customers and depots (d_{ij}), capacity of vehicles (VL_{vp}), and capacity of depots (cap_p_{ip}) with respect to products are reported in Tables 5.2, 5.3 and 5.4.

The transferring fuzzy cost per unit of distance for vehicles in depots ($\tilde{C}T_{ijv}$) is presented in Table 5.5.

The supplying fuzzy cost for products in depots at each period ($\tilde{C}s_{ipt}$) is presented in Table 5.6.

The holding fuzzy cost for products in depots ($\tilde{C}h_{ip}$) is presented in Table 5.7.

To facilitate the computations and present the results effectively, the LINGO 9.0 package is utilized. Table 5.8 shows forward flow decision variables (Yager method). This table displays the forward flow of products between depots and customers, as determined by the Yager ranking method. The values represent the optimized

Table 5.3 The capacity of vehicles

Capacity of vehicle	Product 1	Product 2	Product 3
Vehicle 1	30	50	20
Vehicle 2	10	15	8

Table 5.4 The capacity of depots

Depot capacity	Product 1	Product 2	Product 3
Depot 1	100	85	90
Depot 2	90	80	70
Depot 3	80	75	70
Depot 4	90	100	70
Depot 5	85	65	75
Depot 6	80	70	60
Depot 7	100	70	80

Table 5.5 The transferring fuzzy cost per unit of distance for vehicles in the first depot

Transferring cost per unit of distance	Depot 1							
	Vehicle 1	Yager	MOM	Principle	Vehicle 2	Yager	MOM	Principle
Customer 1	(45, 55, 4, 6)	51	50	50	(25, 32, 3, 6)	30	29	30
Customer 2	(48, 52, 3, 6)	51	50	50	(28, 31, 3, 4)	30	30	30
Customer 3	(49, 57, 4, 5)	54	53	50	(27, 36, 4, 6)	32	32	30

flow of products to meet customer demands while minimizing costs. The Yager method ensures efficient defuzzification, converting fuzzy numbers into crisp values for practical decision-making.

Table 5.9 shows received product (TH) from supplier. This table shows the amount of product received from the supplier at each depot during each period. The values reflect the optimized supply quantities to meet depot requirements ensuring that depots are adequately stocked to fulfill customer demands without overstocking.

Table 5.10 shows storage (S) of products. This table represents the amount of product stored in each depot at the end of each period. The values indicate the stored status of products, with the numbers representing the storage quantities helping to monitor inventory levels and ensure that storage capacities are not exceeded. The "halts" status indicates periods where no storage changes occur, maintaining the described status.

The use of LINGO 9.0 ensures accurate and efficient computations for the fuzzy mathematical model. The results demonstrate optimized product flow, supply, and storage, aligning with the model's objectives. The Yager method and defuzzification process provide practical, real-world solutions by converting fuzzy data into actionable insights.

Whereas activation of depot necessitates returning of vehicles, with a tradeoff between corresponding costs, the model has made decision to activate new depots (1 and 3) at the first of period four. The output of forward flow for the decision variables of principle method is presented in Table 5.11.

The output of forward flow for the decision variables of MOM method is presented in Table 5.12.

The best objectives of three methods are given in Table 5.13.

The objective values of the two fuzzy ranking methods are very close to each other. Therefore, to verify the differences, a sensitivity analysis is suggested.

Table 5.6 The supplying fuzzy cost for products in depots at the first period

Supplying cost	First period											
	Product 1				Product 2				Product 3			
	Fuzzy number	Yager	MOM	Principle	Fuzzy number	Yager	MOM	Principle	Fuzzy number	Yager	MOM	Principle
Depot 1	(10, 17, 2, 2)	14	14	15	(8, 9, 2, 1)	9	9	8	(9, 16, 3, 5)	13	13	10
Depot 2	(8, 14, 3, 6)	12	11	8	(7, 8, 2, 3)	8	8	7	(9, 16, 5, 2)	12	13	15
Depot 3	(7, 14, 5, 3)	10	11	12	(7, 11, 5, 2)	9	9	10	(15, 19, 2, 2)	17	17	15
Depot 4	(10, 16, 3, 4)	14	13	14	(13, 16, 3, 5)	15	15	15	(8, 10, 3, 3)	9	9	9
Depot 5	(12, 16, 1, 4)	15	14	13	(6, 10, 2, 3)	9	8	7	(9, 16, 4, 5)	13	13	16
Depot 6	(9, 16, 3, 6)	14	13	15	(10, 15, 3, 5)	13	13	12	(13, 17, 2, 6)	16	15	13
Depot 7	(5, 8, 2, 2)	7	7	5	(10, 14, 3, 4)	13	12	10	(7, 15, 4, 3)	11	11	7

Table 5.7 The holding fuzzy cost for products in depots

Holding cost	Product 1				Product 2				Product 3			
	Fuzzy number	Yager	MOM	Principle	Fuzzy number	Yager	MOM	Principle	Fuzzy number	Yager	MOM	Principle
Depot 1	(10, 15, 2, 3)	13	13	12	(6, 9, 2, 2)	8	8	6	(11, 16, 3, 3)	14	14	11
Depot 2	(10, 14, 4, 6)	13	12	13	(3, 8, 2, 3)	6	6	4	(9, 11, 5, 3)	10	10	10
Depot 3	(9, 14, 3, 4)	12	12	13	(7, 9, 5, 2)	8	8	7	(13, 19, 2, 2)	16	16	14
Depot 4	(10, 16, 3, 4)	14	13	12	(10, 11, 3, 2)	11	11	10	(8, 10, 3, 3)	9	9	9
Depot 5	(12, 15, 1, 2)	14	14	13	(6, 9, 2, 3)	8	8	7	(11, 16, 4, 5)	14	14	15
Depot 6	(9, 14, 3, 4)	12	12	12	(12, 15, 3, 5)	14	14	14	(11, 17, 5, 6)	15	14	16
Depot 7	(4, 8, 2, 2)	6	6	7	(10, 14, 3, 4)	13	12	13	(7, 10, 4, 3)	9	9	7

Table 5.8 The forward path output using Yager method

X	Depot	Customer	Product	Vehicle	Period	f	$\tilde{Q}P$	QP
1	2	2	1	1	1	3	(53.8, 53.8, 1, 53.8)	67
1	2	2	2	1	1	1	(1, 57, 1, 1)	29
1	2	2	3	1	1	3	(40.2, 40.2, 1, 40.2)	50
1	2	3	3	1	1	1	(1, 9, 1, 1)	5
1	7	1	1	1	1	2	(41, 41, 1, 1)	41
1	7	1	2	2	1	3	(35.4, 35.4, 1, 35.4)	44
1	7	1	3	1	1	3	(1, 109, 1, 1)	55
1	7	3	2	2	1	2	(16.2, 16.2, 1, 16.2)	20
1	7	3	3	1	1	2	(1, 49, 1, 1)	25
1	2	1	1	1	2	1	(21, 21, 1, 1)	21
1	2	1	3	1	2	1	(15.4, 15.4, 1, 15.4)	19
1	2	2	3	1	2	1	(11.4, 11.4, 1, 11.4)	14
1	2	3	2	2	2	2	(13.8, 13.8, 1, 13.8)	17
1	2	3	3	2	2	2	(1, 29, 1, 1)	15
1	7	3	1	1	2	1	(11.4, 11.4, 1, 11.4)	14
1	2	1	1	2	3	4	(1, 61, 1, 1)	31
1	2	1	2	2	3	2	(1, 51, 1, 1)	26
1	2	2	1	2	3	1	(1, 9, 1, 1)	5
1	2	2	2	1	3	1	(17.8, 17.8, 1, 17.8)	22
1	2	3	1	1	3	1	(1, 47, 1, 1)	24
1	2	3	2	1	3	1	(1, 49, 1, 1)	25
1	7	1	3	2	3	3	(15.4, 15.4, 1, 15.4)	19
1	7	2	1	2	3	1	(13, 13, 13, 1)	10
1	7	2	3	2	3	3	(14.6, 14.6, 1, 14.6)	18
1	7	3	3	2	3	3	(16.2, 16.2, 1, 16.2)	20

Table 5.9 The amount of received product in each depot

TH	Product 1	Product 2	Product 3
First period			
Depot 2	67	29	55
Depot 7	41	64	80
Second period			
Depot 2	21	17	48
Depot 7	14	1	0
Third period			
Depot 2	60	73	0
Depot 7	10	0	57

Table 5.10 The amount of stored product in each depot

Legend	
┈┈▶	Continue
Halt ┈┈▶┘	

Depot / product	Period 1	Period 2	Period 3	Period 4
Depot 1				
Depot 2				
Depot 3				
Depot 4				
Depot 5				
Depot 6				
Depot 7 / 2		1 ┈┈▶	1 ┈┈▶	1 ┈┈▶┘

5.5 Conclusions

To address the complexity of real-life environments and existing constraints in supply chain management (SCM), we propose a fuzzy mathematical program as a more effective alternative to deterministic models. Deterministic mathematical programs often fail to capture uncertainties and constraints in SCM, such as fluctuating demand, supply variability, and operational limitations. A fuzzy mathematical program is introduced to better model these uncertainties and provide more flexible and practical solutions. Incorporates fuzzy logic to handle uncertainties in parameters such

Table 5.11 The forward path output for principal method

X	Depot	Customer	Product	Vehicle	Period	*f*	QP
1	2	1	2	1	**1**	1	45
1	2	1	3	1	**1**	3	60
1	2	3	2	2	**1**	1	15
1	2	3	3	2	**1**	1	6
1	3	2	1	1	**1**	3	70
1	3	2	2	1	**1**	1	30
1	3	2	3	1	**1**	3	50
1	4	1	1	1	**1**	2	40
1	4	3	2	2	**1**	1	5
1	4	3	3	2	**1**	3	24
1	2	1	3	1	2	1	18
1	2	3	3	2	2	1	1
1	4	1	1	2	2	2	19
1	4	2	3	1	2	1	13
1	4	3	1	2	2	1	3
1	4	3	2	2	2	1	15
1	4	3	3	2	2	2	16
1	7	3	1	2	2	1	10
1	2	2	1	2	**3**	1	6
1	2	3	1	2	**3**	3	26
1	2	3	2	2	**3**	2	25
1	2	3	3	2	**3**	3	20
1	4	1	1	1	**3**	1	30
1	4	1	2	2	**3**	1	10
1	4	1	3	1	**3**	1	17
1	4	2	1	2	**3**	1	10
1	5	1	2	2	**3**	1	15
1	5	2	2	2	**3**	1	15
1	7	2	2	2	**3**	1	5
1	7	2	3	1	**3**	1	18

as demand, supply, and costs. Uses ranking functions to defuzzify fuzzy numbers and convert them into crisp values for decision-making. The solutions of the fuzzy program are compared with those of deterministic models to highlight its advantages. The fuzzy mathematical program demonstrates greater flexibility in adapting to real-life uncertainties and constraints. Compared to deterministic models, the fuzzy program provides more robust and practical solutions. Sensitivity analysis reveals the model's resilience to changes in key decision variables. The fuzzy mathematical

Table 5.12 The forward path output using MOM method

X	Depot	Customer	Product	Vehicle	Period	f	$\tilde{Q}P$	QP
1	1	1	2	2	**1**	1	(14, 14, 1, 1)	14
1	1	2	3	1	**1**	3	(51, 51, 1, 1)	51
1	1	3	3	2	**1**	1	(8, 8, 1, 1)	8
1	2	1	2	2	**1**	2	(1, 59, 1, 1)	30
1	2	1	3	2	**1**	5	(1, 79, 1, 1)	40
1	2	2	1	1	**1**	3	(1, 133, 1, 1)	67
1	2	2	2	1	**1**	1	(1, 57, 1, 1)	29
1	2	3	2	2	**1**	2	(20, 20, 1, 1)	20
1	2	3	3	2	**1**	1	(8, 8, 1, 1)	8
1	3	1	1	2	**1**	1	(10, 10, 1, 1)	10
1	3	1	3	2	**1**	2	(14, 14, 1, 1)	14
1	3	3	3	2	**1**	1	(1, 1, 1, 1)	1
1	6	1	1	2	**1**	3	(1, 59, 1, 1)	30
1	6	3	3	1	**1**	1	(1, 23, 1, 1)	12
1	1	1	3	1	*2*	1	(11, 11, 1, 1)	11
1	1	3	1	2	*2*	1	(1, 7, 1, 1)	4
1	2	1	1	1	*2*	1	(1, 39, 1, 1)	20
1	2	1	3	2	*2*	1	(1, 15, 1, 1)	8
1	2	3	1	2	*2*	1	(1, 19, 1, 1)	10
1	3	2	3	2	*2*	1	(1, 11, 1, 1)	6
1	3	3	2	2	*2*	1	(1, 29, 1, 1)	15
1	6	1	1	2	*2*	1	(1, 1, 1, 1)	1
1	6	2	3	2	*2*	1	(8, 8, 1, 1)	8
1	6	3	2	1	*2*	1	(1, 3, 1, 1)	2
1	6	3	3	1	*2*	1	(1, 27, 1, 1)	14
1	2	1	1	1	**3**	1	(1, 59, 1, 1)	30
1	2	1	2	1	**3**	1	(1, 51, 1, 1)	26
1	2	1	3	1	**3**	1	(1, 35, 1, 1)	18
1	2	2	2	2	**3**	1	(1, 29, 1, 1)	15
1	2	2	3	2	**3**	1	(1, 15, 1, 1)	8
1	2	3	1	2	**3**	1	(10, 10, 1, 1)	10
1	2	3	2	1	**3**	1	(1, 47, 1, 1)	24
1	3	2	1	1	**3**	1	(1, 25, 1, 1)	13
1	3	2	2	2	**3**	1	(1, 11, 1, 1)	6
1	3	3	2	2	**3**	1	(1, 1, 1, 1)	1
1	3	3	3	1	**3**	1	(1, 39, 1, 1)	20
1	6	2	1	2	**3**	1	(1, 1, 1, 1)	1

(continued)

Table 5.12 (continued)

X	Depot	Customer	Product	Vehicle	Period	f	$\tilde{Q}P$	QP
1	6	2	3	2	**3**	2	(1, 17, 1, 1)	9
1	6	3	1	1	**3**	1	(1, 27, 1, 1)	14

Table 5.13 The best objectives of three methods

	Principle	Yager	MOM
Best objective	**61,081**	**67,924**	**67,853**

program offers a more realistic representation of SCM challenges. The model's flexibility and robustness make it suitable for real-world SCM implementations. The use of ranking functions and sensitivity analysis provides a comprehensive framework for solving fuzzy optimization problems. The proposed fuzzy mathematical program addresses the limitations of deterministic models in SCM by incorporating uncertainties and constraints. Its flexibility, improved solutions, and validated ranking methods make it a valuable tool for optimizing supply chain operations in real-life environments.

References and Further Reading

Akbari, F., Fazlollahtabar, H., & Mahdavi, I. (2013). An uncertain decision making process considering customers and services in evaluating banks: A case study. *International Journal of Strategic Decision Sciences (IJSDS), 4*(2), 48–78.

Esmaeilpour, R., Fazlollahtabar, H., & Aghasi, E. (2013). Multi-objective mathematical model as a decision support for customer service marketing. *International Journal of Applied Operational Research, 3*(2), 27–41.

Fazlollahtabar, H., Aghasi, E., & Forte, P. (2012). Bi-objective two-stage decision-making process for service marketing. *International Journal of Strategic Decision Sciences (IJSDS), 3*(3), 24–39.

Fazlollahtabar, H., Mahdavi, I., Ashoori, M. T., Kaviani, S., & Mahdavi-Amiri, N. (2011). A multi-objective decision-making process of supplier selection and order allocation for multi-period scheduling in an electronic market. *The International Journal of Advanced Manufacturing Technology, 52*, 1039–1052.

Fazlollahtabar, H., Mahdavi, I., & Mohajeri, A. (2013). Applying fuzzy mathematical programming approach to optimize a multiple supply network in uncertain condition with comparative analysis. *Applied Soft Computing, 13*(1), 550–562.

Goetschalckx, M., Vidal, C. J., & Dogan, K. (2002). Modeling and design of global logistics systems: A review of integrated strategic and tactical models and design algorithms. *European Journal of Operational Research, 143*, 1–18.

Guillén, G., Mele, F. D., Bagajewicz, M. J., Espuña, A., & Puigjaner, L. (2005). Multi objective supply chain design under uncertainty. *Chemical Engineering Science, 60*, 1535–1553.

Landeghen, H. V., & Vanmaele, H. (2002). Robust planning: A new paradigm for demand chain planning. *Journal of Operations Management, 20*, 769–783.

Maleki, H. R., Tata, M., & Mashinchi, M. (2000). Linear programming with fuzzy variables. *Fuzzy Sets and Systems, 109*, 21–33.

Min, H., & Zhou, G. (2002). Supply chain modeling: Past, present and future. *Computers and Industrial Engineering, 43*, 231–249.

Nattier, W. (2001). Random fuzzy variables of second order and applications to statistical inference. *Information Sciences, 133*, 69–88.

Yager, R. R. (1981). A procedure for ordering fuzzy subsets of the unit interval. *Information Science, 24*, 143–161.

Yan, H., Yu, Z., & Cheng, T. C. E. (2003). A strategic model for supply chain design with logical constraints: Formulation and solution. *Computers and Operations Research, 30*, 2135–2155.

Chapter 6
Vague Multi-objective Optimization by Branch and Bound Method

Abstract Modern production systems require sophisticated scheduling approaches that simultaneously address machine-task allocation and production planning while accounting for real-world uncertainties. Flexible manufacturing scheduling plays a pivotal role in optimizing production efficiency through shortest processing time sequencing and deadlock prevention. Recognizing the critical impact of uncertainty in manufacturing systems, this study adopts a probabilistic vague set approach, specifically utilizing trapezoidal fuzzy numbers to model process variability through distance-based vague sets. To solve this complex scheduling problem, we develop a hybrid optimization methodology combining an enhanced branch and bound algorithm with genetic algorithm components, validated through robust ranking techniques. The proposed framework is implemented and tested through practical case studies that demonstrate its functionality and analyze the sensitivity of the vague modeling approach. This integrated solution provides decision-makers with powerful tools to quantify operational uncertainties, maintain scheduling flexibility, and effectively respond to dynamic production demands, while the modified hybrid algorithm's performance is systematically verified through both theoretical analysis and practical validation. The results offer manufacturers a mathematically rigorous yet practical approach to optimize scheduling in uncertain environments.

Keywords Vagueness · Flexible manufacturing system scheduling · Branch and bound algorithm (B&B) · Genetic algorithm (GA) · Ranking methods

6.1 Introduction

Flexible Manufacturing Systems (FMS) represent computer-controlled production environments capable of processing multiple part types using constrained shared resources, with applications ranging from eyeglass production to semiconductor manufacturing (Abbasian et al., 2015). These systems exhibit high resource sharing and routing flexibility, particularly in Flexible Job Shop Scheduling Problems

H. Fazlollahtabar, *Uncertain Multi-objective Decision Making: Methods and Industrial Models*, https://doi.org/10.1007/978-981-95-6035-6_6

(FJSSP) where operations can be processed on alternative machines, requiring simultaneous optimization of both machine allocation and operation sequencing (Abbasian et al., 2015). However, real-world implementations face significant uncertainty from internal and external factors that often render deterministic scheduling solutions unreliable (Abedi et al., 2015).

To address these challenges, we employ spaced-gap-based vague sets—an advanced uncertainty modeling technique extending traditional fuzzy sets through interval-based membership functions. This approach builds upon successful applications in scheduling problems with vague due dates and processing times (Fazlollahtabar & Hassanli, 2018). Our solution methodology combines a modified Branch and Bound (B&B) algorithm, which systematically explores solution spaces while pruning suboptimal branches based on estimated bounds (Morrison et al., 2016), with Genetic Algorithm (GA) components for enhanced global search capability in complex solution spaces.

The hybrid B&B-GA approach incorporates robust ranking techniques for fuzzy numbers (Morrison et al., 2016) to validate solutions under uncertainty. This study represents a pioneering quantitative investigation of vague process modeling in FMS, specifically targeting long-term decision-making in dynamic scheduling environments. The framework advances FMS scheduling through (1) novel application of spaced-gap-based vague sets, (2) enhanced optimization via algorithmic hybridization, and (3) comprehensive uncertainty management—addressing the critical need for reliable scheduling in flexible manufacturing systems facing operational variability (Abbasian et al., 2015; Abedi et al., 2015).

The Branch and Bound (B&B) algorithm is a fundamental method for solving discrete and combinatorial optimization problems in mathematical programming. This systematic approach organizes potential solutions through a state space search, conceptualized as a rooted tree structure where the complete solution set resides at the root node (Morrison et al., 2016). The algorithm progressively explores branches representing solution subsets, employing bounding mechanisms to eliminate suboptimal paths. Specifically, each branch is evaluated against estimated upper and lower bounds of the optimal solution, and is pruned if it cannot improve upon the current best solution identified by the algorithm.

In the context of fuzzy systems, robust ranking techniques for fuzzy numbers play a pivotal role in various applications including decision-making, optimization, and forecasting (Morrison et al., 2016). These techniques are particularly valuable when dealing with vague numerical representations of alternative options in decision analysis problems. The ability to accurately rank fuzzy numbers enables demonstration of solution validity under uncertainty.

Building upon these foundations, this study presents a novel quantitative approach to addressing Flexible Manufacturing Systems (FMS) scheduling problems under vague process conditions. Our research contributes to the field by developing a hybrid methodology that combines

- A modified Branch and Bound (B&B) algorithm.
- Genetic Algorithm (GA) components.

This integrated approach specifically targets the challenge of long-term decision-making in uncertain FMS environments, focusing on all aspects of scheduling optimization. The primary objectives of this investigation include

- Development of an enhanced B&B-GA hybrid with superior performance characteristics.
- Implementation of vagueness and fuzzy uncertainty modeling in FMS scheduling.
- Comprehensive analysis of flexible manufacturing systems under uncertain conditions.

The proposed methodology advances current practices by introducing innovative techniques for handling the inherent uncertainties in FMS operations while maintaining scheduling efficiency and resource optimization capabilities.

6.2 Literature Review

This study synthesizes contemporary research on Flexible Manufacturing Systems (FMS) scheduling, building upon three critical research streams: FMS architecture, production scheduling optimization, and uncertainty quantification methods. The foundational work of Souier and Sari (2014) establishes flexibility as the defining characteristic of FMS, emphasizing its strategic importance in justifying system investments through optimal utilization. Our research extends this paradigm by investigating scheduling problems in FMS with routing flexibility under uncertainty conditions, particularly addressing random part arrivals and machine failures while incorporating reliability and maintenance constraints—a challenge previously examined by Diop et al. (2019) in their work on failure-prone production systems.

The literature reveals two predominant approaches to FMS scheduling optimization. Metaheuristic methods have demonstrated particular effectiveness, including (1) NSGA-II implementations for multi-objective optimization of deadlock prevention and reliability maximization (Diop et al., 2019), (2) hybrid flow shop scheduling using self-tuning iterated greedy algorithms (Ying & Lin, 2018), and (3) genetic algorithm–Monte Carlo integration for routing flexibility analysis (Dosdoğru et al., 2015). Complementary to these, heuristic-based solutions have shown promise in specific contexts, such as the two-stage assembly scheduling with variable local search developed by Framinan and Perez-Gonzalez (2019), and dynamic routing strategies for part launching optimization proposed by Wang et al. (2016).

Critical to our methodology is the robust ranking of fuzzy numbers, which Fazlollahtabar and Gholizadeh (2020) have shown to satisfy essential properties of compensation, linearity, and additivity in decision-making under uncertainty. This approach builds upon earlier uncertainty modeling frameworks, including Mohammadi et al.'s (2020) fuzzy-based scheduling for wind-integrated multi-energy systems and Gaula and Sharma's (2015) hybrid framework for failure modeling in flexible manufacturing cells. The present study advances this research trajectory through three key innovations: First, we introduce the novel application of vague set theory to FMS scheduling problems, incorporating trapezoidal vague numbers with robust ranking. Second, we develop a hybrid modified branch and bound algorithm enhanced with GA-assisted population generation and constraint-satisfying heuristic solutions. Third, we implement practical improvements including accelerated convergence and reduced solution space exploration, addressing the human error impacts on production planning identified by Gholizadeh and Fazlollahtabar (2019).

- Our integrated approach specifically targets three operational challenges in FMS: (1) routing flexibility optimization under machine reliability constraints, (2) workload balancing and utilization optimization, and (3) deadlock prevention in uncertain production environments. This builds upon while significantly extending the reliability-focused work of Gaula and Sharma (2015) and the dynamic routing strategies of Wang et al. (2016), offering a comprehensive framework for vague environment scheduling that demonstrates both theoretical rigor and practical implementability.

6.3 Vague Environment

Definition 1 (*Fuzzy Set*) A fuzzy set $\tilde{A} = \{< u, \mu_{\tilde{A}}(u) > | u \in U\}$ in a universe of discourse U is characterized by a membership function, $\mu_{\tilde{A}}$, as follows (Gonzalez-Lopez et al., 2016):

$$\mu_{\tilde{A}} = U \rightarrow [0, 1].$$

Definition 2 A vague set $\tilde{A}$ in a universe of discourse U is characterized. It is expressed by two membership functions, which is a true membership, $T_{\tilde{A}}$, function and a false membership function, $F_{\tilde{A}}$, as follows: $T_{\tilde{A}} : U \rightarrow [0, 1]$, $F_{\tilde{A}} : U \rightarrow [0, 1]$, and $T_{\tilde{A}}(u) + F_{\tilde{A}}(u) \leq 1$, where $T_{\tilde{A}}(u)$ is a lower bound on the grade of membership of u derived from the evidence for u and $F_{\tilde{A}}(u)$ is a lower bound on the grade of membership of the negation of u derived from the evidence against u. Suppose $U = \{u_1, u_2, \ldots, u_n\}$. A vague set $\tilde{A}$ of the universe of discourse U can be represented by $\tilde{A} = \frac{\sum_{i=1}^{N}[T_{\tilde{A}}(u_i), 1-F_{\tilde{A}}(u_i)]}{u_i}$, where $0 \leq T(u_i) \leq 1 - F(u_i) \leq 1$ and $1 \leq i \leq n$. In this case, the degree of membership in u_i will be limited to a period of time, $\left[T_{\tilde{A}}(u_i), 1 - F_{\tilde{A}}(u_i)\right]$ of [0, 1].

Therefore, generalized vague collections have fuzzy sets of membership degrees $\mu_{\tilde{A}}(u)$ of u in Definition 1 which may be inexact in a vague set. We now depict a vague set in Fig. 6.1.

Definition 3 The complement of a vague set $\tilde{A}$ is denoted by $\widetilde{A'}$ and is defined by

$$T_{\widetilde{A'}}(u) = F_{\tilde{A}}(u), \quad 1 - F_{\widetilde{A'}}(u) = 1 - F_{\tilde{A}}(u). \tag{6.1}$$

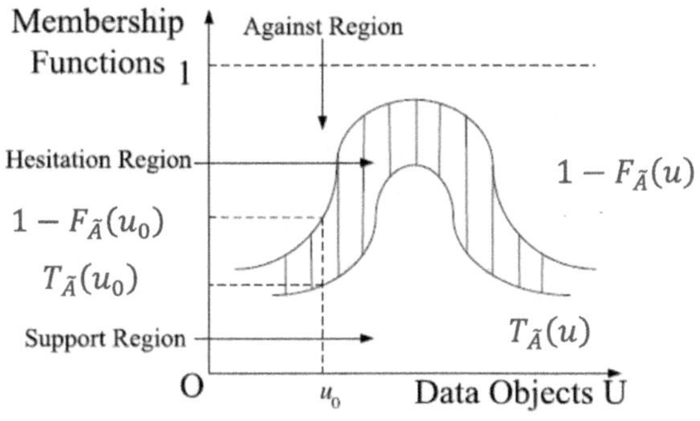

Fig. 6.1 Membership functions of a vague set

Definition 4 Let $\tilde{A}_{T_{\tilde{A}}F_{\tilde{A}}}$ be a vague set of U. Then, we define α_T-cuts and α_F-cuts of $\tilde{A}$ as the crisp sets of U given by

$$\tilde{A}_{\alpha_T} = \{u : T_{\tilde{A}}(u) \geq \alpha_T\} \quad \alpha_T \in [0, 1] \tag{6.2}$$

$$\tilde{A}_{\alpha_F} = \{u : F_{\tilde{A}}(u) \geq \alpha_F\}, \quad \alpha_F \in [0, 1]. \tag{6.3}$$

Definition 5 A vague set $\tilde{A}_{T_{\tilde{A}}F_{\tilde{A}}}$ of $\mathscr{R}$ with continuous membership functions $T_{\tilde{A}}$ and $F_{\tilde{A}}$ is called a vague number if and only if $\tilde{A}_{\alpha_T}$ and $\tilde{A}_{\alpha_F}$, for all $\alpha_T, \alpha_F \in (0, 1]$, are bounded closed intervals, i.e.:

$$\tilde{A}_{\alpha_T} = [\tilde{A}^L_{\alpha_T}, \tilde{A}^U_{\alpha_T}], \quad \tilde{A}_{\alpha_F} = [\tilde{A}^L_{\alpha_F}, \tilde{A}^U_{\alpha_F}]. \tag{6.4}$$

We denote the class of all vague numbers by $A(\mathscr{R})$.

Definition 6 The vague number $\tilde{A}_{T_{\tilde{A}}F_{\tilde{A}}}$ is called a trapezoidal vague number, if

$$T_{\tilde{A}}(u) = \begin{cases} \max\left(0, \frac{((1-\frac{1}{w})a_1+\frac{1}{w}a_2)-u}{\frac{1}{w}(a_2-a_1)}\right) & a_1 \leq u \leq a_2 \\ 0 & a_2 \leq u \leq a_3 \\ \max\left(0, \frac{u-((1-\frac{1}{w})a_4+\frac{1}{w}a_3)}{\frac{1}{w}(a_4-a_3)}\right) & a_3 \leq u \leq a_4 \\ 0 & \text{otherwise} \end{cases} \tag{6.5}$$

$$1 - F_{\tilde{A}}(u) = \begin{cases} \max\left(0, \frac{(a_2-u)}{(a_2-a_1)}\right) & a_1 \leq u \leq a_2 \\ 0 & a_2 \leq u \leq a_3 \\ \max\left(0, \frac{u-a_3}{(a_4-a_3)}\right) & a_3 \leq u \leq a_4 \\ 0 & \text{otherwise,} \end{cases} \tag{6.6}$$

where $w \in [1, \infty)$. We denote such a vague number by $\tilde{A} = (a_1, a_2, a_3, a_4, w)_T$.

6.3.1 Ranking of Trapezoidal Vague Numbers

The operational properties of any system can be effectively modeled using appropriate membership functions. This approach becomes particularly relevant when dealing with imprecise input data, where decision alternatives must be evaluated using vague numerical representations (Gholizadeh et al., 2019). To address this challenge, the robust ranking method serves as a validation mechanism, enabling quantitative comparison between competing options while preserving decision-making integrity. The technique's mathematical foundation ensures reliable performance when handling the inherent ambiguity in complex system evaluations (Gholizadeh et al., 2019). This approach proves particularly valuable in scenarios requiring precise discrimination between alternatives expressed through vague numerical constructs.

Remark: The evaluation metric for a trapezoidal vague number is derived through bilateral area computation, specifically:

1. **Measurement Methodology**
   - Calculates the arithmetic mean of bilateral vague regions.
   - Integrates both left and right boundary areas.
   - Computes relative to the α-cut axis in membership space.
2. **Robust Ranking Characteristics**
   - Maintains three fundamental mathematical properties:
     - **Compensation**: Balances extreme values.
     - **Linearity**: Preserves proportional relationships.
     - **Additivity**: Enables component aggregation.
   - Delivers outcomes that align with
     - Cognitive decision-making processes.
     - Practical judgment expectations.
     - Intuitive interpretation requirements.

This dual-aspect approach (Gholizadeh et al., 2019) provides a mathematically rigorous yet practically applicable framework for comparing and ordering vague numerical values in uncertain environments. The technique's design specifically addresses the need for reliable quantification while maintaining conceptual alignment with human reasoning patterns.

Definition 6 Let $\tilde{A}$ be a normal octagonal vague number. The value $R(\tilde{A}_\alpha)$ called the measure of $\tilde{A}$ is calculated as follows:

$$R(\tilde{A}_{\alpha_t}) = \int_0^1 (0.5)\left[\tilde{l}_1(r)^L_{\alpha_t}, \tilde{l}_1(r)^U_{\alpha_t}\right]d\alpha_t + \int_0^1 (0.5)\left[\tilde{s}_1(t)^L_{\alpha_t}, \tilde{s}_1(t)^U_{\alpha_t}\right]d\alpha_t \quad (6.7)$$

$$R(\tilde{A}_{\alpha_t}) = \int_0^1 (0.5)\left[\tilde{l}_2(r)^L_{\alpha_t}, \tilde{l}_2(r)^U_{\alpha_t}\right]d\alpha_t + \int_0^1 (0.5)\left[\tilde{s}_2(t)^L_{\alpha_t}, \tilde{s}_2(t)^U_{\alpha_t}\right]d\alpha_t \tag{6.8}$$

$$R(\tilde{A}_{\alpha_f}) = \int_0^1 (0.5)\left[\tilde{l}_1(r)^L_{\alpha_f}, \tilde{l}_1(r)^U_{\alpha_f}\right]d\alpha_f + \int_0^1 (0.5)\left[\tilde{s}_1(t)^L_{\alpha_f}, \tilde{s}_1(t)^U_{\alpha_f}\right]d\alpha_f \tag{6.9}$$

$$R(\tilde{A}_{\alpha_f}) = \int_0^1 (0.5)\left[\tilde{l}_2(r)^L_{\alpha_f}, \tilde{l}_2(r)^U_{\alpha_f}\right]d\alpha_f + \int_0^1 (0.5)\left[\tilde{s}_2(t)^L_{\alpha_f}, \tilde{s}_2(t)^U_{\alpha_f}\right]d\alpha_f. \tag{6.10}$$

6.4 Vague B&B Algorithm

The Branch and Bound (B&B) method serves as an exact optimization algorithm for discrete and combinatorial problems. This systematic approach organizes the solution space through a tree structure, where

1. The root node contains the complete solution set.
2. Branches represent progressively constrained solution subsets.
3. Node evaluation incorporates bounding techniques to prune suboptimal paths.

Key operational components (Fazlollahtabar & Gholizadeh, 2019):

- **State space exploration**: Systematically enumerates candidate solutions.
- **Bounding mechanism**: Compares branch potential against current best solution.
- **Pruning criterion**: Eliminates branches that cannot improve the incumbent solution.

For vague environment adaptation (Gholizadeh et al., 2019):

1. **Vague Parameter Handling**
    - Processes vague processing times.
    - Expresses bounds as vague numbers.
    - Implements equation $R(\tilde{A}_\alpha)$ for performance evaluation.
2. **Modified Algorithm Architecture**
    - Incorporates vague number comparison metrics.
    - Maintains exact solution guarantees.
    - Preserves computational efficiency through:
        - Intelligent branching strategies.
        - Adaptive bounding techniques.
        - Effective pruning mechanisms.

The enhanced algorithm demonstrates particular effectiveness for

- Scheduling problems with uncertain durations.
- Resource allocation under vagueness.

Combinatorial optimization with imprecise parameters.

Vague B&B Algorithm

Step 1: Calculate vague lower bound for the two machine vague makespan problem:

$$\widetilde{\mathrm{LB}}(S_r)_{\alpha_t} = \mathrm{Max}\begin{cases} \widetilde{\mathrm{CT1}}(S_r)_{\alpha_t}(+)(+)\tilde{P}_{i1_{\alpha_t}}(+)\min\left(\tilde{P}_{i2_{\alpha_t}}\right) \\ \widetilde{\mathrm{CT2}}(S_r)_{\alpha_t}(+)(+)\tilde{P}_{i2_{\alpha_t}} \end{cases}, \tag{6.11}$$

$$\widetilde{\mathrm{LB}}(S_r)_{\alpha_f} = \mathrm{Max}\begin{cases} \widetilde{\mathrm{CT1}}(S_r)_{\alpha_f}(+)(+)\tilde{P}_{i\alpha_f}(+)\min\left(\tilde{P}_{i2\alpha_f}\right) \\ \widetilde{\mathrm{CT2}}(S_r)_{\alpha_f}(+)(+)\tilde{P}_{i2_{\alpha_f}} \end{cases}, \tag{6.12}$$

where

$\tilde{P}_{ij}$ is the vague processing time of the ith job in the jth machine.

S_r' is the set of $(n - r)$ jobs yet to be assigned to the machines.

$\widetilde{\mathrm{CTK}}$ is the vague completion time of the last job in the sequence S_r at the machine K.

$(+)$ refers to fuzzy addition.

$\widetilde{\mathrm{LB}}$ is the lower bounds.

Step 2: After vague lower bound values for nodes are calculated, branching is done from the lowest bound to form new nodes for all unscheduled jobs. This process is continued till all the jobs are scheduled. Calculate the lower bounds $\widetilde{\mathrm{LB}} = \{\widetilde{\mathrm{LB}}(1), \widetilde{\mathrm{LB}}(2), \widetilde{\mathrm{LB}}(3)\}$.

Step 3: To find the minimum of the vague numbers, the generalized mean value (GMV) of these numbers is calculated and the vague number with the smallest GMV is considered the smallest. If two vague numbers have the same GMVs, to break the tie, the spread is calculated for each vague number as given in step 2 and the number with the smaller spread is adjudged the smallest.

Step 4: After calculating the maximum, now to determine which node to branch from, we have to find the minimum of $\widetilde{\mathrm{LB}}(1), \widetilde{\mathrm{LB}}(2), \widetilde{\mathrm{LB}}(3)$ using the GMVs.

Step 5: Calculate the new completion times of the partial sequence S_r.

Step 6: The vague completion time is computed for the last job in the sequence S_r at machine1, machine 2, and $\widetilde{\mathrm{CTK}}\ (s_r)$.

Step 7: Using the new completion time, fuzzy lower bounds for partial sequences following the steps above are repeated and recalculated.

Step 8: The vague makespan must be calculated as the maximum of the job completion times using a ranking method.

The vague waiting, processing, and completion times for each of the job sequences are calculated. Negative portions of the vague numbers are deleted since it is not realistic.

6.5 Solution Method

For complex scheduling problems of substantial scale, heuristic methods like genetic algorithms (GAs) offer an effective solution framework. This study implements a GA-based approach to address the formulated large-scale model, leveraging its stochastic search capabilities. The methodology involves several key components (Gholizadeh et al., 2019):

1. **Solution Representation**
   - Dual-part chromosome structure:
     - Part 1: Encodes machine-job processing priorities:

       Machine sequencing for each B&B step.
       Job sequencing per machine per gene.
     - Part 2: Incorporates B&B subsections:

       Contains lower bound values with ranked vague numbers.
       Indicates machine operation precedence via ith sub-elements.
2. **Genetic Operators**
   - **Selection**: Randomly pairs parents from top 50% fitness pool.
   - **Crossover**: Single-point exchange:
     - Random crossover point selection.
     - Right-segment swapping between parent chromosomes.
     - Solution repair mechanism for feasibility maintenance.
   - **Mutation**: Pairwise gene swapping
     - Operation schedule interchange.
     - Random machine reassignment (illustrated in Fig. 6.3).
3. **Feasibility Maintenance**
   - Repair process for constraint satisfaction:
     - Identifies and removes infeasible job assignments.
     - Replaces with valid alternatives through

       Random selection from feasible options.
       Example case resolution (Fig. 6.2 demonstration).

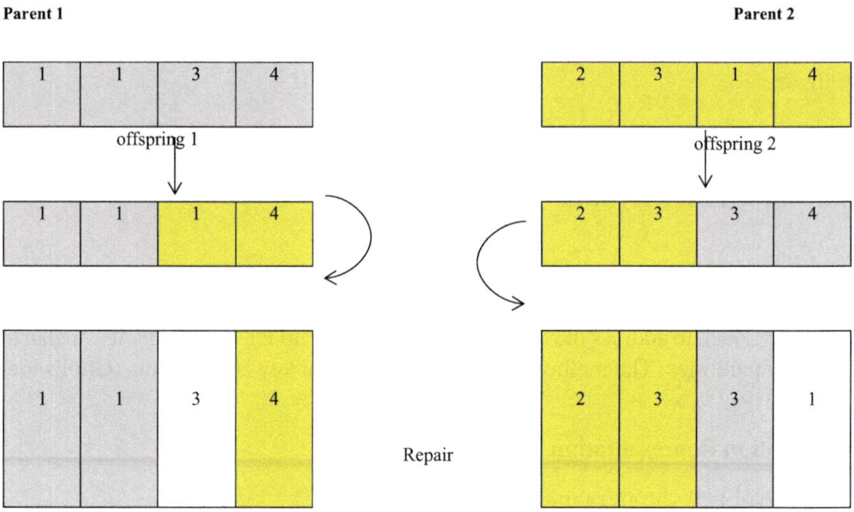

Fig. 6.2 Crossover operator display for two jobs and two machines

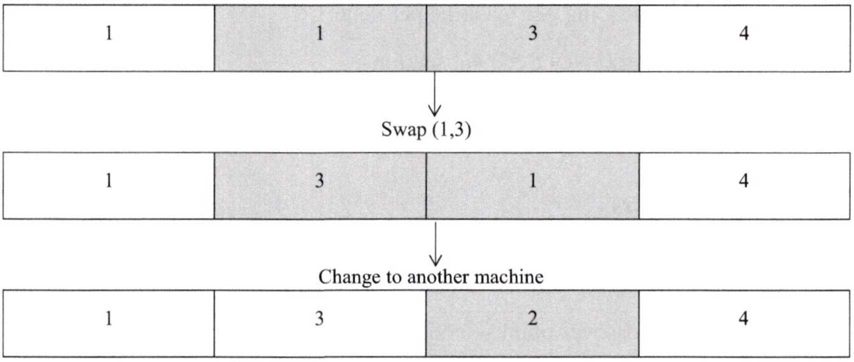

Fig. 6.3 Mutation operator display for two jobs and two machines

- Balances solution quality with computational complexity.

4. **Branch Operator Enhancement**

    - Node organization via vague membership function distance metrics.
    - Integrates B&B principles with evolutionary computation (Figs. 6.3 and 6.4).

The approach demonstrates particular effectiveness for

- Very large problem instances.
- Scenarios where exact methods prove computationally prohibitive.
- Systems requiring solution diversity maintenance.

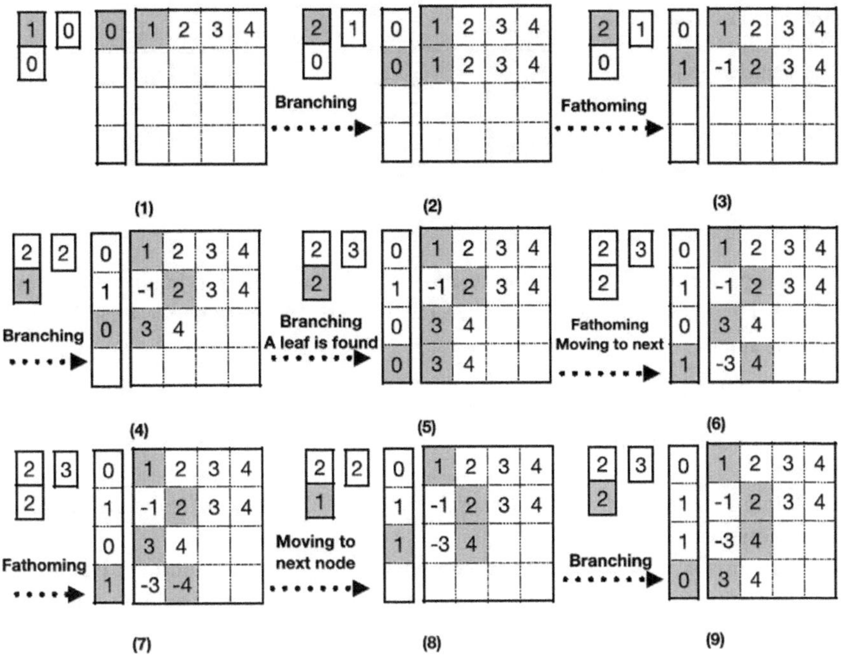

Fig. 6.4 Branch operator display for two jobs and two machines

Parameter Setting

The effectiveness of metaheuristic algorithms is highly influenced by the appropriate configuration of their control parameters. Proper parameter tuning significantly enhances solution quality and computational efficiency. In this study, we employ an orthogonal experimental design to determine the optimal settings for the genetic algorithm (GA) parameters, evaluating their individual and combined effects on performance.

Key parameters and tuning process:

1. Controlled parameters:
   - Crossover probability ($\mathcal{P}_c$).
   - Mutation probability ($\mathcal{P}_m$).
   - Population size (N_{pop}).
   - Maximum iterations (M_{axIt}).
2. Orthogonal experimental design:
   - Evaluates multiple parameter combinations at different levels.
   - Identifies the optimal configuration for balancing exploration and exploitation.
   - Determines the relative significance of each parameter via main effect analysis.

Table 6.1 Levels and optimal values of proposed GA parameters

Parameter	Best level	Value
$\mathcal{P}_c$	3	0.8
$\mathcal{P}_m$	2	0.2
$(N_{\text{pop}}, M_{\text{axIt}})$	1	(100, 200)

3. Optimal parameter selection:

The best performing combination is selected based on solution quality and convergence behavior. Results are summarized in Table 6.1, specifying the ideal levels for each parameter.

Impact of Parameter Tuning

- Crossover probability ($\mathcal{P}_c$): Controls diversity maintenance in offspring generation.
- Mutation probability ($\mathcal{P}_m$): Ensures sufficient exploration to avoid premature convergence.
- Population size (N_{pop}): Balances computational effort and solution diversity.
- Maximum iterations (M_{axIt}): Defines termination criteria based on convergence stability.

6.6 Numerical Example

Consider a manufacturing system comprising M machines and N jobs, where processing times are represented as trapezoidal vague numbers to account for uncertainty. The proposed methodology emphasizes:

1. Data transformation for decision-making:
   - Converts interactive data logs into structured analytical inputs.
   - Addresses challenges in large-scale data management that may impact optimization.
   - Ensures computational tractability while preserving decision quality.
2. Implementation framework:
   - Software: Implemented in MATLAB™ 2013.
   - Hardware: Executed on an *Intel® Core™ i5-5200U @2.20 GHz* (Windows 10).
   - Validation: Tested on 15 benchmark problems of varying scales (small to large).
3. Key contributions:
   - Vague number integration: Trapezoidal fuzzy sets model uncertain processing times.

- Robust optimization: Maintains solution validity despite data complexity.
- Scalability: Handles diverse problem sizes efficiently.

Performance evaluation:

- Test problems assess
  - Computational efficiency.
  - Solution accuracy under uncertainty.
  - Algorithm convergence.

Results confirm the method's effectiveness and reliability for flexible scheduling. Input data are randomly generated with ambiguous features. This section devotes to present the obtained results from solving the research problem using B&B method, traditional GA, and the proposed hybrid B&B-GA. Table 6.2 shows the results of solving the problem in small-sized instances. In Table 6.2, the results achieved by algorithms are compared to the results obtained by B&B. To do this, percentage relative error (PRE) factor is applied. PRE is calculated using following formula:

$$\mathrm{PRE} = \frac{\mathrm{Sol}_{\mathrm{Alg}} - \mathrm{Sol}_{\mathrm{B\&B}}}{\mathrm{Sol}_{\mathrm{B\&B}}} \times 100, \tag{6.13}$$

where $\mathrm{Sol}_{\mathrm{Alg}}$ is solution which is achieved by the algorithms and $\mathrm{Sol}_{\mathrm{B\&B}}$ denotes solution which is obtained by B&B.

According to Table 6.2, the developed hybrid algorithms can achieved high-quality solutions in reasonable time. Figure 6.5 has compared the values of PRE (criteria of solution's quality). As can be seen in Fig. 6.6, in all instances, the proposed hybrid B&B-GA has better performance than traditional GA. Also, Fig. 6.6 compares the algorithms in CPU time criteria which results show that performance of the GA is better than hybrid algorithm in terms of CPU time.

6.7 Conclusions

This study presents an advanced optimization framework for flow shop scheduling problems under uncertainty by integrating a modified Branch and Bound (B&B) algorithm with Genetic Algorithm (GA) components, where job processing times are modeled using vague octagonal numbers to better capture real-world imprecision. The proposed hybrid approach enhances traditional B&B methods through three key innovations: (1) incorporation of vague octagonal numbers for robust uncertainty quantification, enabling decision-makers to process incomplete or ambiguous data while maintaining solution quality; (2) adaptive pruning strategies based on vague number ranking to efficiently eliminate suboptimal branches; and (3) GA-assisted exploration of the solution space to prevent premature convergence. Computational

Table 6.2 Results of comparison between B&B and GA and hybrid B&B-GA solutions for problems possessing

Problem	(*n*, *m*)	B&B		GA				Hybrid B&B-GA			
		Min Cmax	CPU time	Best value	Average value	CPU time	PRE	Best value	Average value	CPU time	PRE
1	(5, 2)	384.5	2.1	384.5	423.0	16.4	0.0	384.5	417.2	21.5	0.0
2	(7, 2)	438.6	3.8	438.6	482.5	16.8	0.0	438.6	475.9	23.4	0.0
3	(7, 3)	405.2	8.5	413.7	455.1	17.5	2.1	405.2	439.6	26.1	0.0
4	(8, 3)	493.7	14.3	507.5	558.3	18.1	2.8	493.7	535.7	29.2	0.0
5	(9, 4)	544.3	75.6	562.8	619.1	18.6	3.4	544.3	590.6	34.6	0.0
6	(10, 4)	607.1	149.5	629.6	692.5	20.4	3.7	607.1	658.7	37.5	0.0
7	(11, 5)	665.9	376.1	693.9	763.3	23.7	4.2	665.9	722.5	44.2	0.0
8	(12, 5)	713.2	508.0	746.0	820.6	25.2	4.6	713.2	773.8	51.9	0.0
9	(12, 6)	688.5	713.6	722.9	795.2	26.5	5.0	688.5	747.0	60.5	0.0
10	(13, 6)	708.4	1348.5	745.9	820.5	28.0	5.3	718.3	779.4	68.3	1.4
11	(14, 7)	795.6	2720.8	844.1	928.5	31.4	6.1	810.7	879.6	75.6	1.9
12	(15, 7)	848.5	3902.2	903.7	994.0	33.6	6.5	868.0	941.8	84.1	2.3
13	(16, 8)	923.7	6859.4	986.5	1085.2	37.2	6.8	949.6	1030.3	92.5	2.8
14	(18, 8)	–	> 10,000	1075.6	1183.2	41.5	–	1024.8	1111.9	107.3	–
15	(20, 9)	–	> 10,000	1283.5	1411.9	46.9	–	1196.3	1298.0	118.4	–

experiments demonstrate the method's effectiveness in improving scheduling decisions under uncertainty, particularly in precision manufacturing sectors like semiconductor production and automotive assembly. For future research, the framework can be extended to incorporate additional vague parameters such as dynamic failure rates, transportation delays, and machine setup times, while parallel computing implementations could further enhance scalability for industrial-scale problems. This work establishes a foundation for more sophisticated uncertainty-aware scheduling systems that balance computational efficiency with practical decision-making needs.

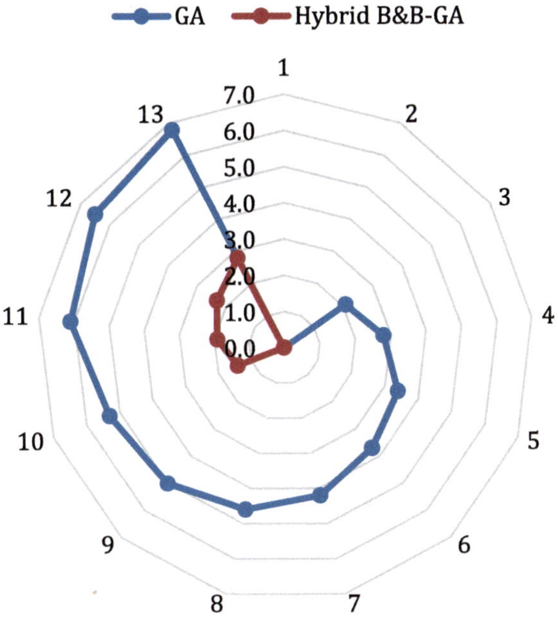

Fig. 6.5 Comparison of the PRE of the algorithms

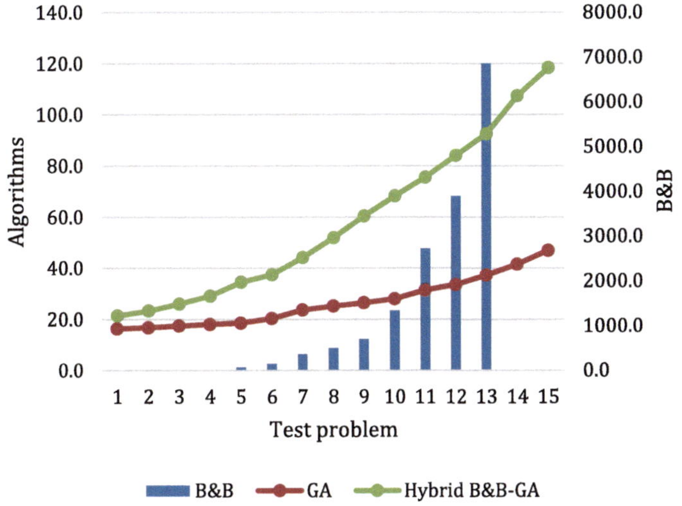

Fig. 6.6 Comparison of the CPU time of the algorithms

References and Further Reading

Abbasian, M., Nosratabadi, H., & Fazlollahtabar, H. (2015). Applying an intelligent dynamic genetic algorithm for solving a multi-objective flexible job shop scheduling problem with maintenance considerations. *Journal of Applied & Computational Mathematics, 4*(4).

Abedi, M., Seidgar, H., Fazlollahtabar, H., & Bijani, R. (2015). Bi-objective optimisation for scheduling the identical parallel batch-processing machines with arbitrary job sizes, unequal job release times and capacity limits. *International Journal of Production Research, 53*(6), 1680–1711.

Diop, I., Nadeau, S., & Emami-Mehrgani, B. (2019). A mathematical model: A flexible manufacturing system, prone to error, making two products each with stochastic demand schedules. *American Journal of Industrial and Business Management, 9*(1), 139–168.

Dosdoğru, A. T., Göçken, M., & Geyik, F. (2015). Integration of genetic algorithm and Monte Carlo to analyze the effect of routing flexibility. *International Journal of Advanced Manufacturing Technology, 81*(5), 1379–1389.

Fazlollahtabar, H., & Gholizadeh, H. (2019). Economic analysis of the M/M/1/N queuing system cost model in a vague environment. *International Journal of Fuzzy Logic and Intelligent Systems, 19*(3), 192–203.

Fazlollahtabar, H., & Gholizadeh, H. (2020). Fuzzy possibility regression integrated with fuzzy adaptive neural network for predicting and optimizing electrical discharge machining parameters. *Computers & Industrial Engineering, 140*, Article 106225.

Fazlollahtabar, H., & Hassanli, S. (2018). Hybrid cost and time path planning for multiple autonomous guided vehicles. *Applied Intelligence, 48*, 482–498.

Framinan, J. M., Perez-Gonzalez, P., & Fernandez-Viagas, V. (2019). Deterministic assembly scheduling problems: A review and classification of concurrent-type scheduling models and solution procedures. *European Journal of Operational Research, 273*(2), 401–417.

Gaula, A. K., & Sharma, R. K. (2015). Analyzing the effect of maintenance strategies on throughput of a typical fmc (3-m, 1-r). *International Journal of Systems Assurance Engineering and Management, 6*(2), 183–190.

Gholizadeh, H., & Fazlollahtabar, H. (2019). Production control process using integrated robust data envelopment analysis and fuzzy neural network. *International Journal of Mathematical, Engineering and Management Sciences, 4*(3), 580.

Gholizadeh, H., Fazlollahtabar, H., & Gholizadeh, R. (2019). A modified branch and bound algorithm for a vague flow-shop scheduling problem. *Iranian Journal of Fuzzy Systems, 16*(4), 55–64.

Gonzalez-Lopez, V. A., Gholizadeh, R., & Shirazi, A. M. (2016). Optimization of queuing theory based on vague environment. *International Journal of Fuzzy System Applications, 5*(1), 1–26.

Mohammadi, M., Noorollahi, Y., & Mohammadi-ivatloo, B. (2020). Fuzzy-based scheduling of wind integrated multi-energy systems under multiple uncertainties. *Sustainable Energy Technologies and Assessments, 37*, Article 100602.

Morrison, D. R., Jacobson, S. H., Sauppe, J. J., & Sewell, E. C. (2016). Branch-and-bound algorithms: A survey of recent advances in searching, branching, and pruning. *Discrete Optimization, 19*, 79–102.

Souier, M., & Sari, Z. (2014). Impacts of scheduling decisions based on PSO algorithm and dispatching rules on FMS performances. *International Journal of Applied Metaheuristic Computing (IJAMC), 5*(2), 22–38.

Wang, Y. C., Chen, T., Chiang, H., & Pan, H. C. (2016). A simulation analysis of part launching and order collection decisions for a flexible manufacturing system. *Simulation Modelling Practice and Theory, 69*, 80–91.

Ying, K.-C., & Lin, S.-W. (2018). Minimizing makespan for the distributed hybrid flowshop scheduling problem with multiprocessor tasks. *Expert Systems with Applications, 92*, 132–141.

Chapter 7
Possible Vague Multi-objective Optimization in Queue System

Abstract In many real-world systems, the vague information leads to inappropriate analysis. Generally, lack of sufficient knowledge of a process in random environment and vagueness of the whole system are two important challenges for modeling a system. Therefore, this chapter considers parallel queues systems with encouraged arrivals shared server. The various arrival rates, service rates, and processing times of jobs/tasks usually assumed to be uncertain. Hence, we describe the theory of queuing in a possible vague environment where encouraged arrival rates and service rates are considered to be uncertain numbers. Then, novel measures to handle vague information are employed. A bi-serial and two parallel servers linked in series are considered having vague parameters. Modeling such a system using specific techniques with respect to industrial application is purposed. In implementation, system characteristics are modeled using vague numbers and sensitivity analysis indicates the validity and effectiveness of the result. The proposed vague queue model is employed for a healthcare system to improve decision-making.

Keywords Encouraged arrivals rate · Possible vague · Parallel systems · Robust ranking technique

7.1 Introduction

The queuing systems are an important tool in the field of performance in various systems. Planning in parallel systems is widespread in computer science, telecommunications, manufacturing, and even healthcare systems. Some organizations work with several parallel queues with each server, while few have a separate server. The chapter highlights the significant attempt made to find various characteristics of queue network in which a bi-serial and two parallel servers are linked in series with a common server in a possible vague environment with the aim to solve problems associated with a variety of fields such as industry, education, banking, insurance, travel and tours, and more significantly in data communication and processing. Therefore, studying on parallel systems of queuing is of special importance. On the other

H. Fazlollahtabar, *Uncertain Multi-objective Decision Making: Methods and Industrial Models*, https://doi.org/10.1007/978-981-95-6035-6_7

hand, organizations consider discounts and offer that customers are more likely to encourage more profits and more discounts provided by organizations. Therefore, attracted customers are considered as encouraged arrivals. Due to uncertainty in process of the queue, fuzzy set theory is employed. From the combination of fuzzy and random uncertainty factors, a new variable can be obtained as a random fuzzy variable. In this chapter, we use spaced-gap-based sets, which, instead of point-based membership in fuzzy sets, use spaced-gap-based membership in an obscure set. Also, encouraged arrival rates and the service rate are possible vague parameters.

We first outline the assumptions of the proposed problem:

- The customers arrive based on the Poisson distribution by parameter $(1 + k)$, where "k" represents the percentage of change in number of customers calculated from past or observed data. That is λ_i, vague with "q_i" vague possibility.
- Service rate is shown by μ that is vague with "w_i" vague possibility.
- P_{ij} : Possibility that the system is in the state (i, j) at any time t.
- n: Number of customers/jobs arriving.

7.2 Literature Review

The theoretical framework for uncertainty modeling in queuing systems has evolved significantly since Zadeh's (1978) seminal work on fuzzy set theory and possibility theory. Early contributions to queuing analysis include Maggu's (1970) introduction of the Bibendum concept in queuing theory and Singh's (1979) investigation of serial/bi-serial channel networks. The application of fuzzy methodologies to queuing systems gained momentum with Prade's (1980) development of possibilistic queuing models, while Nagoorgani and Retha (2006) specifically examined tandem queue configurations in fuzzy environments.

Later advancements in network analysis include Singh and Pardeep's (2009) work on transient behavior in parallel bi-serial server networks, and Seema et al.'s (2013) characterization of queue networks connecting parallel bi-serial servers to a common series server under fuzzy conditions. Kao et al. (1999) contributed significantly to this field through their parametric linear programming approach for constructing membership functions of system characteristics in fuzzy queues.

The theoretical progression from fuzzy to more sophisticated uncertainty models is marked by Gau and Buehrer's (1993) introduction of vague sets, which generalize traditional fuzzy sets by replacing point-valued membership with interval-based membership. This advancement was subsequently applied to queuing optimization by Gonzalez-Lopez et al. (2016). The conceptual spectrum of arrival dynamics has been further expanded through investigations of encouraged arrival phenomena, notably Som and Seth's (2017) analysis of M/M/1/N systems with encouraged arrivals, representing the conceptual inverse of discouraged arrival scenarios.

Chaleshigar Kordasiabi et al. (2020) modeled the situation in which the manager wishes to attract customers and to maximize profit using the queuing theory. The service time of the only server of the facilities behaves with uncertainty, which may be represented by means of a fuzzy number. The objective was to determine the rate to be paid by every customer for his service and the level of publicity which the manager must utilize to maximize his profits.

Fazlollahtabar and Gholizadeh (2020a) conducted a comparative analysis of M/M/1 queuing systems with observable versus unobservable queue lengths, examining how these different information structures influence both system properties and performance metrics under optimal customer behavior strategies. Their work provides important insights into queue management under varying information availability conditions.

In the domain of scheduling algorithms, conventional multilevel feedback queue approaches face significant challenges including.

- Resource starvation for long-duration tasks.
- Inflexible queue quantity limitations.
- Fixed time quantum allocations.
- Performance degradation due to imprecise task attributes.

Addressing manufacturing system uncertainties, Fazlollahtabar and Gholizadeh (2020b) developed an innovative approach using wafer lots transfer probability (WLTP) to model the stochastic flow of wafer lots across production bays. Their open queuing network framework enables evaluation of average work-in-process (WIP) levels, providing crucial metrics for production capacity planning in semiconductor fabrication environments characterized by diverse wafer types and fluctuating quantities.

The application of semi-open queuing networks (SOQNs) has expanded significantly across various sectors including.

- Manufacturing operations.
- Logistics and supply chain management.
- Telecommunication systems.
- Service industries (healthcare, hospitality).

However, as Gholizadeh et al. (2019) demonstrate, traditional SOQN solution methods prove inadequate for systems experiencing time-dependent arrival rate variability, highlighting the need for more sophisticated analytical approaches capable of handling non-homogeneous arrival patterns.

7.3 Possible Vague Environment

Before discussing vague modeling, some basics are explained.

Definition 1 A vague set $\tilde{V}$ in a universe of discourse U is characterized by a true membership function, $t_{\tilde{V}}$, and a false membership function, $f_{\tilde{V}}$, as follows: $t_{\tilde{V}} : U \rightarrow [0, 1]$, $f_{\tilde{V}} : U \rightarrow [0, 1]$, and $t_{\tilde{V}}(u) + f_{\tilde{V}}(u) \leq 1$, where $t_{\tilde{V}}(u)$ is a lower bound on the grade of membership of u derived from the evidence for u, and $f_{\tilde{V}}(u)$ is a lower

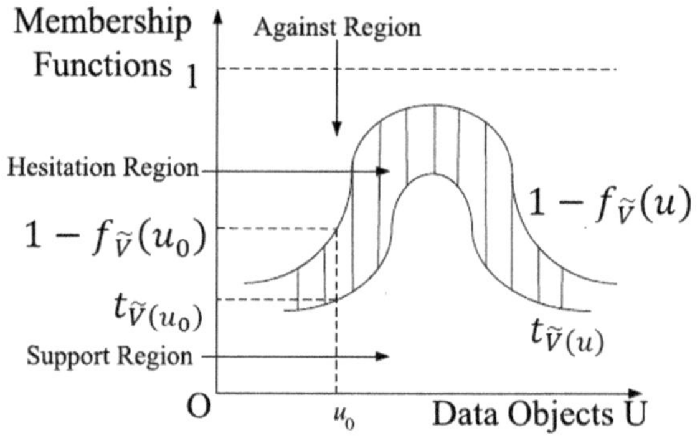

Fig. 7.1 Membership functions of a vague set

bound on the grade of membership of the negation of u derived from the evidence against u (Fazlollahtabar & Gholizadeh, 2019).

Suppose $U = \{u_1, u_2, \ldots u_n\}$. A vague set $\tilde{V}$ of the universe of discourse U can be represented by $\tilde{V} = \sum_{i=1}^{n}\left[t_{\tilde{V}}(u_i), 1 - f_{\tilde{V}}(u_i)\right]/u_i$, where $0 \le t(u_i) \le 1 - f(u_i) \le 1$ and $1 \le i \le n$. In other words, the grade of membership of u_i is bounded to a subinterval $\left[t(u_i), 1 - f_{\tilde{V}}(u_i)\right]$ of [0, 1]. Thus, vague sets are generalization of fuzzy sets, since the grade of membership $\mu_{\tilde{V}}(u)$ of u in Definition 1 may be inexact in a vague set. We now depict a vague set in Fig. 7.1.

Definition 2 The complement of a vague set $\tilde{V}$ is denoted by $\tilde{V}'$ and is defined by

$$t_{\tilde{V}'}(u) = f_{\tilde{V}}(u),$$

$$1 - f_{\tilde{V}'}(u) = 1 - f_{\tilde{V}'}(u).$$

Definition 3 Let $\tilde{V}_{t_{\tilde{V}}, f_{\tilde{V}}}$ be a vague set of U. Then, we define α_t-cuts and α_f-cuts of $\tilde{V}$ as the crisp sets of U given by

$$\tilde{V}_{\alpha_t} = \left\{u : t_{\tilde{V}}(u) \ge \alpha_t\right\}, \quad \alpha_t \in [0, 1],$$

$$\tilde{V}_{\alpha_f} = \left\{u : 1 - f_{\tilde{V}}(u) \ge \alpha_f\right\}, \quad \alpha_f \in [0, 1].$$

Definition 4 A vague set $\tilde{V}_{t_{\tilde{V}}, f_{\tilde{V}}}$ of $\mathcal{R}$ with continuous membership functions $t_{\tilde{V}}$ and $f_{\tilde{V}}$ is called a vague number if and only if $\tilde{V}_{\alpha_t}$ and $\tilde{V}_{\alpha_f}$, for all $\alpha_t, \alpha_f \in (0, 1]$, are bounded closed intervals, i.e., (Fazlollahtabar & Gholizadeh, 2019):

$$\tilde{V}_{\alpha_t} = \left[\tilde{V}^L_{\alpha_t}, \tilde{V}^U_{\alpha_t}\right]$$

$$\tilde{V}_{\alpha_f} = \left[\tilde{V}^L_{\alpha_f}, \tilde{V}^U_{\alpha_f}\right].$$

We denote the class of all vague numbers by $V(\mathcal{R})$.

Definition 5 The vague number $\tilde{V}_{t_{\tilde{V}},f_{\tilde{V}}}$ is called a triangular vague number, if

$$t_{\tilde{V}}(u) = \begin{cases} \frac{u-a_1}{w(a_2-a_1)} & a_1 \le u \le a_2 \\ \frac{a_3-u}{w(a_3-a_2)} & a_2 \le u \le a_3 \\ 0 & \text{otherwise} \end{cases}$$

$$1 - f_{\tilde{V}}(u) = \begin{cases} \frac{u-a_1}{a_2-a_1} & a_1 \le u \le a_2 \\ \frac{a_3-u}{a_3-a_2} & a_2 \le u \le a_3 \ , \\ 0 & \text{otherwise} \end{cases}$$

where $w \in [1, \infty)$. We denote such a vague number by $\tilde{V} = (a_1, a_2, a_3, w)_T$ (Fazlollahtabar & Gholizadeh, 2019).

7.4 Proposed Queue Model

The proposed model features a three-server architecture with specialized routing mechanisms, as illustrated in Fig. 7.2 (Fazlollahtabar & Gholizadeh, 2019). The system configuration consists of.

1. Server S_1 structure:
   - Comprises two bi-serial service channels (S_{11} and S_{12}).
   - Operates with exponential service time distributions for all $S_1 ij$ nodes ($i, j = 1, 2$).
2. Server S_2 composition:
   - Contains two parallel processing units (S_{21} and S_{22}).
   - Maintains identical exponential service time characteristics.
3. Network topology:
   - Server S_3 functions as a common serial node.
   - Connects downstream with both S_1 and S_2 servers.
   - Completes phase-specific service requirements from either parallel branch.
4. Queue formation dynamics:
   - Dedicated queues develop at Q_1 and Q_2 when respective servers are occupied.
   - Queue management follows vague set principles for uncertainty handling.
   - System behavior adheres to the proposed vague theoretical framework.

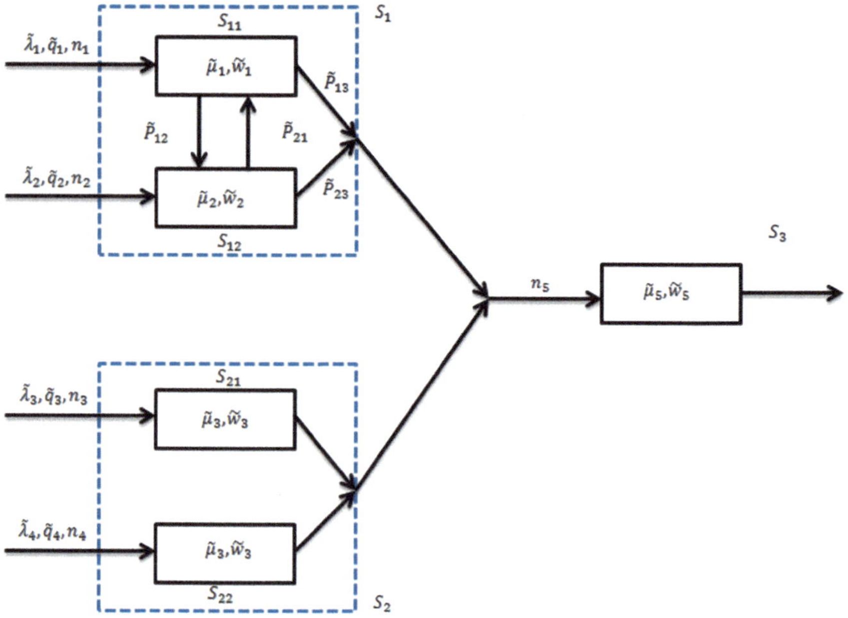

Fig. 7.2 Vague queue model

The integration of these components creates a hybrid serial–parallel service network capable of processing multiple service phases through alternative routing paths while maintaining probabilistic service time distributions.

The customers arrived possibilities $\tilde{q}_1$ with rate $\tilde{\lambda}_1$ after completion of service at S_{11} with possibilities $\tilde{w}_1$ with service rate $\tilde{\mu}_1$ will enter the network of the servers $S_{11} \rightarrow S_{12}$ or $S_{11} \rightarrow S_3$ with possibilities $\tilde{P}_{12}$ or $\tilde{P}_{13}$ such that $\tilde{P}_{12} + \tilde{P}_{13} = 1$.

The customers arrived possibilities $\tilde{q}_2$ with rate $\tilde{\lambda}_2$ after completion of service at S_{12} with possibilities $\tilde{w}_2$ with service rate $\tilde{\mu}_2$ will enter the network of the servers $S_{12} \rightarrow S_{11}$ or $S_{12} \rightarrow S_3$ with possibilities $\tilde{P}_{21}$ or $\tilde{P}_{23}$ such that $\tilde{P}_{21} + \tilde{P}_{23} = 1$.

The customers arrived possibilities $\tilde{q}_3$ with rate $\tilde{\lambda}_3$ after completion of service at S_{21} with possibilities $\tilde{w}_3$ with service rate $\tilde{\mu}_3$ will enter the network of the servers $S_{21} \rightarrow S_3$.

The customers arrived possibilities $\tilde{q}_4$ with rate $\tilde{\lambda}_4$ after completion of service at S_{22} with possibilities $\tilde{w}_4$ with service rate $\tilde{\mu}_4$ will enter the network of the servers $S_{22} \rightarrow S_3$.

The objective is to develop an algorithm to find various queue characteristics for the proposed vague queue model.

7.4.1 Adapted Vague Queue Model

This chapter introduces an advanced queuing model that incorporates parallel processing capabilities while accounting for key uncertainties in system dynamics. The proposed framework specifically addresses.

Uncertain parameter integration:

- Source-encouraged arrival rates ($\tilde{\lambda}$) as vague variables.
- Service rates ($\tilde{\mu}$) represented through vague set theory.
- Percentage change in customer numbers (k) modeled as uncertain parameters.
- Entry probabilities ($\tilde{q}$).
- Service probabilities ($\tilde{w}$).

Model characteristics:

- Parallel server architecture with vague processing capabilities.
- Dynamic arrival patterns influenced by encouragement effects.
- Adaptive service mechanisms for fluctuating demand.

Theoretical foundation:

- Extends traditional queuing theory through vague set incorporation.
- Captures both probabilistic and possibilistic uncertainties.
- Enables robust performance evaluation under incomplete information.

Practical applications:

- Suitable for systems with self-regulating arrival patterns.
- Applicable to service environments with variable throughput.
- Provides decision support for capacity planning under uncertainty.

The model development process includes:

- Formalization of vague arrival and service processes.
- Derivation of system stability conditions.
- Performance metric computation under uncertainty.
- Validation through numerical simulation.

Key innovations include:

- Simultaneous handling of multiple uncertainty sources.
- Integration of encouragement effects in vague environments.
- Development of parallel processing metrics with imprecise parameters.

Implementation considerations:

- Requires specification of vague number parameters.
- Dependent on proper characterization of encouragement factors.
- Demands robust computational methods for solution derivation.

This framework advances queuing theory by providing analytical tools for systems where both service mechanisms and customer behavior exhibit inherent vagueness.

Possibility that the system is in the state (i, j) is $(\tilde{P}_{ij})$ which are not known precisely; therefore, we use triangular vague numbers to represent them. Let

$$\tilde{\lambda} = (\lambda_1, \lambda_2, \lambda_3), \tilde{\mu} = (\mu_1, \mu_2, \mu_3),$$

$$\tilde{K} = (k_1, k_2, k_3), \tilde{q} = (q_1, q_2, q_3),$$

$$\tilde{W} = (w_1, w_2, w_3), \tilde{P} = (p_1, p_2, p_3),$$

where

$$\lambda_1 \leq \lambda_2 \leq \lambda_3,$$

$$\mu_1 \leq \mu_2 \leq \mu_3,$$

$$k_1 \leq k_2 \leq k_3,$$

$$q_1 \leq q_2 \leq q_3,$$

$$w_1 \leq w_2 \leq w_3,$$

$$p_1 \leq p_2 \leq p_3$$

based on subjective judgment. The membership of $t_{\tilde{\lambda}}(\lambda)$, $1-f_{\tilde{\lambda}}(\lambda)$ and $t_{\tilde{\mu}}(\mu)$, $1-f_{\tilde{\mu}}(\mu)$ and $t_{\tilde{k}}(k)$, $1-f_{\tilde{k}}(k)$ and $t_{\tilde{q}}(q)$, $1-f_{\tilde{q}}(q)$ and $t_{\tilde{w}}(w)$, $1-f_{\tilde{w}}(w)$ and $t_{\tilde{p}}(p)$, $1-f_{\tilde{p}}(p)$ are defined as follows:

$$t_{\tilde{\lambda}}(\lambda) = \begin{cases} \frac{\lambda-\lambda_1}{w(\lambda_2-\lambda_1)} & \lambda_1 \leq \lambda \leq \lambda_2 \\ \frac{\lambda_3-\lambda}{w(\lambda_3-\lambda_2)} & \lambda_2 \leq \lambda \leq \lambda_3 \end{cases}, \tag{7.1}$$

$$1-f_{\tilde{\lambda}}(\lambda) = \begin{cases} \frac{\lambda-\lambda_1}{(\lambda_2-\lambda_1)} & \lambda_1 \leq \lambda \leq \lambda_2 \\ \frac{\lambda_3-\lambda}{(\lambda_3-\lambda_2)} & \lambda_2 \leq \lambda \leq \lambda_3 \end{cases} \tag{7.2}$$

$$t_{\tilde{\mu}}(\mu) = \begin{cases} \frac{\mu-\mu_1}{w(\mu_2-\mu_1)} & \mu_1 \leq \mu \leq \mu_2 \\ \frac{\mu_3-\mu}{w(\mu_3-\mu_2)} & \mu_2 \leq \mu \leq \mu_3 \end{cases} \tag{7.3}$$

$$1-f_{\tilde{\mu}}(\mu) = \begin{cases} \frac{\mu-\mu_1}{(\mu_2-\mu_1)} & \mu_1 \leq \mu \leq \mu_2 \\ \frac{\mu_3-\mu}{(\mu_3-\mu_2)} & \mu_2 \leq \mu \leq \mu_3 \end{cases} \tag{7.4}$$

$$t_{\tilde{k}}(k) = \begin{cases} \frac{k-k_1}{w(k_2-k_1)} & k_1 \le k \le k_2 \\ \frac{k_3-k}{w(k_3-k_2)} & k_2 \le k \le k_3 \end{cases} \tag{7.5}$$

$$1 - f_{\tilde{k}}(k) = \begin{cases} \frac{k-k_1}{(k_2-k_1)} & k_1 \le k \le k_2 \\ \frac{k_3-k}{(k_3-k_2)} & k_2 \le k \le k_3 \end{cases} \tag{7.6}$$

$$t_{\tilde{w}}(w) = \begin{cases} \frac{w-w_1}{w(w_2-w_1)} & w_1 \le w \le w_2 \\ \frac{w_3-w}{w(w_3-w_2)} & w_2 \le w \le w_3 \end{cases} \tag{7.7}$$

$$1 - f_{\tilde{w}}(w) = \begin{cases} \frac{w-w_1}{(w_2-w_1)} & w_1 \le w \le w_2 \\ \frac{w_3-w}{(w_3-w_2)} & w_2 \le w \le w_3 \end{cases} \tag{7.8}$$

$$t_{\tilde{q}}(q) = \begin{cases} \frac{q-q_1}{w(q_2-q_1)} & q_1 \le q \le q_2 \\ \frac{q_3-q}{w(q_3-q_2)} & q_2 \le q \le q_3 \end{cases} \tag{7.9}$$

$$1 - f_{\tilde{q}}(q) = \begin{cases} \frac{q-q_1}{(q_2-q_1)} & q_1 \le q \le q_2 \\ \frac{q_3-q}{(q_3-q_2)} & q_2 \le q \le q_3 \end{cases} \tag{7.10}$$

$$t_{\tilde{p}}(p) = \begin{cases} \frac{p-p_1}{w(p_2-p_1)} & p_1 \le p \le p_2 \\ \frac{p_3-p}{w(p_3-p_2)} & p_2 \le p \le p_3 \end{cases} \tag{7.11}$$

$$1 - f_{\tilde{p}}(p) = \begin{cases} \frac{p-p_1}{(p_2-p_1)} & p_1 \le p \le p_2 \\ \frac{p_3-p}{(p_3-p_2)} & p_2 \le p \le p_3 \end{cases} \tag{7.12}$$

$$\tilde{\lambda}_{\alpha_t} = [w\alpha_t(\lambda_2 - \lambda_1) + \lambda_1, \lambda_3 - w\alpha_t(\lambda_3 - \lambda_2)] \tag{7.13}$$

$$\tilde{\lambda}_{\alpha_f} = \left[\alpha_f(\lambda_2 - \lambda_1) + \lambda_1, \lambda_3 - \alpha_f(\lambda_3 - \lambda_2)\right] \tag{7.14}$$

$$\tilde{\mu}_{\alpha_t} = [w\alpha_t(\mu_2 - \mu_1) + \mu_1, \mu_3 - w\alpha_t(\mu_3 - \mu_2)] \tag{7.15}$$

$$\tilde{\mu}_{\alpha_f} = \left[\alpha_f(\mu_2 - \mu_1) + \mu_1, \mu_3 - \alpha_f(\mu_3 - \mu_2)\right] \tag{7.16}$$

$$\tilde{k}_{\alpha_t} = [w\alpha_t(k_2 - k_1) + k_1, k_3 - w\alpha_t(k_3 - k_2)] \tag{7.17}$$

$$\tilde{k}_{\alpha_f} = \left[\alpha_f(k_2 - k_1) + k_1, k_3 - \alpha_f(k_3 - k_2)\right] \tag{7.18}$$

$$\tilde{w}_{\alpha_t} = [w\alpha_t(w_2 - w_1) + w_1, w_3 - w\alpha_t(w_3 - w_2)] \tag{7.19}$$

$$\tilde{w}_{\alpha_f} = \left[\alpha_f(w_2 - w_1) + w_1, w_3 - \alpha_f(w_3 - w_2)\right] \tag{7.20}$$

$$\tilde{q}_{\alpha_t} = \left[w\alpha_t(q_2 - q_1) + q_1, q_3 - w\alpha_t(q_3 - q_2)\right] \tag{7.21}$$

$$\tilde{q}_{\alpha_f} = \left[\alpha_f(q_2 - q_1) + q_1, q_3 - \alpha_f(q_3 - q_2)\right] \tag{7.22}$$

$$\tilde{p}_{\alpha_t} = \left[w\alpha_t(p_2 - p_1) + p_1, p_3 - w\alpha_t(p_3 - p_2)\right] \tag{7.23}$$

$$\tilde{p}_{\alpha_f} = \left[\alpha_f(p_2 - p_1) + p_1, p_3 - \alpha_f(p_3 - p_2)\right] \tag{7.24}$$

$$\tilde{\lambda}_{\alpha_t} = \left[\tilde{\lambda}^L_{\alpha_t}, \tilde{\lambda}^U_{\alpha_t}\right], \tilde{\lambda}_{\alpha_f} = \left[\tilde{\lambda}^L_{\alpha_f}, \tilde{\lambda}^U_{\alpha_f}\right] \tag{7.25}$$

$$\tilde{\mu}_{\alpha_t} = \left[\tilde{\mu}^L_{\alpha_t}, \tilde{\mu}^U_{\alpha_t}\right], \tilde{\mu}_{\alpha_f} = \left[\tilde{\mu}^L_{\alpha_f}, \tilde{\mu}^U_{\alpha_f}\right] \tag{7.26}$$

$$\tilde{k}_{\alpha_t} = \left[\tilde{k}^L_{\alpha_t}, \tilde{k}^U_{\alpha_t}\right], \tilde{k}_{\alpha_f} = \left[\tilde{k}^L_{\alpha_f}, \tilde{k}^U_{\alpha_f}\right] \tag{7.27}$$

$$\tilde{q}_{\alpha_t} = \left[\tilde{q}^L_{\alpha_t}, \tilde{q}^U_{\alpha_t}\right], \tilde{q}_{\alpha_f} = \left[\tilde{q}^L_{\alpha_f}, \tilde{q}^U_{\alpha_f}\right] \tag{7.28}$$

$$\tilde{w}_{\alpha_t} = \left[\tilde{w}^L_{\alpha_t}, \tilde{w}^U_{\alpha_t}\right], \tilde{w}_{\alpha_f} = \left[\tilde{w}^L_{\alpha_f}, \tilde{w}^U_{\alpha_f}\right] \tag{7.29}$$

$$\tilde{p}_{\alpha_t} = \left[\tilde{p}^L_{\alpha_t}, \tilde{p}^U_{\alpha_t}\right], \tilde{p}_{\alpha_f} = \left[\tilde{p}^L_{\alpha_f}, \tilde{p}^U_{\alpha_f}\right]. \tag{7.30}$$

Based on established queue theory principles (Fazlollahtabar & Gholizadeh, 2020a), this study focuses on two fundamental performance measures:

1. Average queue length (L):
   - Represents the mean number of customers/jobs in the system.
   - Calculated considering both visible and hidden queue components.
   - Incorporates vague set representations for uncertain arrivals.
2. Expected waiting time (W):
   - Measures the average time customers/jobs spend in the system.
   - Accounts for parallel processing capabilities.
   - Reflects service rate uncertainties through vague parameters.

Theoretical foundation:

- Builds upon classical M/M/1 queue analysis.
- Extends traditional formulas to accommodate:
  - Vague number arithmetic.
  - Encouraged arrival dynamics.
  - Parallel server configurations.

Calculation methodology:

1. Derive steady-state probabilities using vague set operations.
2. Apply Little's law adaptations for uncertain systems: $L = \lambda W$ (with vague parameter interpretation).
3. Incorporate parallel processing effects through server-specific performance modifiers and resource-sharing factors.

Implementation considerations:

- Requires robust ranking of vague numbers.
- Must account for nonlinear effects of encouragement.
- Demands specialized computational approaches.

Validation approach:

- Compare with classical queue models at certainty limits.
- Verify through Monte Carlo simulation.
- Test sensitivity to vague parameter ranges.

This analytical framework provides decision-makers with crucial insights into system behavior while properly accounting for the inherent uncertainties in real-world queuing scenarios.

Now consider, if

$$\rho_1 = \frac{(\lambda_1 q_1 + \lambda_2 q_2 p_{21})(1+k)}{\mu_1 w_1 (1 - p_{12} p_{21})} \tag{7.31}$$

$$\rho_2 = \frac{(\lambda_2 q_2 + \lambda_1 q_1 p_{12})(1+k)}{\mu_2 w_2 (1 - p_{12} p_{21})} \tag{7.32}$$

$$\rho_3 = \frac{(\lambda'_1 q'_1)(1+k)}{\mu'_1 w'_1} \tag{7.33}$$

$$\rho_4 = \frac{(\lambda'_2 q'_2)(1+k)}{\mu'_2 w'_2} \tag{7.34}$$

$$\rho_5 = \left[\frac{[(\lambda'_1 q'_1) + (\lambda'_2 q'_2)](1+k)}{\mu_3 w_3} + \frac{[p_{13}(\lambda_1 q_1 + \lambda_2 q_2 p_{21}) + p_{23}(\lambda_2 q_2 + \lambda_1 q_1 p_{12})](1+k)}{\mu_2 w_2 (1 - p_{12} p_{21})} \right] \tag{7.35}$$

$$\lambda'_1 = \lambda_{11}, \lambda'_2 = \lambda_{21}, \mu'_1 = \mu_{11}, \mu'_2 = \mu_{21}, w'_1 = w_{11},$$
$$w'_2 = w_{21}, q'_1 = q_{11}, q'_2 = q_{21}.$$

The performance metrics for the proposed hybrid queuing system—comprising two bi-serial servers (S_{11}, S_{12}), two parallel servers (S_{21}, S_{22}), and a common serial server (S_3)—are derived as follows:

$$\text{Mean queue length } L = \frac{\rho_1}{1-\rho_1} + \frac{\rho_2}{1-\rho_2} + \frac{\rho_3}{1-\rho_3} + \frac{\rho_4}{1-\rho_4} + \frac{\rho_5}{1-\rho_5}$$
$$= L_1 + L_2 + L_3 + L_4 + L_5. \quad (7.36)$$

$$\text{Expected waiting time} = E(W) = \frac{L}{\lambda}, \text{ where } \lambda = \lambda_1 + \lambda_2 + \lambda_1' + \lambda_2'. \quad (7.37)$$

This research investigates queue system behavior under two fundamental types of uncertainty, employing vague set theory as the primary analytical framework (Gholizadeh et al., 2019):

1. Nature of ambiguity in the proposed model:
   - First-type uncertainty: Arises from inherent imprecision and insufficient system specialization.
   - Second-type uncertainty: Stems from inadequate sample data for reliable statistical estimation.
2. Methodological approach:
   - Adoption of vague parameters instead of deterministic values.
   - Development of possibility distributions for system characteristics.
   - Implementation of α-cut-based analysis for performance evaluation.
3. Theoretical justification:
   - Vague sets provide superior uncertainty representation compared to
     - Traditional fuzzy sets (single membership values).
     - Probabilistic approaches (requiring complete data).
   - Capability to simultaneously capture
     - Membership degrees (supporting evidence).
     - Non-membership degrees (opposing evidence).
4. Practical implementation:
   - Parameter specification through trapezoidal vague numbers.
   - System analysis using robust ranking techniques.
   - Performance metric derivation under incomplete information.

Let

$$\tilde{\rho}_{1\alpha_t}^{L,U} = \frac{\left(\lambda_{1\alpha_t}^{L,U} q_{1\alpha_t}^{L,U} + \lambda_{2\alpha_t}^{L,U} q_{2\alpha_t}^{L,U} p_{21\alpha_t}^{L,U}\right)\left(1 + k_{\alpha_t}^{L,U}\right)}{\mu_{1\alpha_t}^{L,U} w_{1\alpha_t}^{L,U}\left(1 - p_{12\alpha_t}^{L,U} p_{21\alpha_t}^{L,U}\right)} \quad (7.38)$$

$$\tilde{\rho}_{1\alpha_f}^{L,U} = \frac{\left(\lambda_{1\alpha_f}^{L,U} q_{1\alpha_f}^{L,U} + \lambda_{2\alpha_f}^{L,U} q_{2\alpha_f}^{L,U} p_{21\alpha_f}^{L,U}\right)\left(1 + k_{\alpha_f}^{L,U}\right)}{\mu_{1\alpha_f}^{L,U} w_{1\alpha_f}^{L,U}\left(1 - p_{12\alpha_f}^{L,U} p_{21\alpha_f}^{L,U}\right)} \tag{7.39}$$

$$\tilde{\rho}_{2\alpha_t}^{L,U} = \frac{\left(\lambda_{2\alpha_t}^{L,U} q_{2\alpha_t}^{L,U} + \lambda_{1\alpha_t}^{L,U} q_{1\alpha_t}^{L,U} p_{12\alpha_t}^{L,U}\right)\left(1 + k_{\alpha_t}^{L,U}\right)}{\mu_{2\alpha_t}^{L,U} w_{2\alpha_t}^{L,U}\left(1 - p_{12\alpha_t}^{L,U} p_{21\alpha_t}^{L,U}\right)} \tag{7.40}$$

$$\tilde{\rho}_{2\alpha_f}^{L,U} = \frac{\left(\lambda_{2\alpha_f}^{L,U} q_{2\alpha_f}^{L,U} + \lambda_{1\alpha_f}^{L,U} q_{1\alpha_f}^{L,U} p_{12\alpha_f}^{L,U}\right)\left(1 + k_{\alpha_f}^{L,U}\right)}{\mu_{2\alpha_f}^{L,U} w_{2\alpha_f}^{L,U}\left(1 - p_{12\alpha_f}^{L,U} p_{21\alpha_f}^{L,U}\right)} \tag{7.41}$$

$$\tilde{\rho}_{3\alpha_t}^{L,U} = \frac{\left(\lambda_{1\alpha_t}^{L,U\prime} q_{1\alpha_t}^{L,U\prime}\right)\left(1 + k_{\alpha_t}^{L,U}\right)}{\mu_{1\alpha_t}^{\prime L,U} w_{1\alpha_t}^{\prime L,U}} \tag{7.42}$$

$$\tilde{\rho}_{3\alpha_f}^{L,U} = \frac{\left(\lambda_{1\alpha_f}^{L,U\prime} q_{1\alpha_f}^{L,U\prime}\right)\left(1 + k_{\alpha_f}^{L,U}\right)}{\mu_{1\alpha_f}^{\prime L,U} w_{1\alpha_f}^{\prime L,U}} \tag{7.43}$$

$$\tilde{\rho}_{4\alpha_t}^{L,U} = \frac{\left(\lambda_{2\alpha_t}^{L,U\prime} q_{2\alpha_t}^{L,U\prime}\right)\left(1 + k_{\alpha_t}^{L,U}\right)}{\mu_{2\alpha_t}^{\prime L,U} w_{2\alpha_t}^{\prime L,U}} \tag{7.44}$$

$$\tilde{\rho}_{4\alpha_f}^{L,U} = \frac{\left(\lambda_{2\alpha_f}^{L,U\prime} q_{2\alpha_f}^{L,U\prime}\right)\left(1 + k_{\alpha_f}^{L,U}\right)}{\mu_{2\alpha_f}^{\prime L,U} w_{2\alpha_f}^{\prime L,U}} \tag{7.45}$$

$$\tilde{\rho}_{5\alpha_t}^{L,U} = \left[\frac{\left[\left(\lambda_{1\alpha_t}^{L,U\prime} q_{1\alpha_t}^{L,U\prime}\right) + \left(\lambda_{2\alpha_t}^{L,U\prime} q_{2\alpha_t}^{L,U\prime}\right)\right]\left(1 + k_{\alpha_t}^{L,U}\right)}{\mu_{3\alpha_t}^{L,U} w_{3\alpha_t}^{L,U}} + \frac{\left[p_{13\alpha_t}^{L,U}\left(\lambda_{1\alpha_t}^{L,U} q_{1\alpha_t}^{L,U} + \lambda_{2\alpha_t}^{L,U} q_{2\alpha_t}^{L,U} p_{21\alpha_t}^{L,U}\right) + p_{23\alpha_t}^{L,U}\left(\lambda_{2\alpha_t}^{L,U} q_{2\alpha_t}^{L,U} + \lambda_{1\alpha_t}^{L,U} q_{1\alpha_t}^{L,U} p_{12\alpha_t}^{L,U}\right)\right]\left(1 + k_{\alpha_t}^{L,U}\right)}{\mu_{2\alpha_t}^{L,U} w_{2\alpha_t}^{L,U}\left(1 - p_{12\alpha_t}^{L,U} p_{21\alpha_t}^{L,U}\right)}\right] \tag{7.46}$$

$$\tilde{\rho}_{5\alpha_f}^{L,U} = \left[\frac{\left[\left(\lambda_{1\alpha_f}^{L,U\prime} q_{1\alpha_f}^{L,U\prime}\right) + \left(\lambda_{2\alpha_f}^{L,U\prime} q_{2\alpha_f}^{L,U\prime}\right)\right]\left(1 + k_{\alpha_f}^{L,U}\right)}{\mu_{3\alpha_t}^{L,U} w_{3\alpha_t}^{L,U}} + \frac{\left[p_{13\alpha_f}^{L,U}\left(\lambda_{1\alpha_f}^{L,U} q_{1\alpha_f}^{L,U} + \lambda_{2\alpha_f}^{L,U} q_{2\alpha_f}^{L,U} p_{21\alpha_f}^{L,U}\right) + p_{23\alpha_f}^{L,U}\left(\lambda_{2\alpha_f}^{L,U} q_{2\alpha_f}^{L,U} + \lambda_{1\alpha_f}^{L,U} q_{1\alpha_f}^{L,U} p_{12\alpha_f}^{L,U}\right)\right]\left(1 + k_{\alpha_f}^{L,U}\right)}{\mu_{2\alpha_f}^{L,U} w_{2\alpha_f}^{L,U}\left(1 - p_{12\alpha_f}^{L,U} p_{21\alpha_f}^{L,U}\right)}\right] \tag{7.47}$$

$$\tilde{L}_{1\alpha_t}^{L,U} = \frac{\left[\frac{\left(\lambda_{1\alpha_t}^{L,U} q_{1\alpha_t}^{L,U} + \lambda_{2\alpha_t}^{L,U} q_{2\alpha_t}^{L,U} p_{21\alpha_t}^{L,U}\right)\left(1 + k_{\alpha_t}^{L,U}\right)}{\mu_{1\alpha_t}^{L,U} w_{1\alpha_t}^{L,U}\left(1 - p_{12\alpha_t}^{L,U} p_{21\alpha_t}^{L,U}\right)}\right]}{1 - \left[\frac{\left(\lambda_{1\alpha_t}^{L,U} q_{1\alpha_t}^{L,U} + \lambda_{2\alpha_t}^{L,U} q_{2\alpha_t}^{L,U} p_{21\alpha_t}^{L,U}\right)\left(1 + k_{\alpha_t}^{L,U}\right)}{\mu_{1\alpha_t}^{L,U} w_{1\alpha_t}^{L,U}\left(1 - p_{12\alpha_t}^{L,U} p_{21\alpha_t}^{L,U}\right)}\right]} \tag{7.48}$$

$$\tilde{L}_{1\alpha_f}^{L,U} = \frac{\left[\frac{\left(\lambda_{1\alpha_f}^{L,U} q_{1\alpha_f}^{L,U} + \lambda_{2\alpha_f}^{L,U} q_{2\alpha_f}^{L,U} p_{21\alpha_f}^{L,U}\right)\left(1+k_{\alpha_f}^{L,U}\right)}{\mu_{1\alpha_f}^{L,U} w_{1\alpha_f}^{L,U}\left(1-p_{12\alpha_f}^{L,U} p_{21\alpha_f}^{L,U}\right)}\right]}{1-\left[\frac{\left(\lambda_{1\alpha_f}^{L,U} q_{1\alpha_f}^{L,U} + \lambda_{2\alpha_f}^{L,U} q_{2\alpha_f}^{L,U} p_{21\alpha_f}^{L,U}\right)\left(1+k_{\alpha_f}^{L,U}\right)}{\mu_{1\alpha_f}^{L,U} w_{1\alpha_f}^{L,U}\left(1-p_{12\alpha_f}^{L,U} p_{21\alpha_f}^{L,U}\right)}\right]} \tag{7.49}$$

$$\tilde{L}_{2\alpha_t}^{L,U} = \frac{\left[\frac{\left(\lambda_{2\alpha_t}^{L,U} q_{2\alpha_t}^{L,U} + \lambda_{1\alpha_t}^{L,U} q_{1\alpha_t}^{L,U} p_{12\alpha_t}^{L,U}\right)\left(1+k_{\alpha_t}^{L,U}\right)}{\mu_{2\alpha_t}^{L,U} w_{2\alpha_t}^{L,U}\left(1-p_{21\alpha_t}^{L,U}\right)}\right]}{1-\left[\frac{\left(\lambda_{2\alpha_t}^{L,U} q_{2\alpha_t}^{L,U} + \lambda_{1\alpha_t}^{L,U} q_{1\alpha_t}^{L,U} p_{12\alpha_t}^{L,U}\right)\left(1+k_{\alpha_t}^{L,U}\right)}{\mu_{2\alpha_t}^{L,U} w_{2\alpha_t}^{L,U}\left(1-p_{12\alpha_t}^{L,U} p_{21\alpha_t}^{L,U}\right)}\right]} \tag{7.50}$$

$$\tilde{L}_{2\alpha_f}^{L,U} = \frac{\left[\frac{\left(\lambda_{2\alpha_f}^{L,U} q_{2\alpha_f}^{L,U} + \lambda_{1\alpha_f}^{L,U} q_{1\alpha_f}^{L,U} p_{12\alpha_f}^{L,U}\right)\left(1+k_{\alpha_f}^{L,U}\right)}{\mu_{2\alpha_f}^{L,U} w_{2\alpha_f}^{L,U}\left(1-p_{12\alpha_f}^{L,U} p_{21\alpha_f}^{L,U}\right)}\right]}{1-\left[\frac{\left(\lambda_{2\alpha_f}^{L,U} q_{2\alpha_f}^{L,U} + \lambda_{1\alpha_f}^{L,U} q_{1\alpha_f}^{L,U} p_{12\alpha_f}^{L,U}\right)\left(1+k_{\alpha_f}^{L,U}\right)}{\mu_{2\alpha_f}^{L,U} w_{2\alpha_f}^{L,U}\left(1-p_{12\alpha_f}^{L,U} p_{21\alpha_f}^{L,U}\right)}\right]} \tag{7.51}$$

$$\tilde{L}_{3\alpha_t}^{L,U} = \frac{\left[\frac{\left(\lambda_{1\alpha_t}^{L,U'} q_{1\alpha_t}^{L,U'}\right)\left(1+k_{\alpha_t}^{L,U}\right)}{\mu_{1\alpha_t}^{\prime L,U} w_{1\alpha_t}^{\prime L,U}}\right]}{1-\left[\frac{\left(\lambda_{1\alpha_t}^{L,U'} q_{1\alpha_t}^{L,U'}\right)\left(1+k_{\alpha_t}^{L,U}\right)}{\mu_{1\alpha_t}^{\prime L,U} w_{1\alpha_t}^{\prime L,U}}\right]} \tag{7.52}$$

$$\tilde{L}_{3\alpha_f}^{L,U} = \frac{\left[\frac{\left(\lambda_{1\alpha_f}^{L,U'} q_{1\alpha_f}^{L,U'}\right)\left(1+k_{\alpha_f}^{L,U}\right)}{\mu_{1\alpha_f}^{\prime L,U} w_{1\alpha_f}^{\prime L,U}}\right]}{1-\left[\frac{\left(\lambda_{1\alpha_f}^{L,U'} q_{1\alpha_f}^{L,U'}\right)\left(1+k_{\alpha_f}^{L,U}\right)}{\mu_{1\alpha_f}^{\prime L,U} w_{1\alpha_f}^{\prime L,U}}\right]} \tag{7.53}$$

$$\tilde{L}_{4\alpha_t}^{L,U} = \frac{\left[\frac{\left(\lambda_{2\alpha_t}^{L,U'} q_{2\alpha_t}^{L,U'}\right)\left(1+k_{\alpha_t}^{L,U}\right)}{\mu_{2\alpha_t}^{\prime L,U} w_{2\alpha_t}^{\prime L,U}}\right]}{1-\left[\frac{\left(\lambda_{2\alpha_t}^{L,U'} q_{2\alpha_t}^{L,U'}\right)\left(1+k_{\alpha_t}^{L,U}\right)}{\mu_{2\alpha_t}^{\prime L,U} w_{2\alpha_t}^{\prime L,U}}\right]} \tag{7.54}$$

$$\tilde{L}_{4\alpha_f}^{L,U} = \frac{\left[\frac{\left(\lambda_{2\alpha_f}^{L,U'} q_{2\alpha_f}^{L,U'}\right)\left(1+k_{\alpha_f}^{L,U}\right)}{\mu_{2\alpha_f}^{\prime L,U} w_{2\alpha_f}^{\prime L,U}}\right]}{1-\left[\frac{\left(\lambda_{2\alpha_f}^{L,U'} q_{2\alpha_f}^{L,U'}\right)\left(1+k_{\alpha_f}^{L,U}\right)}{\mu_{2\alpha_f}^{\prime L,U} w_{2\alpha_f}^{\prime L,U}}\right]} \tag{7.55}$$

$$\tilde{L}_{5\alpha_t}^{L,U} = \frac{\left[\frac{\left[\left(\lambda_{1\alpha_t}^{L,U\prime} q_{1\alpha_t}^{L,U\prime}\right) + \left(\lambda_{2\alpha_t}^{L,U\prime} q_{2\alpha_t}^{L,U\prime}\right)\right]\left(1 + k_{\alpha_t}^{L,U}\right)}{\mu_{3\alpha_t}^{L,U} w_{3\alpha_t}^{L,U}} + \frac{\left[p_{13\alpha_t}^{L,U}\left(\lambda_{1\alpha_t}^{L,U} q_{1\alpha_t}^{L,U} + \lambda_{2\alpha_t}^{L,U} q_{2\alpha_t}^{L,U} p_{21\alpha_t}^{L,U}\right) + p_{23\alpha_t}^{L,U}\left(\lambda_{2\alpha_t}^{L,U} q_{2\alpha_t}^{L,U} + \lambda_{1\alpha_t}^{L,U} q_{1\alpha_t}^{L,U} p_{12\alpha_t}^{L,U}\right)\right]\left(1 + k_{\alpha_t}^{L,U}\right)}{\mu_{2\alpha_t}^{L,U} w_{2\alpha_t}^{L,U}\left(1 - p_{12\alpha_t}^{L,U} p_{21\alpha_t}^{L,U}\right)}\right]}{1 - \left[\frac{\left[\left(\lambda_{1\alpha_t}^{L,U\prime} q_{1\alpha_t}^{L,U\prime}\right) + \left(\lambda_{2\alpha_t}^{L,U\prime} q_{2\alpha_t}^{L,U\prime}\right)\right]\left(1 + k_{\alpha_t}^{L,U}\right)}{\mu_{3\alpha_t}^{L,U} w_{3\alpha_t}^{L,U}} + \frac{\left[p_{13\alpha_t}^{L,U}\left(\lambda_{1\alpha_t}^{L,U} q_{1\alpha_t}^{L,U} + \lambda_{2\alpha_t}^{L,U} q_{2\alpha_t}^{L,U} p_{21\alpha_t}^{L,U}\right) + p_{23\alpha_t}^{L,U}\left(\lambda_{2\alpha_t}^{L,U} q_{2\alpha_t}^{L,U} + \lambda_{1\alpha_t}^{L,U} q_{1\alpha_t}^{L,U} p_{12\alpha_t}^{L,U}\right)\right]\left(1 + k_{\alpha_t}^{L,U}\right)}{\mu_{2\alpha_t}^{L,U} w_{2\alpha_t}^{L,U}\left(1 - p_{12\alpha_t}^{L,U} p_{21\alpha_t}^{L,U}\right)}\right]} \tag{7.56}$$

$$\tilde{L}_{5\alpha_f}^{L,U} = \frac{\left[\frac{\left[\left(\lambda_{1\alpha_f}^{L,U\prime} q_{1\alpha_f}^{L,U\prime}\right) + \left(\lambda_{2\alpha_f}^{L,U\prime} q_{2\alpha_f}^{L,U\prime}\right)\right]\left(1 + k_{\alpha_f}^{L,U}\right)}{\mu_{3\alpha_t}^{L,U} w_{3\alpha_t}^{L,U}} + \frac{\left[p_{13\alpha_f}^{L,U}\left(\lambda_{1\alpha_f}^{L,U} q_{1\alpha_f}^{L,U} + \lambda_{2\alpha_f}^{L,U} q_{2\alpha_f}^{L,U} p_{21\alpha_f}^{L,U}\right) + p_{23\alpha_f}^{L,U}\left(\lambda_{2\alpha_f}^{L,U} q_{2\alpha_f}^{L,U} + \lambda_{1\alpha_f}^{L,U} q_{1\alpha_f}^{L,U} p_{12\alpha_f}^{L,U}\right)\right]\left(1 + k_{\alpha_f}^{L,U}\right)}{\mu_{2\alpha_f}^{L,U} w_{2\alpha_f}^{L,U}\left(1 - p_{12\alpha_f}^{L,U} p_{21\alpha_f}^{L,U}\right)}\right]}{1 - \left[\frac{\left[\left(\lambda_{1\alpha_f}^{L,U\prime} q_{1\alpha_f}^{L,U\prime}\right) + \left(\lambda_{2\alpha_f}^{L,U\prime} q_{2\alpha_f}^{L,U\prime}\right)\right]\left(1 + k_{\alpha_f}^{L,U}\right)}{\mu_{3\alpha_t}^{L,U} w_{3\alpha_t}^{L,U}} + \frac{\left[p_{13\alpha_f}^{L,U}\left(\lambda_{1\alpha_f}^{L,U} q_{1\alpha_f}^{L,U} + \lambda_{2\alpha_f}^{L,U} q_{2\alpha_f}^{L,U} p_{21\alpha_f}^{L,U}\right) + p_{23\alpha_f}^{L,U}\left(\lambda_{2\alpha_f}^{L,U} q_{2\alpha_f}^{L,U} + \lambda_{1\alpha_f}^{L,U} q_{1\alpha_f}^{L,U} p_{12\alpha_f}^{L,U}\right)\right]\left(1 + k_{\alpha_f}^{L,U}\right)}{\mu_{2\alpha_f}^{L,U} w_{2\alpha_f}^{L,U}\left(1 - p_{12\alpha_f}^{L,U} p_{21\alpha_f}^{L,U}\right)}\right]} \tag{7.57}$$

and

$$\text{Mean queue length} = \tilde{L} = \tilde{L}_1 + \tilde{L}_2 + \tilde{L}_3 + \tilde{L}_4 + \tilde{L}_5. \tag{7.58}$$

$$\text{Mean arrival rate} = \tilde{\lambda} = \tilde{\lambda}_1 + \tilde{\lambda}_2 + \tilde{\lambda}_1' + \tilde{\lambda}_2'. \tag{7.59}$$

$$\text{Expected waiting time} = E\left(\tilde{W}\right) = \frac{\tilde{L}}{\tilde{\lambda}}. \tag{7.60}$$

The membership functions for $\tilde{L}$ are

$$t_{\tilde{L}}(L) = \begin{cases} \frac{L-L_1}{w(L_2-L_1)} & L_1 \le L \le L_2 \\ \frac{L_3-L}{w(L_3-L_2)} & L_2 \le L \le L_3 \end{cases} \tag{7.61}$$

$$1 - f_{\tilde{L}}(L) = \begin{cases} \frac{L-L_1}{(L_2-L_1)} & L_1 \le L \le L_2 \\ \frac{L_3-L}{(L_3-L_2)} & L_2 \le L \le L_3 \end{cases} \tag{7.62}$$

$$\tilde{L}_{\alpha_t} = [w\alpha_t(l_2 - l_1) + l_1, l_3 - w\alpha_t(l_3 - l_2)] \tag{7.63}$$

$$\tilde{L}_{\alpha_f} = \left[\alpha_f(l_2 - l_1) + l_1, l_3 - \alpha_f(l_3 - l_2)\right] \tag{7.64}$$

$$\tilde{L}_{\alpha_t} = \left[\tilde{L}^L_{\alpha_t}, \tilde{L}^U_{\alpha_t}\right], \tilde{L}_{\alpha_f} = \left[\tilde{L}^L_{\alpha_f}, \tilde{L}^U_{\alpha_f}\right] \tag{7.65}$$

$$\left(\tilde{W}\right)^L_{\alpha_t} = \frac{\left(\tilde{L}\right)^L_{\alpha_t}}{\tilde{\lambda}^L_{\alpha_t}}, \left(\tilde{W}\right)^U_{\alpha_t} = \frac{\left(\tilde{L}\right)^U_{\alpha_t}}{\tilde{\lambda}^U_{\alpha_t}} \tag{7.66}$$

$$\left(\tilde{W}\right)^L_{\alpha_f} = \frac{\left(\tilde{L}\right)^L_{\alpha_f}}{\tilde{\lambda}^L_{\alpha_f}}, \left(\tilde{W}\right)^U_{\alpha_f} = \frac{\left(\tilde{L}\right)^U_{\alpha_f}}{\tilde{\lambda}^U_{\alpha_f}}. \tag{7.67}$$

The characteristics of a system can be described by its membership function. Since input information is ambiguous, and in decision-making issues, the value and importance of different options are expressed as vague numbers. Therefore, the robust ranking technique is used to demonstrate the validity of the evaluation.

If $\tilde{A} = (a_1, a_2, a_3)$ is a triangular fuzzy number, where

$$\begin{aligned} \tilde{A}_{\alpha_t} &= [w\alpha_t(a_2 - a_1) + a_1, a_3 - w\alpha_t(a_3 - a_2)]; \\ \tilde{A}_{\alpha_f} &= \left[\alpha_f(a_2 - a_1) + a_1, a_3 - \alpha_f(a_3 - a_2)\right], \end{aligned} \tag{7.68}$$

$$\tilde{A}_{\alpha_t} = \left[\tilde{a}^L_{\alpha_t}, \tilde{a}^U_{\alpha_t}\right], \tilde{A}_{\alpha_f} = \left[\tilde{a}^L_{\alpha_f}, \tilde{a}^U_{\alpha_f}\right], \tag{7.69}$$

then robust ranking is defined by

$$R\left(\tilde{A}_{\alpha_t}\right) = \int_0^1 (0.5)\left[\tilde{a}^L_{\alpha_t}, \tilde{a}^U_{\alpha_t}\right] d\alpha_t, \tag{7.70}$$

$$R\left(\tilde{A}_{\alpha_f}\right) = \int_0^1 (0.5)\left[\tilde{a}^L_{\alpha_f}, \tilde{a}^U_{\alpha_f}\right] d\alpha_f. \tag{7.71}$$

As stated, due to the uncertainty of fuzzy and vague systems, we use the vague system because it is more accurate than fuzzy systems and covers various scenarios. Therefore, making decisions is easier for organizations.

7.5 Numerical Example

This study examines a clinical diagnostic process as a queuing network, where patients flow through sequential testing stages with probabilistic routing decisions. The system architecture consists of (1) an initial diagnostic stage (S_1) performing lab tests and basic examinations; (2) parallel secondary testing paths including electrocardiogram (S_{21}) and chest X-ray (S_{22}) services; and (3) a final physician consultation stage (S_3) for result interpretation. Patient routing follows conditional probabilities based on preliminary test outcomes, where cases may either proceed directly from S_1 to S_3 for normal results or divert to specialized testing (S_{21}/S_{22}) before final evaluation. The model incorporates vague set theory to capture inherent uncertainties in service times (due to test complexity variations) and routing decisions (reflecting diagnostic uncertainty), with complete operational parameters specified in Table 7.1. This framework enables analysis of key performance metrics including average patient throughput time, resource utilization rates, and bottleneck identification, while accommodating real-world healthcare complexities such as priority cases, equipment sharing between parallel paths, and potential retesting requirements. The implementation demonstrates how vague queuing models can optimize multi-stage service systems with flexible routing under uncertainty, particularly valuable for clinical process improvement and diagnostic center design.

According to the proposed formulations, the queue length is computed in Table 7.2.

Meanwhile, the waiting time is shown in Table 7.3.

As a result, the final ranking of average queue length and waiting times for different alfa levels are presented in Table 7.4.

As can be seen, given the sensitivity analysis, the higher the alpha value, the lower the mean queue length and the average wait time are. Also according to the rankings,

Table 7.1 Input data

Number	Number of customers (n)	Arrival rate ($\tilde{\lambda}$)	Possibility of arrive ($\tilde{q}$)	Service rate ($\tilde{\mu}$)	Possibility of service ($\tilde{w}$)	Possibilities ($\tilde{P}_{ij}$)	Percentage change in number of customers ($\tilde{K}$)
1	6	[2, 3, 4]	[0.1, 0.2, 0.3]	[9, 10, 11]	[0.3, 0.4, 0.5]	$\tilde{P}_{12}$ [0.3, 0.4, 0.5]	[0.25, 0.5, 0.75]
2	5	[3, 4, 5]	[0.2, 0.3, 0.4]	[10, 11, 12]	[0.2, 0.3, 0.4]	$\tilde{P}_{13}$ [0.2, 0.3, 0.4]	[0.25, 0.5, 0.75]
3	5	[4, 5, 6]	[0.3, 0.4, 0.5]	[12, 13, 14]	[0.4, 0.5, 0.6]	$\tilde{P}_{21}$ [0.1, 0.2, 0.3]	[0.25, 0.5, 0.75]
4	4	[5, 6, 7]	[0.5, 0.6, 0.7]	[8, 9, 10]	[0.5, 0.6, 0.7]	$\tilde{P}_{23}$ [0.4, 0.5, 0.6]	[0.25, 0.5, 0.75]
5	18			[14, 15, 16]	[0.1, 0.2, 0.3]		

Table 7.2 Computed queue length

		$\widetilde{L_1}$		$\widetilde{L_2}$		$\widetilde{L_3}$		$\widetilde{L_4}$		$\widetilde{L_5}$		$\tilde{L}$	
α_t	w	$(\widetilde{L_1})^L_{\alpha_t}$	$(\widetilde{L_1})^U_{\alpha_t}$	$(\widetilde{L_2})^L_{\alpha_t}$	$(\widetilde{L_2})^U_{\alpha_t}$	$(\widetilde{L_3})^L_{\alpha_t}$	$(\widetilde{L_3})^U_{\alpha_t}$	$(\widetilde{L_4})^L_{\alpha_t}$	$(\widetilde{L_4})^U_{\alpha_t}$	$(\widetilde{L_5})^L_{\alpha_t}$	$(\widetilde{L_5})^U_{\alpha_t}$	$(\tilde{L})^L_{\alpha_t}$	$(\tilde{L})^U_{\alpha_t}$
0.15	1	0.1737	1.6553	0.8321	2.1670	0.1217	0.5668	0.6463	2.4495	0.8052	4.3286	2.579	11.1672
	2	0.2096	1.3543	0.8680	1.8660	0.1424	0.5190	0.6870	2.4017	0.8411	4.0276	2.7481	10.1686
	3	0.2496	1.1242	0.9080	1.6359	0.1642	0.4712	0.7905	2.3539	0.8811	3.7975	2.9934	9.3827
	4	0.2986	0.9425	0.9570	1.450	0.1871	0.4227	0.8134	2.3054	0.9301	3.6116	3.1862	8.7322
	5	0.3452	0.7925	1.0036	1.30	0.213	0.3716	0.9363	2.2543	0.9767	3.4686	3.4748	8.1870
0.30	1	0.2096	1.3543	0.8680	1.8660	0.1424	0.5190	0.6870	2.4017	0.8411	4.0276	2.7481	10.1686
	2	0.2986	0.9425	0.9570	1.45	0.1871	0.4227	0.8134	2.3054	0.9301	3.6116	3.1862	8.7322
	3	0.4026	0.7740	1.0610	1.2815	0.2364	0.3964	0.8627	2.2091	1.0341	3.3431	3.5968	8.0041
	4	0.4937	0.6488	1.0921	1.1863	0.2903	0.3279	0.9166	2.1106	1.1752	3.2179	4.0379	7.4915
	5	0.5848	0.5936	1.1251	1.1511	0.3042	0.3159	1.0705	2.0086	1.3663	3.0927	4.2709	7.0519
0.45	1	0.2496	1.3242	0.9080	1.6359	0.1642	0.4712	0.7905	2.3539	0.8811	3.7975	2.9934	9.5827
	2	0.4026	1.0040	1.0610	1.3815	0.2364	0.3564	0.8627	2.2091	1.0341	3.3431	3.5941	8.2941
	3	0.5315	0.9055	1.0899	1.2313	0.2895	0.3216	0.9349	2.0743	1.263	2.8929	4.1088	7.4256
	4	0.6072	0.8803	1.1656	1.1961	0.3035	0.3168	1.418	1.9695	1.6387	2.7677	4.3330	7.1004
	5	0.7342	0.7551	1.1926	1.1999	0.3107	0.3128	1.612	1.9407	1.3657	2.6425	4.8152	6.2510
α_f	w	$(\widetilde{L_1})^L_{\alpha_f}$	$(\widetilde{L_1})^U_{\alpha_f}$	$(\widetilde{L_2})^L_{\alpha_f}$	$(\widetilde{L_2})^U_{\alpha_f}$	$(\widetilde{L_3})^L_{\alpha_f}$	$(\widetilde{L_3})^U_{\alpha_f}$	$(\widetilde{L_4})^L_{\alpha_f}$	$(\widetilde{L_4})^U_{\alpha_f}$	$(\widetilde{L_5})^L_{\alpha_f}$	$(\widetilde{L_5})^U_{\alpha_f}$	$(\tilde{L})^L_{\alpha_f}$	$(\tilde{L})^U_{\alpha_f}$
0.55	1	0.2790	0.9986	1.1585	1.8781	0.1794	0.4109	0.6655	1.9304	1.8515	3.2917	4.1339	8.5097
0.75	1	0.3452	0.7954	1.3880	1.8382	0.2456	0.3877	0.8594	1.7231	2.0447	2.9357	4.8829	7.6801
0.95	1	0.4235	0.6381	1.5894	1.6403	0.3239	0.3302	0.9424	1.5158	2.2551	2.5837	5.5343	6.7081

Table 7.3 Calculated waiting time

		$\widetilde{w_1}$		$\widetilde{w_2}$		$\widetilde{w_3}$		$\widetilde{w_4}$		$\tilde{W}$	
α_t	w	$(\widetilde{w_1})^L_{\alpha_t}$	$(\widetilde{w_1})^U_{\alpha_t}$	$(\widetilde{w_2})^L_{\alpha_t}$	$(\widetilde{w_2})^U_{\alpha_t}$	$(\widetilde{w_3})^L_{\alpha_t}$	$(\widetilde{w_3})^U_{\alpha_t}$	$(\widetilde{w_4})^L_{\alpha_t}$	$(\widetilde{w_4})^U_{\alpha_t}$	$(\tilde{W})^L_{\alpha_t}$	$(\tilde{W})^U_{\alpha_t}$
0.15	1	0.0807	0.4299	0.2641	0.4468	0.0293	0.0968	0.12	0.3575	0.1766	0.5218
	2	0.0911	0.3660	0.2630	0.3970	0.03311	0.0910	0.1296	0.3584	0.1807	0.4888
	3	0.1018	0.3166	0.2631	0.3595	0.03689	0.0849	0.1450	0.3593	0.1894	0.4644
	4	0.1148	0.2772	0.2658	0.3295	0.0406	0.0782	0.1452	0.3602	0.1942	0.4455
	5	0.1255	0.2438	0.2676	0.3058	0.0448	0.0707	0.1628	0.3606	0.2044	0.4308
0.30	1	0.0911	0.3660	0.2630	0.3970	0.03311	0.0910	0.1296	0.3584	0.1807	0.4888
	2	0.1148	0.2772	0.2658	0.3295	0.0406	0.0782	0.1452	0.3602	0.1942	0.4455
	3	0.1388	0.2496	0.2720	0.3125	0.0482	0.0777	0.1462	0.3621	0.2043	0.4350
	4	0.1542	0.2387	0.2600	0.3121	0.0630	0.0683	0.1478	0.3638	0.2147	0.4355
	5	0.1670	0.2374	0.2500	0.3211	0.0553	0.0702	0.1646	0.3652	0.2235	0.4207
0.45	1	0.1018	0.5404	0.2631	0.3595	0.0368	0.0849	0.1450	0.3593	0.1894	0.4743
	2	0.1388	0.3464	0.2720	0.3369	0.0482	0.0698	0.1462	0.3621	0.2042	0.4507
	3	0.1586	0.2702	0.2505	0.3373	0.0541	0.06916	0.1472	0.3671	0.2117	0.4473
	4	0.1597	0.2316	0.2428	0.3737	0.0523	0.0754	0.2085	0.3787	0.2421	0.4217
	5	0.1727	0.1776	0.2271	0.4363	0.0497	0.0834	0.2223	0.4085	0.2267	0.4070
α_f	w	$(\widetilde{w_1})^L_{\alpha_f}$	$(\widetilde{w_1})^U_{\alpha_f}$	$(\widetilde{w_2})^L_{\alpha_f}$	$(\widetilde{w_2})^U_{\alpha_f}$	$(\widetilde{w_3})^L_{\alpha_f}$	$(\widetilde{w_3})^U_{\alpha_f}$	$(\widetilde{w_4})^L_{\alpha_f}$	$(\widetilde{w_4})^U_{\alpha_f}$	$(\tilde{W})^L_{\alpha_f}$	$(\tilde{W})^U_{\alpha_f}$
0.55	1	0.1094	0.2894	0.3263	0.4220	0.0394	0.0570	0.1462	0.3542	0.2551	0.4297
0.75	1	0.1255	0.2447	0.3701	0.4325	0.0517	0.0357	0.1494	0.3282	0.2872	0.4042
0.95	1	0.1435	0.2092	0.4276	0.3803	0.0654	0.0257	0.1583	0.2505	0.3109	0.3685

Table 7.4 Ranking of the average queue length and calculated waiting time

		$\tilde{W}$		$\tilde{L}$		$R(\tilde{W})$	$R(\tilde{L})$
α_t	w	$(\tilde{W})^L_{\alpha_t}$	$(\tilde{W})^u_{\alpha_t}$	$(\tilde{L})^L_{\alpha_t}$	$(\tilde{L})^U_{\alpha_t}$	$R(\tilde{W})_{\alpha_t}$	$R(\tilde{L})_{\alpha_t}$
0.15	1	0.1766	0.5218	2.579	11.1672	0.3492	6.8731
	2	0.1807	0.4888	2.7481	10.1686	0.3347	6.4583
	3	0.1894	0.4644	2.9934	9.3827	0.3269	6.1880
	4	0.1942	0.4455	3.1862	8.7322	0.3198	5.9592
	5	0.2044	0.4308	3.4748	8.1870	0.3176	5.8309
0.30	1	0.1807	0.4888	2.7481	10.1686	0.3347	6.4583
	2	0.1942	0.4455	3.1862	8.7322	0.3198	5.9592
	3	0.2043	0.4350	3.5968	8.0041	0.3196	5.8004
	4	0.2147	0.4355	4.0379	7.4915	0.3251	5.7647
	5	0.2235	0.4207	4.2709	7.0519	0.3221	5.6614
0.45	1	0.1894	0.4743	2.9934	9.5827	0.3318	6.2880
	2	0.2042	0.4507	3.5941	8.2941	0.3274	5.9441
	3	0.2117	0.4473	4.1088	7.4256	0.3295	5.7672
	4	0.2421	0.4217	4.333	7.1004	0.3319	5.7167
	5	0.2267	0.4070	4.8152	6.2510	0.3168	5.5331
α_f	w	$(\tilde{W})^L_{\alpha_f}$	$(\tilde{W})^U_{\alpha_f}$	$(\tilde{L})^L_{\alpha_f}$	$(\tilde{L})^U_{\alpha_f}$	$R(\tilde{W})_{\alpha_f}$	$R(\tilde{L})_{\alpha_f}$
0.55	1	0.2551	0.4297	4.1339	8.5097	0.3424	6.3218
0.75	2	0.2872	0.4042	4.8829	7.6801	0.3457	6.2815
0.95	3	0.3109	0.3685	5.5343	6.7081	0.3397	6.1212

the upper and lower limits of queue length and waiting time cannot be outside the calculated range. By increasing the weight of the numbers, vague numbers converge to the real numbers. That is, the amount of the upper limit of the length of the queue is reduced and the lower limit of the queue length is increased, indicating that the upper limit and the lower limit are closer to each other, which approaches the real number.

7.6 Conclusions

In this chapter, the vague analysis of synergistic collaboration of bi-serial and parallel servers with a common server has been conducted using server queuing model with encouraged arrivals. Further, optimization was performed such that the arrival rate and service rate are vague numbers showing with possibility vague. The system can also be studied for heterogeneous service. A clinical physical test procedure where the patients have to pass through a series of stages (lab tests, electrocardiogram, chest X-ray, etc.) was considered as an experimental study. And with the numerical example and sensitivity analysis, organizations can make the best decisions. In decision-making issues, the value and importance of different options are expressed as vague numbers. Finally, by using the robust ranking technique validity test for evaluation was proposed.

References and Further Reading

Chaleshigar Kordasiabi, M., Gholizadeh, H., Fazlollahtabar, H., & Javadian, N. (2020). Analysis of cost model with queuing system under uncertainty. *Journal of Industrial and Production Engineering, 37*(6), 292–304.

Fazlollahtabar, H., & Gholizadeh, H. (2019). Economic analysis of the M/M/1/N queuing system cost model in a vague environment. *International Journal of Fuzzy Logic and Intelligent Systems, 19*(3), 192–203.

Fazlollahtabar, H., & Gholizadeh, H. (2020a). Fuzzy possibility regression integrated with fuzzy adaptive neural network for predicting and optimizing electrical discharge machining parameters. *Computers and Industrial Engineering, 140*, Article 106225.

Fazlollahtabar, H., & Gholizadeh, H. (2020b). Application of queuing theory in quality control of multi-stage flexible flow shop. *Yugoslav Journal of Operations Research, 30*(1), 101–119.

Gau, W. L., & Buehrer, D. J. (1993). Vague sets. *IEEE Transaction Systems Man Cybernet, 23*(2), 610–614.

Gholizadeh, H., Fazlollahtabar, H., & Gholizadeh, R. (2019). A modified branch and bound algorithm for a vague flow-shop scheduling problem. *Iranian Journal of Fuzzy Systems, 16*(4), 55–64.

Gonzalez-Lopez, V. A., Gholizadeh, R., & Shirazi, A. M. (2016). Optimization of queuing theory based on vague environment. *International Journal of Fuzzy System Applications, 5*(1), 1–26.

Kao, C., Li, C., & Chen, S. (1999). Parametric programming to the analysis of fuzzy queues. *Fuzzy Sets and Systems, 107*, 93–100.

Maggu, P. L. (1970). Phase type service queues with two servers in biseries. *Journal of Operational Research Society of Japan, 13*(1), 1–6.

Nagoorgani, A., & Retha, W. (2006). Fuzzy tandem queues. *Acta Cienicia Indica, XXXII*(M), 257–263.

Prade, H. M. (1980). An outline of fuzzy or possiblistic models for queuing systems. In P. P. Wang & S. K. Chang (Eds.), *Fuzzy sets: Theory and applications to policy analysis and information systems* (pp. 147–153). Plenum Press.

Seema, Gupta, D., & Sharma, S. (2013). Analysis of biserial servers linked to a common server in fuzzy environment. *International Journal of Computer Applications, 68*(6), 26–32.

Singh, M. (1979). *On certain queuing networks of serial/bi-serial channels*. Ph. D Thesis, Meerut University.

Singh, T. P., & Pardeep. (2009). Serial queue model with blocking under fuzzy environment. *JMASS, 5*(2), 86–94.

Som, B. K., & Seth, S. (2017). An M/M/1/N queuing system with encouraged arrivals. *Global Journal of Pure and Applied Mathematics, 13*(7), 3443–3453.

Zadeh, L. A. (1978). Fuzzy sets as a basis for a theory of possibility. *Fuzzy Sets and Systems, 1*, 3–28.

Chapter 8
Rough Multi-objective Optimization Using Best-Worst Method

Abstract This chapter delves into the multifaceted aspects of sustainability within the biofuel supply chain, aiming to assess its overall value. The significance of sustainability dimensions in generating value has been widely acknowledged. Employing an appropriate framework of key performance indicators ensures precise valuation of the chain. Initially, the chapter outlines the relevance and advantages of biofuels, followed by an exploration of their various types. A novel approach is introduced by conceptualizing an energy chain model that integrates sustainability dimensions and metrics. Uncertainty, particularly of the rough type, is a critical factor in this analysis. Rough set theory stands out as an effective mathematical method for addressing such uncertainties. Subsequently, the rough linear mathematical programming model is optimized using advanced tools, yielding the sustainable value of the chain.

Keywords Biofuel supply chain · Multiple objective decision making · Rough set

8.1 Introduction

Energy supply has long been a critical issue for humanity. Simultaneously, ensuring a healthy environment for material prosperity and mental well-being remains a priority for society. Beyond environmental concerns, the integration of economic growth and social advancement has given rise to the unified concept of sustainability (Ebadi et al., 2024). This principle outlines responsible human behavior in resource provision and utilization, ensuring that current actions do not compromise the planet's integrity or deprive future generations of access to these resources. In recent times, there has been a concerted effort to replace traditional resources with alternatives that are environmentally friendly, cost-effective, and readily accessible. This shift has led to the adoption of renewable or innovative energy sources. Equally crucial is the process of converting these resources into usable energy for diverse applications. In the global energy sector, renewable energy and sustainable supply chains have become pivotal. Biofuels, derived from biomass, encompass a broad spectrum of

H. Fazlollahtabar, *Uncertain Multi-objective Decision Making: Methods and Industrial Models*, https://doi.org/10.1007/978-981-95-6035-6_8

fuel types (Ren et al., 2019). Presently, biofuels are categorized into four generations. First-generation biofuels consist of ethanol and biodiesel produced from food crops, while second, third, and fourth generations are derived from non-food sustainable feedstocks and agricultural residues (Abbaspour et al., 2022).

This chapter presents a three-tier energy supply chain framework, emphasizing that each layer independently addresses sustainability dimensions based on its unique attributes. Aligned with the three pillars of sustainability, the model is multi-objective, designed to quantify the sustainable value for every indicator within each dimension across the supply chain layers. By analyzing the derived values and considering the interdependencies among the layers, conditional uncertainty is introduced using the Bayes Rough method. This approach enables the calculation of the overall sustainable value for the entire chain.

8.2 Literature Review

This chapter examines the biofuel supply chain from a sustainability perspective, highlighting critical gaps in existing research. Based on the summarized studies, no prior work has explored biofuel supply chain modeling that integrates sustainability dimensions, their indicators, and rough uncertainty to derive the chain's value (Fazlollahtabar & Ebadi, 2023). The identified research gaps can be categorized as follows. Existing studies fail to comprehensively define and explain indicators for each sustainability dimension (economic, environmental, and social) in alignment with sustainable development goals. This omission hinders the ability to measure and evaluate sustainability performance effectively. Current research does not incorporate sustainable goal rates as a basis for calculating the value of the biofuel supply chain. This limits the ability to assess how well the chain aligns with broader sustainability objectives. Rough uncertainty, a significant factor in supply chain modeling, has not been addressed using appropriate rough set theory tools. This oversight reduces the robustness of models in handling real-world uncertainties. There is a lack of mathematical frameworks specifically designed to quantify the sustainable value of the biofuel supply chain. This gap prevents a systematic and precise evaluation of the chain's overall sustainability.

The absence of these elements in prior research underscores the need for a novel approach that integrates sustainability indicators, goal rates, rough uncertainty tools, and mathematical modeling. Addressing these gaps will provide a more comprehensive and accurate framework for evaluating the biofuel supply chain's sustainability and value. A summary of the researches already studied are given in Table 8.1.

Table 8.1 Summary of past works

Research	Description						
	Model type		Sustainability			Uncertainty	Value evaluate
	Quantitative	Qualitatively	Environmental	Economic	Social		
Sharma et al., (2013)		✓				✓	
Yue et al., (2014)		✓	✓	✓	✓	✓	
Zaimes et al., (2015)		✓	✓	✓	✓	✓	
Yılmaz Balaman and Selim (2016)	✓		✓	✓		✓	
Fattahi and Govindan (2018)	✓		✓	✓	✓	✓	
Nugroho and Zhu (2019)	✓		✓	✓	✓	✓	
Kesharwani et al., (2019)	✓		✓	✓		✓	✓
Kazemitash et al., (2021)	✓		✓	✓			

8.3 Method

The biofuel supply chain model developed in this chapter comprises three integrated layers: (1) upstream biomass suppliers providing raw materials, (2) midstream biorefineries converting biomass into fuel, and (3) downstream distribution networks delivering products to customers, as visually represented in Fig. 8.1. This vertically-structured network coordinates material flows from diverse agricultural sources through centralized processing facilities to end markets, while simultaneously managing reverse flows of information and payments across the supply web. The model specifically addresses critical biofuel industry challenges including biomass seasonality, conversion efficiency optimization, and demand variability by incorporating: flexible collection protocols to accommodate fluctuating feedstock availability, adaptive refining capacities that adjust to input quality variations, and responsive distribution mechanisms that balance inventory costs with service level requirements. Key operational considerations embedded in the framework include sustainability metrics tracking carbon footprints across each tier, quality control checkpoints at layer interfaces, and resilience strategies for mitigating disruptions in

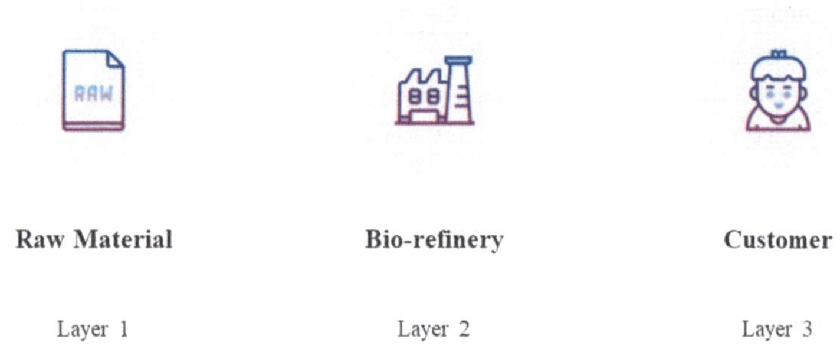

Fig. 8.1 Proposed supply chain

this biologically-dependent supply system. The integrated three-layer architecture enables comprehensive analysis of both conventional supply chain performance indicators (cost, time, quality) and bioenergy-specific metrics (carbon intensity, energy return on investment) while providing decision-support for strategic network design and tactical operations planning in the renewable fuels sector.

Raw Material: Non-food Crops (Second Generation)

Second-generation biofuels are derived from non-food sources such as agricultural residues, wood waste, food waste, and used cooking oil (Fazlollahtabar, 2023). These biofuels are produced from the remnants and byproducts of food crops, offering a sustainable alternative to traditional energy sources. However, the high costs associated with production processes pose significant challenges, limiting the large-scale commercialization of second-generation biofuels (Kazemitash et al., 2021). Despite these hurdles, ongoing research and development efforts highlight the immense potential of second-generation biofuels, positioning them as a promising and innovative energy source for the future.

Bio-refinery: Second Generation

Customer

In this chapter, customers are categorized from two distinct perspectives, reflecting the diverse applications and significance of biofuels in the global energy landscape. Business-to-Business (B2B) Model includes commercial and industrial entities that utilize biofuels as part of their operations. Examples of such customers are

Hotels and Restaurants: For heating, cooking, or power generation.
Factories and Manufacturing Units: As an energy source for machinery and processes.
Corporate Customers: Organizations seeking sustainable energy solutions to reduce their carbon footprint.

The B2B model highlights the role of biofuels in supporting large-scale, energy-intensive operations while promoting sustainability.

Business-to-Consumer (B2C) Model focuses on individual consumers who directly use biofuels. A key application is as an alternative to gasoline, particularly in the automotive sector. Biofuels are increasingly adopted as a cleaner, renewable fuel for cars, reducing reliance on fossil fuels and minimizing environmental impact. The B2C model underscores the growing consumer awareness and demand for eco-friendly energy solutions.

The versatility of biofuels extends beyond these models, encompassing a broad range of uses that demonstrate their global importance. From transportation to industrial processes, biofuels are emerging as a critical component in the transition to sustainable energy systems.

Sustainability

Sustainability is considered within three dimensions as shown in Fig. 8.2.

Sustainability Indicators

In this section, for each dimension of sustainability, a set of key performance indicators in the value creation process is identified.

First layer: raw material		
Environmental	Economic	Social
Greenhouse gases, carbon dioxide emissions	Capacity to create sustainable growth	Food security
Soil quality (erosion, acidification, degradation)	To generate income and employment	Average time spent on volunteering
Land use change	Return on capital related to environmental protection	Equal opportunities
Water use and quality (availability and productivity)	Rate of investment in environmental technology	Social initiative at national and local level
Biodiversity	Gross margin ratio growth	Farm workers
Availability of marginal lands	Effective use of resources	Healthy living and working conditions

Fig. 8.2 Sustainable dimensions

Second layer: bio-refinery		
Environmental	Economic	Social
Renewable resource rates	Increase the flexibility of processes related to the raw materials used	Encourage employees to embrace cultural change
Efficiency material use rates	Reduce energy intensity	Improving the quality of working conditions
Direct or indirect emission rates	Improve overall productivity and process reliability	Health and job safety
Waste reduction rate	Income spent	The rate of social innovation
Recycled material used rate	Dedicated energy production costs	Investor satisfaction
–	Total investment	Employee satisfaction rate

Third layer: customer		
Environmental	Economic	Social
Air quality and pollution	Annual energy cost of urban and rural households	Biofuel Support (Advanced)
Reduce the concentration of the ozone layer	Fuel price fluctuations	Advanced biofuel policy support status
Global warming	Energy efficiency	Customer satisfaction
Living environment	–	Human health
–	–	Next generations
–	–	Social responsibility

Sustainable Goal Rate

The sustainability goals and their associated indicators, as introduced in the previous section, have been evaluated based on expert opinions and reports from global organizations. This evaluation highlights the importance of maintaining strategic alignment between the goals and indicators across the three dimensions of sustainability: economic, environmental, and social.

Social and Economic Dimensions: Achieving social goals, such as employee satisfaction and improved working conditions, directly contributes to the economic dimension. For instance:

Enhanced employee well-being leads to more efficient and reliable production processes.

Improved productivity and process reliability align with economic objectives, such as cost reduction and increased profitability.

Environmental and Economic Dimensions: Environmental goals are strategically linked to economic outcomes. For example: Implementing sustainable practices reduces resource consumption and waste, lowering operational costs. Compliance with environmental regulations minimizes risks and enhances long-term economic stability.

Defining sustainability goals and indicator sets requires a holistic approach to ensure alignment across dimensions. Strategic coherence ensures that progress in one dimension positively influences the others, creating a synergistic effect that amplifies overall sustainability performance. Organizations must adopt an integrated approach to sustainability, recognizing the interdependencies between social, economic, and environmental goals. By aligning strategic objectives with sustainability indicators, organizations can achieve balanced and measurable progress toward their sustainability commitments.

I_{EN11}	88%	I_{EC11}	80%	I_{SO11}	100%
I_{EN12}	100%	I_{EC12}	100%	I_{SO12}	40%
I_{EN13}	100%	I_{EC13}	100%	I_{SO13}	73%
I_{EN14}	100%	I_{EC14}	85%	I_{SO14}	56%
I_{EN15}	96%	I_{EC15}	81%	I_{SO15}	90%
I_{EN16}	79%	I_{EC16}	78%	I_{SO16}	100%

I_{EN21}	100%	I_{EC21}	92%	I_{SO21}	76%
I_{EN22}	100%	I_{EC22}	100%	I_{SO22}	91%
I_{EN23}	72%	I_{EC23}	90%	I_{SO23}	100%
I_{EN24}	64%	I_{EC24}	60%	I_{SO24}	65%
I_{EN25}	64%	I_{EC25}	88%	I_{SO25}	60%
I_{EN26}	–	I_{EC26}	100%	I_{SO26}	44%

I_{EN31}	100%	I_{EC31}	100%	I_{SO31}	100%
I_{EN32}	80%	I_{EC32}	90%	I_{SO32}	95%
I_{EN33}	84%	I_{EC33}	74%	I_{SO33}	40%
I_{EN34}	98%	I_{EC34}	–	I_{SO34}	100%
I_{EN35}	–	I_{EC35}	–	I_{SO35}	99%
I_{EN36}	–	I_{EC36}	–	I_{SO36}	100%

Sustainability goals are central to the evaluation and optimization of mathematical models, particularly in the context of multi-dimensional and uncertain systems. To address these complexities, a comprehensive approach is adopted, integrating sustainability indicators, expert evaluations, and advanced mathematical methods.

Due to the interconnected nature of sustainability dimensions (economic, environmental, and social), all three objective functions must be evaluated simultaneously. This ensures a balanced and holistic assessment of sustainability performance.

Each sustainability indicator is assigned a specific weight based on expert evaluations. These weights reflect the relative importance of each indicator in achieving sustainability goals. The Best-Worst Rough Method is employed to determine these weights, effectively addressing the inherent uncertainty in the model.

By combining sustainability goals with the weighted values of their corresponding indicators, the overall value of sustainability indices is derived. This value is then integrated into a multi-objective mathematical model to evaluate the system's performance.

Using the Rough Weighted Linear Method, the final value of the chain is calculated. This method incorporates the weighted sustainability indices, providing a robust and precise measure of the system's sustainability.

This approach enables a systematic and data-driven evaluation of sustainability performance, supporting decision-making in complex systems. It bridges the gap between theoretical sustainability goals and practical implementation, offering a scalable framework for diverse applications.

8.4 Mathematical Model

8.4.1 Rough Number

Rough set theory is a suitable method to handle uncertainty which was first introduced in (Pawlak, 1982).

Consider U as a universe including all objects and X is a random object from U. There certainly exist a set with k levels showing decision maker's preferences, $Ro = (J_1, J_2, ..., J_k)$, with condition $J_1 < J_2 < ... < J_k$. Then, $\forall X \in U$, $J_q \in R$, $1 \le q \le k$ lower approximation $\underline{Apro}(J_q)$, upper approximation $\overline{Apro}(J_q)$ and boundary interval $Bndr(J_q)$ are determined, respectively, as follows:

$$\underline{Apro}(J_q) = \cup\{X \in U/Ro(X) \le J_q\} \tag{8.1}$$

$$\overline{Apro}(J_q) = \cup\{X \in U/Ro(X) \ge J_q\} \tag{8.2}$$

$$\begin{aligned} Bndr(J_q) &= \cup\{X \in U/Ro(X) \ne J_q\} \\ &= \{X \in U/Ro(X) < J_q\} \cup \{X \in U/Ro(X) > J_q\} \end{aligned} \tag{8.3}$$

The object can be presented with rough number (RoN) defined with lower limit $\underline{Lim}(J_q)$ and upper limit $\overline{Lim}(J_q)$, respectively:

$$\underline{Lim}(J_q) = \frac{1}{M_L} \sum Ro(X)|X \in \underline{Apro}(J_q) \tag{8.4}$$

$$\overline{Lim}(J_q) = \frac{1}{M_U} \sum Ro(X)|X \in \overline{Apro}(J_q) \tag{8.5}$$

where M_L and M_U show the sum of objects in the lower and upper object approximation of J_q, respectively. For object J_q, rough boundary interval $IRBnd(J_q)$ presents an interval between the lower and the upper limits as

$$IRBnd(J_q) = \overline{Lim}(J_q) - \underline{Lim}(J_q) \tag{8.6}$$

The rough boundary interval presents measure of uncertainty. The bigger $IRBnd(J_q)$ value represents the differentiation of experts' preferences, while smaller values represent the harmonized opinions of experts without major deviations. In $IRBnd(J_q)$ are comprised all the objects between lower limit $\underline{Lim}(J_q)$ and upper limit $\overline{Lim}(J_q)$ of rough number $RN(J_q)$. That means that $RN(J_q)$ can be presented using $\underline{Lim}(J_q)$ and $\overline{Lim}(J_q)$.

$$RoN(J_q) = [\underline{Lim}(J_q), \overline{Lim}(J_q)] \tag{8.7}$$

Operation with Rough Number:

Operation of two rough numbers $RoN(\alpha) = [\underline{Lim}(\alpha), \overline{Lim}(\alpha)]$ and $RoN(\beta) = [\underline{Lim}(\beta), \overline{Lim}(\beta)]$ according to (Fazlollahtabar et al., 2017) are

Addition (+) of two rough numbers (α) and (β)

$$RoN(\alpha) + RoN(\beta) = \left[\underline{Lim}(\alpha) + \underline{Lim}(\beta), \overline{Lim}(\alpha) + \overline{Lim}(\beta)\right] \tag{8.8}$$

Subtraction (−) of two rough numbers (α) and (β)

$$RoN(\alpha) - RoN(\beta) = \left[\underline{Lim}(\alpha) - \underline{Lim}(\beta), \overline{Lim}(\alpha) - \overline{Lim}(\beta)\right] \tag{8.9}$$

Multiplication (×) of two rough numbers (α) and (β)

$$RoN(\alpha) \times RoN(\beta) = \left[\underline{Lim}(\alpha) \times \underline{Lim}(\beta), \overline{Lim}(\alpha) \times \overline{Lim}(\beta)\right] \tag{8.10}$$

Division (÷) of two rough numbers (α) and (β)

$$RoN(\alpha) \div RoN(\beta) = \left[\underline{Lim}(\alpha) \div \underline{Lim}(\beta), \overline{Lim}(\alpha) \div \overline{Lim}(\beta)\right] \tag{8.11}$$

Scalar multiplication of rough number $RN(\alpha)$, where μ is a nonzero constant

$$\mu \times RoN(\alpha) = \left[\mu \times \underline{Lim}(\alpha), \mu \times \overline{Lim}(\alpha)\right] \tag{8.12}$$

8.4.2 MARDM: R-BWM

According to the (Rezaei, 2015) BWM is an easy-to-understand and easy-to-apply MCDM method. Straightly, the steps of BWM follow here:

Step 1. Determine the criteria including m decision makers and a set of evaluation criteria $C = \{c_1, c_2, ..., c_n\}$, where n is the number of criteria.
Step 2. Determine the most and least significance criteria.
Step 3. Determine the preference analysis on the most significance criteria (determine the degree of significance for the best criteria). The preference of criteria against each other is shown by $a_{Bj}^e (j = 1, 2, ..., n;\ 1 \leq e \leq m)$ being in interval $a_{Bj}^e \in [1, 9]$ determining the Best-to-Others (BO) vector as follows:

$$A_B^e = (a_{B1}^e, a_{B2}^e, ..., a_{Bn}^e);\ 1 \leq e \leq m \tag{8.13}$$

where a_{Bj}^e shows the influence (preference) of the best criterion B over criterion j, whereby $a_{BB}^e = 1$. This is how we obtain BO matrices $A_B^1, A_B^2, ..., A_B^m$ for each expert.
Step 4. Repeat Step 3 for the worst criterion and call the preference value as a_{jW}^e, $(j = 1, 2, ..., n;\ 1 \leq e \leq m)$ being in interval $a_{jW}^e \in [1, 9]$. As a result, an Others-to-Worst (OW) vector is obtained:

$$A_W^e = (a_{1W}^e, a_{2W}^e, ..., a_{nW}^e);\ 1 \leq e \leq m \tag{8.14}$$

where a_{jW}^e represents the influence (preference) of criterion j in relation to criterion W, whereby $a_{WW}^e = 1$. This is how we obtain OW matrices $A_W^1, A_W^2, ..., A_W^m$ for each expert.
Step 5. Determine the rough BO matrix by computing the average value of all experts' answers leading to form the aggregated sequence matrix:

$$A_B^{*e} = [a_{B1}^m, a_{B2}^m, ..., a_{B1}^k; a_{B2}^1, a_{B2}^2, ..., a_{B2}^m; ...; a_{Bn}^1, a_{Bn}^2, ..., a_{Bn}^m]_{1\times n} \tag{8.15}$$

where $a_{Bj}^e = \{a_{Bj}^1, a_{Bj}^2, ..., a_{Bn}^n\}$ shows sequences of preferences for B criterion. By (8.1)–(8.6), the a_{Bj}^e is changed into a rough form $RoN(a_{Bj}^e) = [\overline{Lim}(a_{Bj}^e), \underline{Lim}(a_{Bj}^e)]$, where $\underline{Lim}(a_{Bj}^e)$ represents the lower limit and $\overline{Lim}(a_{Bj}^e)$ represents upper limit of the rough sequence $RoN(a_{Bj}^e)$ (Vasiljević et al., 2018).

Thus, for sequence $RoN(a_{Bj}^e)$, a BO matrix $A_B^{*1}, A_B^{*2}, ..., A_B^{*m}$ is obtained. Also, the average rough sequence for BO matrix is computed using (8.16) as follows:

$$RN(\overline{a}_{Bj}) = RN(a_{Bj}^1, a_{Bj}^2, \ldots, a_{Bj}^e) = \begin{cases} \overline{a}_{Bj}^L = \frac{1}{m}\sum\limits_{e=1}^{m} a_{Bj}^{eL} \\ \overline{a}_{Bj}^U = \frac{1}{m}\sum\limits_{e=1}^{m} a_{Bj}^{eU} \end{cases} \tag{8.16}$$

where e is the e - th expert ($e = 1, 2, ..., m$), $RoN(a_{Bj}^e)$ is the rough sequences. Then, the average rough BO is obtained as follows:

$$\overline{A}_B = [\overline{a}_{B1}, \overline{a}_{B2}, ..., \overline{a}_{Bn}]_{1\times n} \tag{8.17}$$

Step 6. Determine the rough WO matrix being the average answers of all experts:

$$A_W^{e*} = [a_{1W}^1, a_{1W}^2, ..., a_{1W}^m; a_{2W}^1, a_{2W}^2, ..., a_{2W}^m; ...; a_{nW}^1, a_{nW}^2, ..., a_{nW}^m]_{1\times n} \tag{8.18}$$

where $a_{jW}^e = \{a_{jW}^1, a_{jW}^2, ..., a_{nW}^m\}$ represents the sequence with which the relative significance of criterion j is described in relation to criterion W.

As in Step 5, using Eqs. (8.1)–(8.6), the rough sequence $RoN(a_{jW}^e) = [\overline{Lim}(a_{jW}^e), \underline{Lim}(a_{jW}^e)]$ is in hand and then the averaged OW matrix is formed using experts e, ($1 \le e \le m$) opinions:

$$RoN(\overline{a}_{jW}) = RoN(a_{jW}^1, a_{jW}^2, \ldots, a_{jW}^e) = \begin{cases} \overline{a}_{jW}^L = \frac{1}{m}\sum\limits_{e=1}^{m} a_{jW}^{eL} \\ \overline{a}_{jW}^U = \frac{1}{m}\sum\limits_{e=1}^{m} a_{jW}^{eU} \end{cases} \tag{8.19}$$

$RoN(a_{jW})$ is the rough sequences and the averaged OW matrix is

$$\overline{A}_W = [\overline{a}_{1W}, \overline{a}_{2W}, ..., \overline{a}_{nW}]_{1\times n} \tag{8.20}$$

Step 7. Calculation of the optimal rough values of the weight coefficients of the criteria $[RoN(W_1), RoN(w_2), ..., RoN(w_n)]$ from set C being the difference in the maximum absolute values (8.21):

$$\left|\frac{RoN(w_B)}{RoN(w_j)} - RoN(a_{Bj})\right| and \left|\frac{RoN(w_j)}{RoN(w_W)} - RoN(a_{jW})\right| \tag{8.21}$$

for each value of j is minimized. This occurs when $\sum_{j=1}^{n} w_j^L \le 1$ and $\sum_{j=1}^{n} w_j^U \ge 1$. In this way, the condition is met that the weight coefficients are found at interval $w_j \in [0, 1]$, ($j = 1, 2, ..., n$) that $\sum_{j=1}^{n} w_j = 1$.

The previously defined limits will be presented in the following min–max model:

$$\min \max_{j} \left\{ \left| \frac{RoN(w_B)}{RoN(w_j)} - RoN(a_{Bj}) \right|, \left| \frac{RoN(w_j)}{RoN(w_W)} - RoN(a_{jW}) \right| \right\}$$

s.t.

$$\begin{cases} \sum_{j=1}^{n} w_j^L \leq 1 \\ \sum_{j=1}^{n} w_j^U \geq 1 \\ w_j^L \leq w_j^U,\ \forall j = 1, 2, \ldots, n \\ w_j^L, w_j^U \geq 0,\ \forall j = 1, 2, \ldots, n \end{cases} \tag{8.22}$$

where $RoN(w_j) = [\underline{Lim}(w_j), \overline{Lim}(w_j)] = [w_j^L, w_j^U]$ is the rough weight coefficient of a criterion. The counterpart of Model (8.22) is the following model:

$$\min \xi$$

s.t.

$$\begin{cases} \left| \frac{w_B^L}{w_j^U} - \overline{a}_{Bj}^U \right| \leq \zeta;\ \left| \frac{w_B^U}{w_j^L} - \overline{a}_{Bj}^L \right| \leq \zeta \\ \left| \frac{w_j^L}{w_W^U} - \overline{a}_{jW}^U \right| \leq \zeta;\ \left| \frac{w_j^U}{w_W^L} - \overline{a}_{jW}^L \right| \leq \zeta \\ \sum_{j=1}^{n} w_j^L \leq 1 \\ \sum_{j=1}^{n} w_j^U \geq 1 \\ w_j^L \leq w_j^U, \forall j = 1, 2, \ldots, n \\ w_j^L,\ w_j^U \geq 0, \forall j = 1, 2, \ldots, n \end{cases} \tag{8.23}$$

where $RoN(w_j) = \left[w_j^L, w_j^U\right]$ represents the optimum values of the weight coefficients, $RoN(w_B) = \left[w_B^L, w_B^U\right]$ and $RoN(w_W) = \left[w_W^L, w_W^U\right]$ represents the weight coefficients of the best and worst criterion respectively, while $RoN(\overline{a}_{jW}) = \left[\overline{a}_{jW}^L, \overline{a}_{jW}^U\right]$ and $RoN(\overline{a}_{Bj}) = \left[\overline{a}_{Bj}^L, \overline{a}_{Bj}^U\right]$, respectively, represent the values from the average rough OW and rough BO matrices (see Eqs. (8.17) and (8.20)).

By solving model (8.23) we obtain the optimal values of the weight coefficients for the evaluation criteria $[RoN(w_1), RoN(w_2), \ldots, RoN(w_n)]$ and ξ^*.

8.4.3 MORDM: LWSRM

In this chapter, the inherent uncertainty in the values derived from sustainability indicators. In this chapter, due to the uncertainty in the values obtained from the sustainability indicators, the objective function of the problem is of the rough type and a crisp feasible set is considered. Therefore, in the following model (8), problems with a crisp feasible set and rough objective function are introduced (Abbaspour et al., 2022):

$$\begin{aligned}
&\max(f_1)\\
&\sum_{i=1}^{n}\sum_{j=1}^{n}\left[\overline{a}_{ij},\underline{b}_{ij}\right]x_{ij}\\
&\max(f_2)\\
&\sum_{i=1}^{n}\sum_{j=1}^{n}\left[\overline{c}_{ij},\underline{d}_{ij}\right]x_{ij}\\
&\max(f_3)\\
&\sum_{i=1}^{n}\sum_{j=1}^{n}\left[\overline{e}_{ij},\underline{f}_{ij}\right]x_{ij}
\end{aligned}\tag{8.24}$$

Due to the nature of the problem for the proposed objective functions, the utility functions are defined as follows (Model 9):

$$\begin{aligned}
&\max(U_{f1})\\
&\sum_{i=1}^{n}\sum_{j=1}^{n}\left[\overline{a}_{ij},\underline{b}_{ij}\right]x_{ij}\\
&\max(U_{f2})\\
&\sum_{i=1}^{n}\sum_{j=1}^{n}\left[\overline{c}_{ij},\underline{d}_{ij}\right]x_{ij}\\
&\max(U_{f3})\\
&\sum_{i=1}^{n}\sum_{j=1}^{n}\left[\overline{e}_{ij},\underline{f}_{ij}\right]x_{ij}
\end{aligned}\tag{8.25}$$

Due to the existence of different indicators and their preferences over each other, the utility function has been used. Now the cost limit is added to the introduced model:

$$\begin{aligned}
&\max(U_{f1})\\
&\sum_{i=1}^{n}\sum_{j=1}^{n}\left[\overline{a}_{ij},\underline{b}_{ij}\right]c_{ij}x_{ij}\\
&\text{s.t.}\\
&\sum_{i=1}^{n}c_{ij}\le B_1\ \forall j=\{1,2,\ldots,m\}
\end{aligned}$$

$$x_{ij} \geq 0$$
$$\max(U_{f2})$$
$$\sum_{i=1}^{n}\sum_{j=1}^{n}\left[\overline{c}_{ij}, \underline{d}_{ij}\right]c_{ij}x_{ij}$$
s.t.
$$\sum_{i=1}^{n} c_{ij} \leq B_2 \; \forall j = \{1, 2, \ldots, m\}$$
$$x_{ij} \geq 0$$
$$\max(U_{f3})$$
$$\sum_{i=1}^{n}\sum_{j=1}^{n}\left[\overline{e}_{ij}, \underline{f}_{ij}\right]c_{ij}x_{ij}$$
s.t.
$$\sum_{i=1}^{n} c_{ij} \leq B_3 \; \forall j = \{1, 2, \ldots, m\}$$
$$x_{ij} \geq 0 \tag{8.26}$$

Model (10) is the final model introduced in this chapter. In the next section, a numerical example is solved according to the model. The following table is compiled to better understand the formulated model:

First layer: raw material	f_1	u_{11}	Environmental dimension at the first layer	U_{f1}	$U(F)$
		u_{12}	Economic dimension at the first layer		
		u_{13}	Social dimension at the first layer		
Second layer: bio-refinery	f_2	u_{21}	Environmental dimension at the second layer	U_{f2}	
		u_{22}	Economic dimension at the second layer		
		u_{23}	Social dimension at the second layer		
Third layer: customer	f_3	u_{31}	Environmental dimension at the third layer	U_{f3}	
		u_{32}	Economic dimension at the third layer		
		u_{33}	Social dimension at the third layer		

8.5 Numerical Example

The Best-Worst Rough Algorithm is a structured approach for determining the weights of criteria in multi-dimensional problems, such as sustainability assessments. This section outlines the process of running the algorithm, focusing on the selection of criteria, expert involvement, and the transformation of the vector (BO) into a matrix (BO).

Experts first identify the best and worst criteria within each dimension (e.g., social, economic, environmental). For example, in the social dimension, experts select the most and least important elements from the third layer of the hierarchy.

The process involves multiple layers of experts: First Layer: 5 experts; Second Layer: 6 experts;

Third Layer: 7 experts. These experts can include energy specialists, company managers, professionals in key performance or sustainability indicators, environmental activists, social science professors, or economists.

The vector (BO) represents the preferences of the best criterion over other criteria, as determined by the experts. This vector is then transformed into a matrix (BO) based on the nature of the problem.

The matrix (BO) captures the expert preferences, showing how the best criterion is prioritized over other criteria. This matrix serves as a critical input for calculating the weights of the criteria using the Best-Worst Rough Algorithm.

This method ensures a systematic and transparent approach to weighting criteria, accounting for expert opinions and the inherent uncertainty in sustainability assessments. It is particularly useful in complex decision-making scenarios, such as evaluating sustainability performance, resource allocation, or policy development.

$$A_B^* = \begin{bmatrix} 3\,3\,3\,4\,4\,4\,3 \\ 7\,7\,6\,7\,6\,7\,7 \\ 8\,9\,9\,9\,9\,9\,9 \\ 2\,3\,2\,2\,2\,3\,2 \\ 5\,5\,5\,6\,6\,6\,6 \\ 1\,1\,1\,1\,1\,1\,1 \end{bmatrix}$$

After establishing the matrix (BO), the next step in the Best–Worst Rough Algorithm involves determining the matrix (OW). This matrix represents the preferences of other criteria over the worst criterion, as evaluated by experts. In this specific dimension, the best (most important) criterion is identified as 6, and the worst (least important) criterion is 3. This process is applied in the third level of the chain (customer layer) and the third dimension of sustainability (social dimension).

The matrix (OW) captures the expert evaluations of how other criteria compare to the worst criterion. It complements the matrix (BO), which focuses on the best criterion, ensuring a comprehensive weighting process.

Best Criterion (8.6): Social Responsibility.

This criterion is deemed the most important in the social dimension, reflecting its critical role in sustainability assessments.

Worst Criterion (8.3): Customer Satisfaction.

While still relevant, this criterion is considered the least important in this specific context.

The evaluation is conducted in the third level of the chain, which focuses on the customer layer. This ensures that the weighting process aligns with the specific needs and priorities of the customer dimension within the broader sustainability framework.

The determination of matrix (OW) enhances the robustness of the Best-Worst Rough Algorithm, providing a balanced and nuanced approach to weighting criteria. This method is particularly valuable in sustainability assessments, where multiple dimensions and layers of criteria must be evaluated simultaneously.

$$A_W^* = \begin{bmatrix} 6\,5\,5\,5\,5\,5\,6 \\ 4\,3\,3\,3\,4\,4\,4 \\ 1\,1\,1\,1\,1\,1\,1 \\ 6\,6\,7\,7\,7\,7\,7 \\ 3\,3\,3\,2\,3\,2\,3 \\ 9\,8\,8\,9\,8\,9\,9 \end{bmatrix}$$

Then, using Eqs. (8.1) to (8.6), the definitive evaluation of the experts in the (BO) and (WO) matrices is converted to rough numbers. For example, this operation is performed for criterion (8.1) in the vector (BO) below

$$c_1 = \{3, 3, 3, 4, 4, 4, 3\}$$

$$\underline{Lim}(3) = 3,\ \overline{Lim}(3) = \frac{1}{7}(3+3+3+4+4+4+3) = 3.43$$

$$\underline{Lim}(3) = 3,\ \overline{Lim}(3) = \frac{1}{7}(3+3+3+4+4+4+3) = 3.43$$

$$\underline{Lim}(3) = 3,\ \overline{Lim}(3) = \frac{1}{7}(3+3+3+4+4+4+3) = 3.43$$

$$\underline{Lim}(4) = \frac{1}{7}(3+3+3+4+4+4+3) = 3.43,\ \overline{Lim}(4) = 4$$

$$\underline{Lim}(4) = \frac{1}{7}(3+3+3+4+4+4+3) = 3.43,\ \overline{Lim}(4) = 4$$

$$\underline{Lim}(4) = \frac{1}{7}(3+3+3+4+4+4+3) = 3.43,\ \overline{Lim}(4) = 4$$

$$\underline{Lim}(3) = 3,\ \overline{Lim}(3) = \frac{1}{7}(3+3+3+4+4+4+3) = 3.43$$

Then the upper limit and the lower limit of the whole vector are calculated:

$$\underline{c}_1 = \frac{3+3+3+3.43+3.43+3.43+3}{7} = 3.18$$

$$\overline{c}_1 = \frac{3.43 + 3.43 + 3.43 + 3 + 3 + 3 + 3.43}{7} = 3.67$$

Thus, the rough number of this vector is obtained as [3.18, 3.67]. This operation is performed for all criteria of this vector and vector (OW):

$$BO = \begin{Bmatrix} [3.18, 3, 67] \\ [6.51, 6.92] \\ [8.73, 8.98] \\ [2.08, 2.48] \\ [5.32, 5.81] \\ [1.00, 1.00] \end{Bmatrix}, \quad OW = \begin{Bmatrix} [5.08, 5.48] \\ [3.32, 3.81] \\ [1.00, 1.00] \\ [6.51, 6.92] \\ [2.51, 2.92] \\ [8.32, 8.81] \end{Bmatrix}$$

After obtaining the weight coefficients of the criteria, the best–worst optimization model is solved, which is obtained by solving the optimal values of the weight coefficients of the criteria:

$$SO_3 = \begin{Bmatrix} RoN(w_1) = [0.1542, 0.2698] \\ RoN(w_2) = [0.0750, 0.0925] \\ RoN(w_3) = [0.0430, 0.0450] \\ RoN(w_4) = [0.2186, 0.2260] \\ RoN(w_5) = [0.0602, 0.1225] \\ RoN(w_6) = [0.4491, 0.4491] \end{Bmatrix}$$

This is done for other sustainable dimensions at all layers of the chain, and the weight of the sustainable elements is obtained according to the best–worst rough method as follows:

$$EN_1 = \begin{Bmatrix} RoN(w_1) = [0.3416, 0.4024] \\ RoN(w_2) = [0.1779, 0.2040] \\ RoN(w_3) = [0.0977, 0.1440] \\ RoN(w_4) = [0.2560, 0.2730] \\ RoN(w_5) = [0.0741, 0.1013] \\ RoN(w_6) = [0.0530, 0.0530] \end{Bmatrix}, \quad EC_1 = \begin{Bmatrix} RoN(w_1) = [0.2010, 0.2132] \\ RoN(w_2) = [0.3922, 0.4107] \\ RoN(w_3) = [0.1362, 0.1546] \\ RoN(w_4) = [0.1623, 0.1747] \\ RoN(w_5) = [0.0385, 0.0385] \\ RoN(w_6) = [0.0699, 0.0879] \end{Bmatrix}$$

$$SO_1 = \begin{Bmatrix} RoN(w_1) = [0.4106, 0.4242] \\ RoN(w_2) = [0.0394, 0.0399] \\ RoN(w_3) = [0.1084, 0.1196] \\ RoN(w_4) = [0.0443, 0.0784] \\ RoN(w_5) = [0.1690, 0.2863] \\ RoN(w_6) = [0.2281, 0.2378] \end{Bmatrix}, \quad EN_2 = \begin{Bmatrix} RoN(w_1) = [0.3892, 0.3892] \\ RoN(w_2) = [0.2244, 0.2961] \\ RoN(w_3) = [0.1317, 0.1455] \\ RoN(w_4) = [0.0677, 0.1101] \\ RoN(w_5) = [0.0532, 0.0590] \end{Bmatrix}$$

$$EC_2 = \begin{Bmatrix} RoN(w_1) = [0.1440, 0.1626] \\ RoN(w_2) = [0.2471, 0.3006] \\ RoN(w_3) = [0.0678, 0.1294] \\ RoN(w_4) = [0.0415, 0.0421] \\ RoN(w_5) = [0.0751, 0.0804] \\ RoN(w_6) = [0.4244, 0.4390] \end{Bmatrix}, SO_2 = \begin{Bmatrix} RoN(w_1) = [0.0433, 0.0433] \\ RoN(w_2) = [0.2474, 0.3801] \\ RoN(w_3) = [0.3948, 0.4312] \\ RoN(w_4) = [0.0523, 0.0533] \\ RoN(w_5) = [0.1733, 0.1936] \\ RoN(w_6) = [0.0883, 0.1090] \end{Bmatrix}$$

$$EN_3 = \begin{Bmatrix} RoN(w_1) = [0.4567, 0.4883] \\ RoN(w_2) = [0.2313, 0.2670] \\ RoN(w_3) = [0.1359, 0.1646] \\ RoN(w_4) = [0.0806, 0.0823] \end{Bmatrix}, EC_3 = \begin{Bmatrix} RoN(w_1) = [0.4512, 0.4919] \\ RoN(w_2) = [0.3171, 0.4251] \\ RoN(w_3) = [0.0830, 0.0830] \end{Bmatrix}$$

After obtaining the rough weight coefficients, it is time to combine these coefficients with the relevant sustainable goals to determine the value of each element of the sustainable dimensions at each layer of the chain, for example the social dimension from the third layer according to Eqs. (8.8) to (8.12):

$$I_{EN2} = \begin{Bmatrix} 1 \\ 1 \\ 0.72 \\ 0.64 \\ 0.64 \end{Bmatrix} \times EN_2 = \begin{Bmatrix} RoN(w_1) = [0.3892, 0.3892] \\ RoN(w_2) = [0.2244, 0.2961] \\ RoN(w_3) = [0.1317, 0.1455] \\ RoN(w_4) = [0.0677, 0.1101] \\ RoN(w_5) = [0.0532, 0.0590] \end{Bmatrix}$$

$$= IEN_2^* \begin{Bmatrix} [0.3892, 0.3892] \\ [0.2244, 0.2961] \\ [0.0949, 0.1048] \\ [0.0433, 0.0705] \\ [0.0340, 0.0378] \end{Bmatrix}$$

We perform this operation for all elements in different layers:

$$IEN_1^* = \begin{Bmatrix} [0.3006, 0.3541] \\ [0.1779, 0.2040] \\ [0.0977, 0.1440] \\ [0.2560, 0.2730] \\ [0.0711, 0.0972] \\ [0.0419, 0.0419] \end{Bmatrix}, IEC_1^* = \begin{Bmatrix} [0.1608, 0.1706] \\ [0.3922, 0.4107] \\ [0.1362, 0.1546] \\ [0.1379, 0.1485] \\ [0.0312, 0.0312] \\ [0.0545, 0.0.686] \end{Bmatrix}$$

$$ISO_1^* = \begin{Bmatrix} [0.4106, 0.4242] \\ [0.0157, 0.0160] \\ [0.0791, 0.0873] \\ [0.0242, 0.0439] \\ [0.1521, 0.2577] \\ [0.2281, 0.2378] \end{Bmatrix}, \; IEC_2^* = \begin{Bmatrix} [0.1325, 0.1496] \\ [0.2471, 0.3006] \\ [0.0610, 0.1168] \\ [0.0249, 0.0253] \\ [0.0660, 0.0707] \\ [0.4244, 0.4390] \end{Bmatrix}$$

$$ISO_2^* = \begin{Bmatrix} [0.0330, 0.0330] \\ [0.2251, 0.3459] \\ [0.3948, 0.4312] \\ [0.0399, 0.0346] \\ [0.1039, 0.1161] \\ [0.0389, 0.0479] \end{Bmatrix}, \; IEN_3^* = \begin{Bmatrix} [0.4567, 0.4883] \\ [0.1850, 0.2136] \\ [0.1141, 0.1382] \\ [0.0789, 0.0806] \end{Bmatrix}$$

$$IEC_3^* = \begin{Bmatrix} [0.4512, 0.4919] \\ [0.2854, 0.3826] \\ [0.0614, 0.0614] \end{Bmatrix}, \; ISO_3^* = \begin{Bmatrix} [0.1542, 0.2698] \\ [0.0712, 0.0878] \\ [0.0172, 0.0180] \\ [0.2186, 0.2260] \\ [0.0595, 0.1213] \\ [0.4491, 0.4491] \end{Bmatrix}$$

Objective Function:

In this section, after obtaining the value of sustainable indicators in each dimension of the chain, the optimization model for each layer is formed as follows:

$$F_1 = \begin{cases} f_1 = [0.3006, 0.3541]x_{11} + [0.1779, 0.2040]x_{12} + [0.0977, 0.1440]x_{13} \\ \quad +[0.2560, 0.2730]x_{14} + [0.0711, 0.0972]x_{15} + [0.0419, 0.0419]x_{16} \\ f_2 = [0.1608, 0.1706]x_{21} + [0.3922, 0.4107]x_{22} + [0.1362, 0.1546]x_{23} \\ \quad +[0.1379, 0.1485]x_{24} + [0.0312, 0.0312]x_{25} + [0.0545, 0.0.686]x_{26} \\ f_3 = [0.4106, 0.4242]x_{31} + [0.0157, 0.0160]x_{32} + [0.0791, 0.0873]x_{33} \\ \quad +[0.0242, 0.0439]x_{34} + [0.1521, 0.2577]x_{35} + [0.2281, 0.2378]x_{36} \end{cases}$$

For the second layer:

$$F_2 = \begin{cases} f_1 = [0.3892, 0.3892]x_{11} + [0.2244, 0.2961]x_{12} + [0.0949, 0.1048]x_{13} \\ \quad +[0.0433, 0.0705]x_{14} + [0.0340, 0.0378]x_{15} \\ f_2 = [0.1325, 0.1496]x_{21} + [0.2471, 0.3006]x_{22} + [0.0610, 0.1168]x_{23} \\ \quad +[0.0249, 0.0253]x_{24} + [0.0660, 0.0707]x_{25} + [0.4244, 0.4390]x_{26} \\ f_3 = [0.0330, 0.0330]x_{31} + [0.2251, 0.3459]x_{32} + [0.3948, 0.4312]x_{33} \\ \quad +[0.0399, 0.0346]x_{34} + [0.1039, 0.1161]x_{35} + [0.0389, 0.0479]x_{36} \end{cases}$$

For the third layer:

$$F_3\begin{cases} f_1 = [0.4567, 0.4883]x_{11} + [0.1850, 0.2136]x_{12} + [0.1141, 0.1382]x_{13} \\ +[0.0789, 0.0806]x_{14} \\ f_2 = [0.4512, 0.4919]x_{21} + [0.2854, 0.3826]x_{22} + [0.0614, 0.0614]x_{23} \\ f_3 = [0.1542, 0.2698]x_{31} + [0.0712, 0.0878]x_{32} + [0.0172, 0.0180]x_{33} \\ +[0.2186, 0.2260]x_{34} + [0.0595, 0.1213]x_{35} + [0.4491, 0.4491]x_{36} \end{cases}$$

According to the three layers of the supply chain, three objective functions are formed, each of this layer has its own sub-layers based on the dimensions of sustainability. Due to the nature of the problem and the purpose of the problem, which is to obtain the value of the chain, the above functions are transformed into the form of the utility function and the model constraint, which is of cost type, is added to the model:

$$U_{f1} = \begin{cases} u_{11} = [0.3006, 0.3541]x_{11}c_{11} + [0.1779, 0.2040]x_{12}c_{12} + [0.0977, 0.1440]x_{13}c_{13} \\ +[0.2560, 0.2730]x_{14}c_{14} + [0.0711, 0.0972]x_{15}c_{15} + [0.0419, 0.0419]x_{16}c_{16} \\ u_{12} = [0.1608, 0.1706]x_{21}c_{21} + [0.3922, 0.4107]x_{22}c_{22} + [0.1362, 0.1546]x_{23}c_{23} \\ +[0.1379, 0.1485]x_{24}c_{24} + [0.0312, 0.0312]x_{25}c_{25} + [0.0545, 0.0.686]x_{26}c_{26} \\ u_{13} = [0.4106, 0.4242]x_{31}c_{31} + [0.0157, 0.0160]x_{32}c_{32} + [0.0791, 0.0873]x_{33}c_{33} \\ +[0.0242, 0.0439]x_{34}c_{34} + [0.1521, 0.2577]x_{35}c_{35} + [0.2281, 0.2378]x_{36}c_{36} \\ \sum_{i=1}^{n} c_{ij} \le B_1 \;\; \forall j = \{1, 2, \dots, m\} \end{cases}$$

For the second layer:

$$U_{f2} = \begin{cases} u_{21} = [0.3892, 0.3892]x_{11}c_{11} + [0.2244, 0.2961]x_{12}c_{12} + [0.0949, 0.1048]x_{13}c_{13} \\ +[0.0433, 0.0705]x_{14}c_{14} + [0.0340, 0.0378]x_{15}c_{15} \\ u_{22} = [0.1325, 0.1496]x_{21}c_{21} + [0.2471, 0.3006]x_{22}c_{22} + [0.0610, 0.1168]x_{23}c_{23} \\ +[0.0249, 0.0253]x_{24}c_{24} + [0.0660, 0.0707]x_{25}c_{25} + [0.4244, 0.4390]x_{26}c_{26} \\ u_{23} = [0.0330, 0.0330]x_{31}c_{31} + [0.2251, 0.3459]x_{32}c_{32} + [0.3948, 0.4312]x_{33}c_{33} \\ +[0.0399, 0.0346]x_{34}c_{34} + [0.1039, 0.1161]x_{35}c_{35} + [0.0389, 0.0479]x_{36}c_{36} \\ \sum_{i=1}^{n} c_{ij} \le B_2 \; \forall j = \{1, 2, \dots, m\} \end{cases}$$

For the third layer:

$$U_{f3}\begin{cases} u_{31} = [0.4567, 0.4883]x_{11}c_{11} + [0.1850, 0.2136]x_{12}c_{12} + [0.1141, 0.1382]x_{13}c_{13} \\ +[0.0789, 0.0806]x_{14}c_{14} \\ u_{32} = [0.4512, 0.4919]x_{21}c_{21} + [0.2854, 0.3826]x_{22}c_{22} + [0.0614, 0.0614]x_{23}c_{23} \\ u_{33} = [0.1542, 0.2698]x_{31}c_{31} + [0.0712, 0.0878]x_{32}c_{32} + [0.0172, 0.0180]x_{33}c_{33} \\ +[0.2186, 0.2260]x_{34}c_{34} + [0.0595, 0.1213]x_{35}c_{35} + [0.4491, 0.4491]x_{36}c_{36} \\ \sum_{i=1}^{n} c_{ij} \le B_3 \;\; \forall j = \{1, 2, ..., m\} \end{cases}$$

The following tables specify the cost of selecting each indicator, which is expressed as a constraint in the problem model, based on each layer.

First layer: raw material

Environmental		Economic		Social	
3.5	c_{31}	2	c_{21}	3.5	c_{11}
1	c_{32}	3.5	c_{22}	2	c_{12}
1	c_{33}	1.5	c_{23}	1.5	c_{13}
1	c_{34}	2	c_{24}	2.5	c_{14}
3	c_{35}	1	c_{25}	1	c_{15}
2.5	c_{36}	1	c_{26}	1	c_{16}

Second layer: bio-refinery

Environmental		Economic		Social	
1	c_{31}	1.5	c_{21}	4	c_{11}
3.5	c_{32}	3	c_{22}	3	c_{12}
4	c_{33}	1.5	c_{23}	1.5	c_{13}
1	c_{34}	1	c_{24}	1	c_{14}
2	c_{35}	1	c_{25}	1	c_{15}
1	c_{36}	4	c_{26}	–	–

Third layer: customer

Environmental		Economic		Social	
2	c_{31}	4.5	c_{21}	4.5	c_{11}
1	c_{32}	4	c_{22}	2.5	c_{12}
1	c_{33}	1	c_{23}	1.5	c_{13}
2	c_{34}	–	–	1	c_{14}
1	c_{35}	–	–	–	–
4.5	c_{36}	–	–	–	–

The proposed model is solved by optimization software and the results are given in the table below:

Preferences	Value
If the first layer is a priority	[1.1985, 1.3308]
If the second layer is a priority	[1.6378, 1.6960]
If the third layer is a priority	[2.0320, 2.1011]

8.6 Conclusions

This chapter presents a comprehensive methodological framework for designing and evaluating sustainable biofuel supply chains, with particular emphasis on handling data uncertainty through advanced computational techniques. The proposed approach combines rough set theory with the Best-Worst Method (BWM) to systematically assess and prioritize sustainability indicators, addressing inherent data imprecision in biomass sourcing, conversion processes, and distribution logistics. Building upon this foundation, we develop a multi-objective mathematical model that simultaneously optimizes economic, environmental, and social dimensions of biofuel production networks. The optimization process employs utility-based techniques to generate solution intervals that capture the preference ranges of various stakeholders, including policymakers and supply chain strategists. Key innovations include (1) a novel integration of rough set theory for uncertainty management with BWM for robust indicator weighting, (2) a flexible multi-objective formulation accommodating conflicting sustainability targets, and (3) a utility-theoretic optimization framework that translates qualitative preferences into quantitative design parameters. Computational results demonstrate the model's effectiveness as a decision-support tool, providing policymakers with clear preference intervals for strategic planning while maintaining the necessary flexibility to adapt to regional variations in biomass availability, processing technologies, and market demands. The framework particularly excels in balancing the triple-bottom-line requirements of biofuel systems—optimizing economic viability alongside environmental stewardship and social responsibility—while rigorously accounting for data uncertainties characteristic of renewable energy supply chains.

Rough Set Theory is used to manage the inherent uncertainty in sustainability indicators, ensuring a robust evaluation framework. Best-Worst Method (BWM) integrated with rough set theory to prioritize and weight the indicators effectively. The model is extended to consider multiple objectives, reflecting the complex nature of biofuel energy supply chains. This formulation allows for a comprehensive analysis of trade-offs between different sustainability dimensions (e.g., economic, environmental, social). The optimization process is driven by utility functions, which align the model's outcomes with the preferences of decision and policy makers. This approach ensures that the results are both practical and aligned with strategic goals. The model generates intervals for each preference, providing decision makers with a range of feasible solutions. The approach proves to be a valuable tool for strategists, enabling them to analyze and evaluate the sustainability of biofuel energy supply chains across different dimensions. The study offers a novel framework for designing and evaluating biofuel energy supply chains, addressing critical challenges such as data uncertainty and multi-dimensional sustainability. It provides actionable insights for decision and policy makers, supporting the development of sustainable energy strategies.

References and Further Reading

Abbaspour, M., Fazlollahtabar, H., & Stevic, Z. (2022). Multi-objective rough best-worst method to evaluate sustainability of a biofuel energy supply chain. *International Journal of Industrial Engineering and Production Research, 33*(1), 34–50.

Ebadi, S., Tajdin, A., Mahdavi, I., & Fazlollahtabar, H. (2024). *Rough multi-period network data envelopment analysis for evaluation of supply chain: A case study of skill training in Iran.* Facta Universitatis, Series: Mechanical Engineering.

Fattahi, M., & Govindan, K. (2018). A multi-stage stochastic program for the sustainable design of biofuel supply chain networks under biomass supply uncertainty and disruption risk: A real-life case study. *Transportation Research Part e: Logistics and Transportation Review, 118*, 534–567. https://doi.org/10.1016/j.tre.2018.08.008

Fazlollahtabar, H., Vasiljević, M., Stević, Ž., & Vesković, S. (2017). Evaluation of supplier criteria in automotive industry using rough AHP. In: *The 1st international conference on management, engineering and environment ICMNEE* (pp. 186–197).

Fazlollahtabar, H. (2023). Rough DT2R2ML for renewable energy supplier selection in the presence of big data. *Concurrency and Computation Practice and Experience, 35*(23), Article e7773. https://doi.org/10.1002/cpe.7773

Fazlollahtabar, H., & Ebadi, S. (2023). Skill training supply chain performance evaluation using network data envelopment analysis. *International Journal of Industrial Engineering and Production Research, 34*(1), 1–14.

Kazemitash, N., Fazlollahtabar, H., & Abbaspour, M. (2021). Rough best-worst method for supplier selection in biofuel companies based on green criteria. *Operational Research in Engineering Sciences: Theory and Applications, 4*(2), 1–12.

Kesharwani, R., Sun, Z., Dagli, C., & Xiong, H. (2019). Moving second generation biofuel manufacturing forward: Investigating economic viability and environmental sustainability considering two strategies for supply chain restructuring. *Applied Energy, 242*, 1467–1496. https://doi.org/10.1016/j.apenergy.2019.03.098

Nugroho, Y., & Zhu, L. (2019). Platforms planning and process optimization for biofuels supply chain. *Renewable Energy, 140*, 563–579. https://doi.org/10.1016/j.renene.2019.03.072

Pawlak, Z. (1982). Rough sets. *International Journal of Computer and Information Sciences, 11*(5), 341–356.

Ren, J., Scipioni, A., Manzardo, A., & Liang, H. (2019). *Biofuels for a more sustainable future* (1st ed.). Elsevier.

Rezaei, J. (2015). Best-worst multi-criteria decision-making method. *Omega, 53*, 49–57. https://doi.org/10.1016/j.omega.2014.11.009

Sharma, B., Ingalls, R., Jones, C., & Khanchi, A. (2013). Biomass supply chain design and analysis: Basis, overview, modeling, challenges, and future. *Renewable and Sustainable Energy Reviews, 24*, 608–627. https://doi.org/10.1016/j.rser.2013.03.049

Vasiljević, M., Fazlollahtabar, H., Stević, Ž, & Vesković, S. (2018). A rough multicriteria approach for evaluation of the supplier criteria in automotive industry. *Decision Making: Applications in Management and Engineering, 1*(1), 82–96.

Yılmaz Balaman, Ş, & Selim, H. (2016). Sustainable design of renewable energy supply chains integrated with district heating systems: A fuzzy optimization approach. *Journal of Cleaner Production, 133*, 863–885. https://doi.org/10.1016/j.jclepro.2016.06.001

Yue, D., You, F., & Snyder, S. (2014). Biomass-to-bioenergy and biofuel supply chain optimization: Overview, key issues and challenges. *Computers and Chemical Engineering, 66*, 36–56. https://doi.org/10.1016/j.compchemeng.2013.11.016

Zaimes, G., Vora, N., Chopra, S., Landis, A., & Khanna, V. (2015). Design of sustainable biofuel processes and supply chains: Challenges and opportunities. *Processes, 3*(3), 634–663. https://doi.org/10.3390/pr3030634

Chapter 9
Multi-objective Possibility Theory

Abstract One of the significant challenges in large-scale enterprises is the lack of integrity in operational systems, which hinders effective information flow across departments. This research proposes a decision support system (DSS) to integrate three critical corporate units: purchasing, sales, and production. The process is examined in two steps of information flow within each unit, with a focus on addressing data uncertainty and ensuring organizational integrity. The implementation leverages possibility theory to handle uncertainty and identifies key events that drive organizational integrity. A case study is conducted to validate the system using actual enterprise data, ensuring objective and practical results.

Keywords Decision support system · Possibility theory · Organizational management · Rule base

9.1 Introduction

Decision Support Systems (DSS) are applications designed to assist in decision-making across various conditions. Their nature evolves in parallel with advancements in computer technology, making them inherently dynamic and constantly changing. This dynamism makes it virtually impossible to confine DSS applications to a static set. Notably, Moore's Law, which predicts the doubling of computer power every 18 months, is expected to remain valid for at least the next five years. This suggests that information systems technology will continue to advance in new and diverse directions, further influencing the development of DSS.

DSS applications are not static; they adapt and evolve alongside advancements in computer technology. This continuous evolution ensures that DSS remains relevant and effective in addressing complex decision-making needs. Moore's Law underscores the rapid pace of technological progress, directly impacting the capabilities of DSS. The doubling of computer power every 18 months enables more sophisticated and diverse DSS applications.

H. Fazlollahtabar, *Uncertain Multi-objective Decision Making: Methods and Industrial Models*, https://doi.org/10.1007/978-981-95-6035-6_9

The ongoing advancements in computer technology will drive DSS to explore new functionalities and applications. This progress will likely lead to more integrated, intelligent, and user-friendly DSS solutions. The dynamic nature of DSS necessitates continuous innovation and adaptation to leverage emerging technologies. Organizations must stay updated with technological trends to effectively implement and utilize DSS in their decision-making processes (Mojaveri & Fazlollahtabar, 2012).

Effective decision-making emerges from a synthesis of human expertise, computational resources, and data-driven insights. The evolution of Decision Support Systems (DSS) has been propelled by technological advancements, equipping organizations with tools to process and leverage information efficiently—a key determinant of competitive advantage (Reference, if applicable). In today's volatile and hyper-competitive markets, managerial decisions must be rapid, precise, and optimized to avoid costly missteps. The adage "learning from mistakes" no longer holds; instead, businesses face a "zero-tolerance" reality where errors can lead to irreversible consequences, particularly in an era marked by economic fluctuations, mergers, and acquisitions.

To navigate this high-stakes landscape, organizations must swiftly align with—and surpass—customer expectations, both tangible and perceived. This demands an intelligent, integrated DSS capable of synthesizing diverse analytical models, internal and external data sources, and real-time information to empower managers with actionable insights. Such systems are indispensable for ensuring agility, accuracy, and strategic foresight in decision-making processes, ultimately safeguarding organizational resilience and long-term success.

DSS combines intellectual efforts, computing power, and data to produce decisions that are critical for business success. It enables managers to handle complex information and make informed decisions quickly. The global business environment is highly competitive, fast-paced, and marked by economic instability. Managers face immense pressure to make optimal decisions without the luxury of learning from mistakes. An integrated intelligent DSS leverages diverse models and data resources to support better decision-making. It helps organizations meet customer needs effectively and stay ahead in the competitive market. Marketing managers operate in a rapidly changing and highly competitive environment. DSS provides them with the tools to make better decisions, enhancing their competitiveness. Organizations must invest in advanced DSS to support their managers in navigating the complexities of the modern business environment. The ability to make fast, accurate, and optimal decisions is crucial for survival and success in today's competitive landscape (Delen & Pratt, 2006).

The DSS aims to unify purchasing, sales, and production units by integrating data, applications, and business processes on a single organizational platform. This integration enhances information flow, reduces inefficiencies, and supports better decision-making.

In the first step, Information flow within each unit is analyzed to identify decision-making requirements. In the Second Step, data from the first step is used to implement system rules, addressing uncertainty through possibility theory. Possibility Theory is applied to manage uncertainty in enterprise data, ensuring robust and reliable decision-making. This approach identifies events that significantly impact organizational integrity, allowing enterprises to focus on these areas to achieve their goals. A case study is conducted using actual enterprise data to test the proposed DSS. This ensures the system's practicality, objectivity, and applicability in real-world scenarios.

The proposed DSS addresses a critical gap in large-scale enterprises by improving operational system integrity and information flow. It provides a structured framework for integrating purchasing, sales, and production units, enhancing overall organizational efficiency. The use of possibility theory and case study validation ensures the system's reliability and practical relevance.

9.2 Literature Review

The field of Decision Support Systems (DSS) has seen significant advancements, with researchers expanding the frontiers of computer-supported modeling. One notable development is the concept of 'Brandaid', introduced by Osorio et al. (2008), which is designed to support decisions related to product, promotion, pricing, and advertising. This highlights the application of DSS in marketing and strategic decision-making. Additionally, Power (2008) published a professional book that provides a comprehensive taxonomy of computerized DSS, offering a structured framework to understand and categorize these systems. Furthermore, Mir and Quadri (2009) contributed to the development and refinement of DSS, shaping its evolution and enhancing its capabilities.

Research and practical applications of Decision Support Systems (DSS) highlight their evolution as vital tools for structuring information and enhancing managerial decision-making. Over the years, numerous DSS frameworks have been developed to aid in complex business scenarios. Yet, a persistent challenge remains: decision-making under uncertainty, particularly in contexts requiring prioritization or expert judgment, where incomplete or ambiguous data can critically impact outcomes (Fazlollahtabar et al., 2010).

To mitigate these challenges, contemporary studies advocate for Artificial Intelligence (AI)-driven approaches, which outperform conventional decision-analysis methods. Notably, Fuzzy Logic has emerged as a powerful technique for handling imprecise or incomplete information, making it highly suitable for uncertain environments. Evaluating the efficacy of Fuzzy Logic in DSS requires quantifying the system's robustness—a defining feature that ensures reliable performance despite missing, unreliable, or inherently uncertain inputs (Fazlollahtabar et al., 2010). Robust DSS frameworks must adapt dynamically, maintaining accuracy even when reasoning rules or data quality are compromised.

The sales, production, and procurement departments are critical units in any enterprise, as the decisions made within these units significantly impact the organization's success and growth. Timely and accurate decision-making, along with the integrity of information across these units, is essential. To achieve this, an integrated information system is required, which ensures data consistency and provides accurate information to decision-makers in the shortest possible time. One such system is the Decision Support System (DSS), which supports management decisions effectively. However, a major challenge in information technology organizations is the lack of integration in complex systems, often referred to as the "island view" (Fazlollahtabar, 2022). This research aims to design a DSS process to integrate the purchasing, sales, and production units, leveraging Possibility Theory to handle the inherent uncertainty and vagueness in organizational information.

Some of the importance of Integrated Information Systems are: Ensures data consistency and integrity across departments; Facilitates timely and accurate decision-making. The challenges of the "Island View" are: Lack of integration leads to fragmented data and inefficient processes; Hinders the sharing of efficient and integrated data, applications, and business processes. The Decision Support Systems (DSS) supports management decisions by providing accurate and timely information and enhances the decision-making process in critical units like sales, production, and procurement.

Introduced by Zadeh (1978) as an extension of fuzzy sets and fuzzy logic it provides a mathematical framework for handling uncertainty, offering an alternative to probability theory. It uses Possibility Theory as the computational core to manage vague and uncertain information. An integrated DSS ensures seamless communication and data sharing across departments. Possibility Theory enhances the system's ability to handle uncertainty, improving decision-making accuracy.

9.3 Decision Supports in Industrial Systems

9.3.1 Decision Support Systems in Marketing and Sales

In the modern competitive landscape, marketing and sales are pivotal to the sustainability and expansion of manufacturing firms. A customer-centric approach to product development—where offerings are tailored to meet market demands—proves instrumental in capturing business opportunities. The efficacy of new products heavily relies on accurately interpreting and addressing consumer requirements, underscoring the necessity of Decision Support Systems (DSS). These systems synthesize design specifications and market intelligence, though their prioritization may differ based on designers' strategic focus (Chan & Ip, 2011).

To navigate increasingly intricate market dynamics, specialized tools like Marketing Decision Support Systems (MDSSs) have emerged. These frameworks enhance analytical capabilities, enabling data-driven strategies for segmentation, pricing, and product positioning (Fazlollahtabar, 2022). By leveraging MDSSs, firms can align R&D with real-time consumer insights, minimizing risks and maximizing market responsiveness.

9.3.2 Decision Support Systems in Production

Research on Decision Support Systems (DSS) for manufacturing managers has often focused on using single intelligent tools to address specific decision-making scenarios. Fuzzy Decision Support System by Kulak (2005) was developed to solve material handling selection problems. Utilized fuzzy logic to handle uncertainty and imprecision in decision-making. Knowledge-Based DSS by Ozbayrak and Bell (2003) was designed for managing parts, tools, and information flow in a Flexible Manufacturing System (FMS). Employed rule-based intelligent systems to help managers identify relevant tools and optimize production schedules in complex FMS environments. Intelligent Information System by Wen et al. (2005) focused on analyzing decision situations for enterprise mergers and acquisitions. Provided a more aggregated level of decision support, leveraging intelligent systems to evaluate complex business scenarios.

Narrow Focus of DSS: Most systems are tailored to address specific decision-making challenges, such as material handling, FMS management, or mergers and acquisitions.
Use of Intelligent Tools: Techniques like fuzzy logic, rule-based systems, and intelligent information systems are commonly employed to enhance decision-making accuracy and efficiency.
Application in Manufacturing: These systems are particularly valuable in manufacturing environments, where complex and dynamic decision-making is required.

There is potential for developing integrated DSS that combine multiple intelligent tools to address broader decision-making scenarios. Future systems could focus on cross-functional integration, enabling seamless decision-making across different manufacturing units.

Effective production planning and control (PPC) in modern manufacturing requires coordinated decision-making across multiple operational levels. This complex process encompasses critical determinations regarding production volumes, inventory management, resource procurement, task allocation, and job sequencing—all while navigating competing objectives and operational constraints. To facilitate these multifaceted decisions, specialized decision support systems (DSS) have emerged as vital tools for production planners.

The implementation challenges become particularly pronounced in globally distributed manufacturing networks. Caricato and Grieco (2009) addressed this complexity through their development of an advanced DSS specifically designed for multinational corporations operating production facilities across different geographical regions. Their system provides structured support for

- synchronizing production schedules across locations,
- optimizing resource allocation considering regional constraints, and
- balancing centralized coordination with local operational realities.

Key features of such PPC-oriented DSS include

- multi-criteria optimization capabilities,
- real-time data integration from disparate sources,
- scenario analysis tools for trade-off evaluation, and
- visualization interfaces for complex decision spaces.

These systems enable manufacturers to maintain responsiveness while achieving efficiency targets in increasingly complex global supply networks. The DSS framework proves particularly valuable when dealing with

- fluctuating regional demand patterns,
- varied regulatory environments,
- distributed supplier networks, and
- transportation and logistics variables.

Purpose of the DSS is to provide production planners with a tool that combines constraint programming and mathematical programming, and to enable flexible modeling of production planning problems. The DSS is a software tool composed of multiple modules, each dedicated to a specific task. This modular design allows for customization and scalability in addressing diverse production planning challenges. The DSS developed by Caricato and Grieco (2009) highlights the importance of advanced modeling techniques in production planning. Such systems can help manufacturers optimize resources, reduce costs, and improve efficiency in complex, multi-location production environments.

Production and marketing represent two core functions in most manufacturing firms, operating in a tightly interconnected relationship across critical areas such as sales forecasting, production planning, capacity allocation, inventory control, quality assurance, and distribution logistics. While these functions are inherently interdependent, many organizations continue to manage them through separate, decentralized decision-making structures. This functional siloing frequently leads to misaligned objectives and suboptimal performance outcomes, as production schedules may not fully reflect market demand patterns, inventory levels may exceed actual requirements, and distribution networks may operate below peak efficiency. The lack of integration between these vital functions often manifests in increased operational costs, reduced responsiveness to market changes, and ultimately diminished competitive advantage. This persistent division highlights the need for more unified decision-making frameworks that can bridge the gap between production capabilities and market requirements, ensuring organizational resources are allocated in a manner that maximizes both operational efficiency and market responsiveness.

Marketing decisions are made without full consideration of production constraints, and vice versa. This separation can result in inefficiencies and missed opportunities for optimization. Marketing aims for profit maximization, while production focuses on cost minimization. These conflicting goals can create tension and hinder overall organizational performance. Marketing and production often operate as separate units with distinct action sets and roles. This structural divide makes coordination a complex task.

Many studies suggest combining and jointly making production and marketing decisions to achieve better outcomes. This approach can align objectives and improve overall efficiency. Developing integrated decision support systems (DSS) that consider both production and marketing factors. Such systems can facilitate better communication and coordination between the two functions. Establishing cross-functional teams that include members from both production and marketing. These teams can work together to address shared challenges and align strategies. Coordinated decision-making can lead to more efficient resource allocation and better utilization of production capacity. Aligning production and marketing goals can help achieve a balance between cost minimization and profit maximization. Encouraging collaboration between production and marketing can foster a more cohesive organizational culture.

9.3.3 Decision Support Systems in Procurement

In today's turbulent and unpredictable business environment, the success of an organization heavily depends on its ability to manage and coordinate the complex network of communication within its supply chain. Supply chain management (SCM) involves the integrated management of material and goods flow, data and money flow, and the ability to respond quickly to environmental changes. One of the most critical aspects of SCM is the evaluation, selection, and continuous improvement of suppliers, which directly impacts the efficiency and resilience of the supply chain.

The process of selecting suppliers is often expensive and time-consuming, requiring extensive analysis and evaluation. Organizations must balance multiple criteria, such as cost, quality, reliability, and sustainability. The rise of information and communication technologies (ICT) has transformed traditional supplier selection methods. Organizations are increasingly leveraging decision support systems (DSS) to enhance decision-making processes.

DSS are designed to reduce reliance on intuition and provide data-driven insights for supplier evaluation and selection. These systems integrate advanced tools based on information technology to streamline the decision-making process. Fazlollahtabar and Gholizadeh (2020) developed a DSS to address the challenges of supplier evaluation and selection. Their system incorporates multi-criteria decision-making (MCDM) techniques to ensure a comprehensive and objective evaluation process. DSS enable organizations to automate and optimize supplier evaluation, reducing time and costs. They provide a structured framework for decision-making, ensuring consistency and accuracy. By leveraging data analytics and artificial intelligence, DSS can identify the most suitable suppliers based on predefined criteria. These systems also support continuous improvement by monitoring supplier performance over time. DSS facilitate better communication and collaboration with suppliers, fostering stronger relationships. They help organizations identify and address potential issues proactively, ensuring a more resilient supply chain. The development and implementation of DSS are essential for organizations aiming to achieve a competitive advantage in today's dynamic business environment. By adopting these systems, organizations can enhance their supply chain efficiency, reduce risks, and improve overall performance.

Procurement risk management is a critical component of supply chain management, and its effective implementation is essential for the success of supply management. Decision support systems (DSS) play a pivotal role in enhancing procurement risk management by providing structured frameworks and tools to identify, assess, mitigate, and monitor risks. This approach ensures that organizations can make informed decisions in dynamic and uncertain business environments (Nagali et al, 2008). Many previous studies on supplier risk management have shortcomings in modeling procurement risks and providing decision support for order allocation among multiple suppliers. Traditional methods often rely on estimates and experience, which may not be sufficient in complex and volatile markets.

The presence of spot markets and the need to balance contract suppliers with spot market opportunities add complexity to procurement decisions. Organizations must adapt to changing conditions while minimizing risks and meeting business requirements.

Hong and Lee (2013) developed a comprehensive Decision Support System (DSS) to address procurement risk through a structured four-stage framework for supplier selection and order allocation. Their innovative approach begins with risk identification and assessment, where potential supply chain vulnerabilities are systematically cataloged and evaluated. The second stage focuses on risk quantification, employing advanced modeling techniques to measure the probability and impact of identified risks. In the third phase, the system generates optimized supplier portfolios by balancing cost considerations with risk exposure, using multi-criteria decision analysis. The final stage implements dynamic order allocation strategies that adapt to changing risk profiles, ensuring supply chain resilience. This end-to-end solution provides procurement managers with data-driven insights for making robust sourcing decisions under uncertainty, while maintaining operational efficiency and cost-effectiveness. The framework's staged methodology allows for continuous risk monitoring and strategy adjustment, making it particularly valuable in volatile market conditions where supplier reliability and material availability fluctuate significantly. The phases consist of four stages:

Supply Risk Identification: Identify potential risks in the supply chain, such as supplier reliability, market volatility, and geopolitical factors. This stage lays the foundation for a comprehensive risk management strategy.

Supply Risk Assessment: Use Monte Carlo simulation and Profit–SaR (Profit-at-Risk) maps to quantify and assess risks. These tools provide a probabilistic understanding of potential outcomes and their impact on profitability.

Supply Risk Mitigation: Implement a goal programming model to develop strategies for mitigating identified risks. This stage focuses on optimizing order allocation among multiple suppliers while balancing cost, risk, and business objectives.

Supply Risk Monitoring: Continuously monitor supply risks and adjust strategies as needed. This stage ensures that the organization remains agile and responsive to changing market conditions.

Benefits of the Proposed Framework.

The DSS provides data-driven insights, enabling organizations to make informed decisions on supplier selection and order allocation. It reduces reliance on intuition and estimates, enhancing the accuracy and reliability of procurement decisions. By integrating risk identification, assessment, mitigation, and monitoring, the framework ensures a comprehensive approach to procurement risk management. It helps organizations minimize potential disruptions and maintain supply chain resilience.

The framework allows organizations to leverage both contract suppliers and spot markets to meet business requirements effectively. It ensures that resources are allocated efficiently, balancing cost and risk. The adoption of a decision support system for procurement risk management is essential for organizations operating in dynamic and uncertain environments. This approach not only enhances supply chain efficiency but also strengthens the organization's ability to navigate risks and achieve long-term success.

9.4 Proposed Process and Modeling

Here, the sales, production, and procurement process in an assumed company is described.

Sales Department:

The process of handling customer orders involves several steps, from initial contact to final product dispatch. This workflow ensures efficiency, accuracy, and customer satisfaction. Below is a detailed breakdown of the process:

Step 1: Customer Contact and Background Survey

When customers contact the sales department, an operator initiates a verification process to determine their purchase history. For returning customers, the system retrieves and utilizes their existing customer ID to maintain continuity and streamline transactions. New customers, however, are assigned a unique ID, enabling the creation of a fresh profile for future interactions. This structured approach ensures efficient record-keeping, personalized service, and seamless integration into the company's database for both repeat and first-time buyers.

The sales department uses two forms: Form A (for customers with an existing ID) and Form B (for new customers). If the customer has an ID, Form A is selected. If the customer is new, Form B is selected, and the following information is collected: How the customer knows the organization; Type of customer (retailer or wholesaler) and scope of business activities. After completing the form, the sales operator assigns an ID to the customer.

Step 2: Product Information and Order Submission

The customer provides information about the product type, quantity, delivery date, and technical plan. This information is recorded in the customer database, and an order form is sent to the production unit. The production unit reviews the order for material availability and technical requirements. The order is either admitted or rejected based on this review.

Step 3: Production and Financial Check

Upon order approval, production commences following the predefined process plan. Simultaneously, the system evaluates the customer's economic behavior by querying the financial database. For customers with a history of timely payments, a volume-based discount is automatically applied to incentivize loyalty. Conversely, customers with irregular payment records are charged standard base prices, mitigating financial risk for the organization. This dual-track pricing strategy balances reward for reliability with protection against defaults, ensuring both customer retention and fiscal prudence.

Step 4: Product Dispatch

Once the production and financial checks are complete, the products are dispatched to the customer.

The use of Form A and Form B streamlines the initial contact process, ensuring quick and accurate data collection. The production unit's review ensures that orders are feasible and meet technical requirements. The financial check and discount system reward loyal customers, enhancing satisfaction and retention. The workflow integrates sales, production, and finance departments, ensuring seamless coordination.

This workflow is essential for organizations aiming to optimize operations, reduce errors, and enhance customer experience. By leveraging databases and structured processes, organizations can ensure timely delivery and accurate order fulfillment (Asadi et al., 2015a). The proposed process for sales is diagrammed in Fig. 9.1.

Production Department:

The process of handling customer orders in the production unit involves multiple steps to ensure efficient resource utilization, timely delivery, and quality assurance. Below is a detailed breakdown of the workflow:

Step 1: Order Submission and Initial Check

When a customer order is received, the system automatically accesses the comprehensive production database to assess manufacturing feasibility and schedule operations. This database consolidates real-time data across multiple critical dimensions, including current raw material inventories, equipment availability and maintenance statuses, workforce allocation with skill specifications, and finished goods stock levels. The system evaluates production capacity by analyzing machine utilization

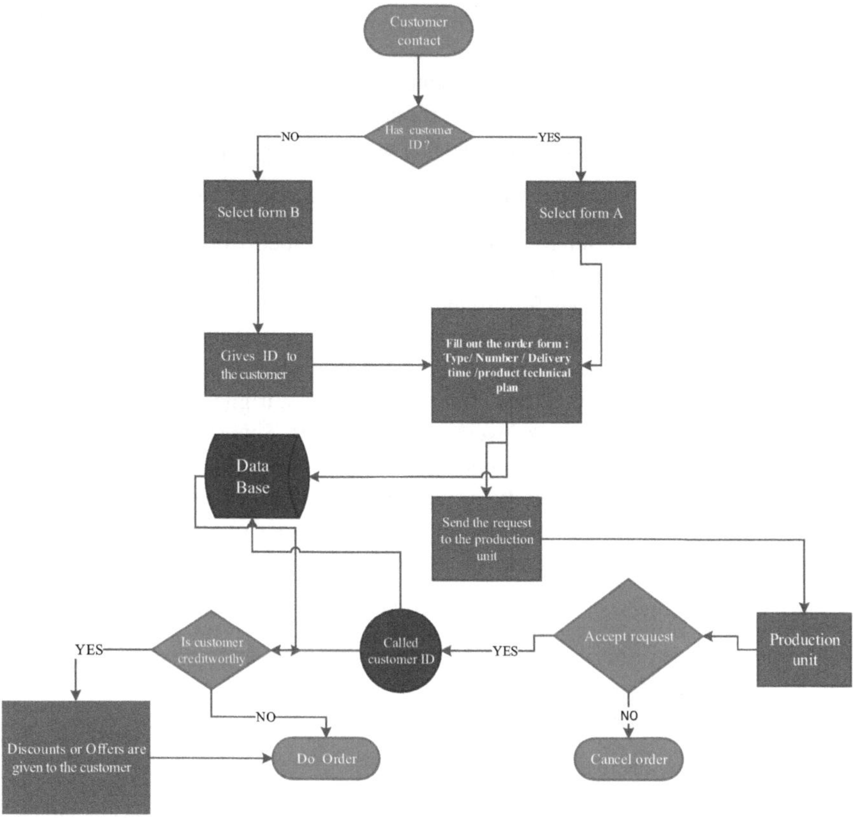

Fig. 9.1 A proposed sales process

rates, distinguishing between scheduled runtime and idle periods, while simultaneously cross-referencing labor availability through shift patterns and attendance records that track both planned and unplanned absences.

For immediate order processing, the system first checks existing finished product inventories, while for made-to-order items it generates optimized production schedules based on material availability, equipment status, and personnel assignments. The database maintains precise personnel-to-machine ratios and tracks all scheduled production line downtime, enabling dynamic adjustments to accommodate urgent orders. This integrated approach facilitates just-in-time material requirements planning while optimizing equipment utilization and workforce deployment. The system features automated alert mechanisms that notify managers of potential bottlenecks, including material shortages with less than 48 h of coverage, critical skill gaps in workforce scheduling, or machine overutilization exceeding 90% capacity threshold. By continuously updating every 15 min, the database ensures production planners have access to the most current operational data for informed decision-making, balancing efficient order fulfillment with sustainable resource management across the manufacturing ecosystem.

The database checks whether the product is in stock. If available, the product is sent directly to the customer. If not, the production supervisor evaluates whether the order can be fulfilled based on available resources and the requested delivery date.

Step 2: Resource Evaluation and Material Requirement Planning (MRP)

Upon receiving a customer order, the production supervisor conducts a comprehensive evaluation of equipment availability and workforce capacity. If production constraints prevent order fulfillment, the request is systematically canceled to maintain operational integrity. For viable orders, the material requirements planning (MRP) data is transmitted to the central database for verification. The system then performs an automated audit of raw material inventories, triggering a purchase order to the procurement unit if stock levels are insufficient. This streamlined workflow ensures that production commitments align with actual resource capabilities while maintaining supply chain continuity through just-in-time procurement protocols. The integrated process minimizes production delays by proactively addressing material shortages and optimizing resource utilization across manufacturing operations.

Step 3: Procurement and Production Planning

The procurement unit prepares the required materials and delivers them to the storage. The production supervisor performs production planning and creates an agenda for manufacturing.

Step 4: Manufacturing and Quality Monitoring

The product is manufactured according to the production plan. The product undergoes quality monitoring to ensure it meets the required standards. If the product meets the standards, it is sent to the warehouse. If not, the manufacturing process is repeated.

The use of the production database ensures quick and accurate resource evaluation. The MRP process ensures that raw materials are procured efficiently, minimizing delays. Quality monitoring ensures that only products meeting the standards are delivered to the customer. The workflow integrates production, procurement, and quality control departments, ensuring seamless coordination.

This workflow is essential for organizations aiming to optimize production processes, reduce lead times, and enhance customer satisfaction. By leveraging databases and structured processes, organizations can ensure timely delivery and high-quality products. The proposed process for production is diagrammed in Fig. 9.2.

Procurement Department:

The procurement process ensures that the production unit receives the necessary materials and components in a timely and efficient manner. Below is a detailed breakdown of the workflow:

Step 1: Purchase Request and Supplier Selection

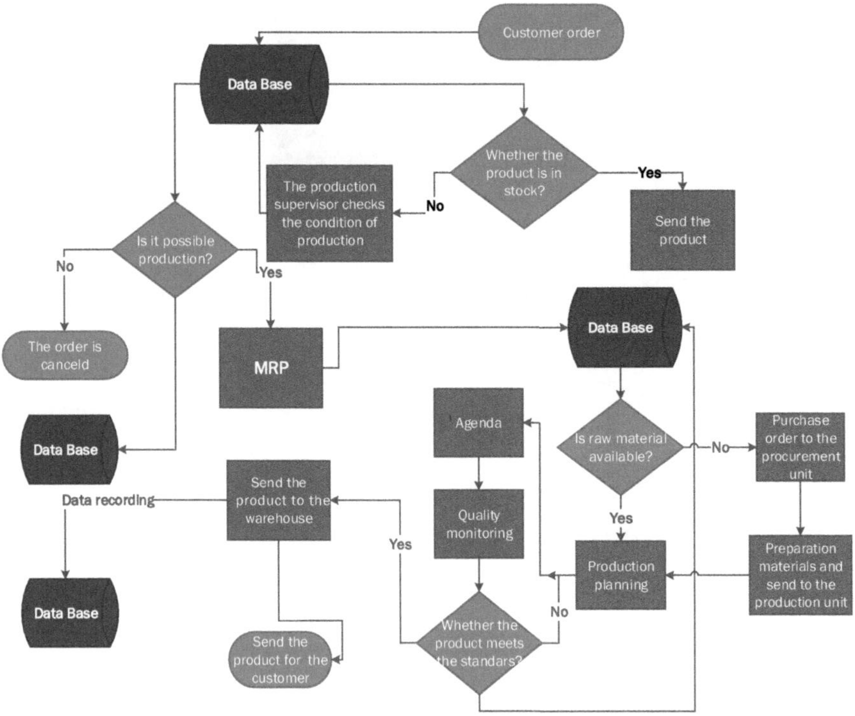

Fig. 9.2 A proposed production process

A purchase request is sent to the procurement unit, specifying the needs to be fulfilled by a supplier. The operator searches the database for a list of potential suppliers. The operator contacts the suppliers, communicates the request, and assesses their circumstances (e.g., availability, pricing, delivery timelines). Based on the assessment, the operator selects the most suitable supplier and records the order in the database.

Step 2: Product Receipt and Compliance Check

Upon receipt of the product, the procurement unit checks whether the product complies with the order. If the product does not comply, it is returned to the supplier. If the product complies, it is sent to the warehouse, and the inventory of the production unit is updated.

The use of a supplier database ensures quick and accurate supplier selection. The compliance check ensures that only products meeting the order requirements are accepted. Updating the production unit's inventory ensures accurate tracking of materials and components. The workflow integrates procurement, warehouse, and production units, ensuring seamless coordination.

This workflow is essential for organizations aiming to optimize procurement processes, reduce lead times, and ensure material quality. By leveraging databases and

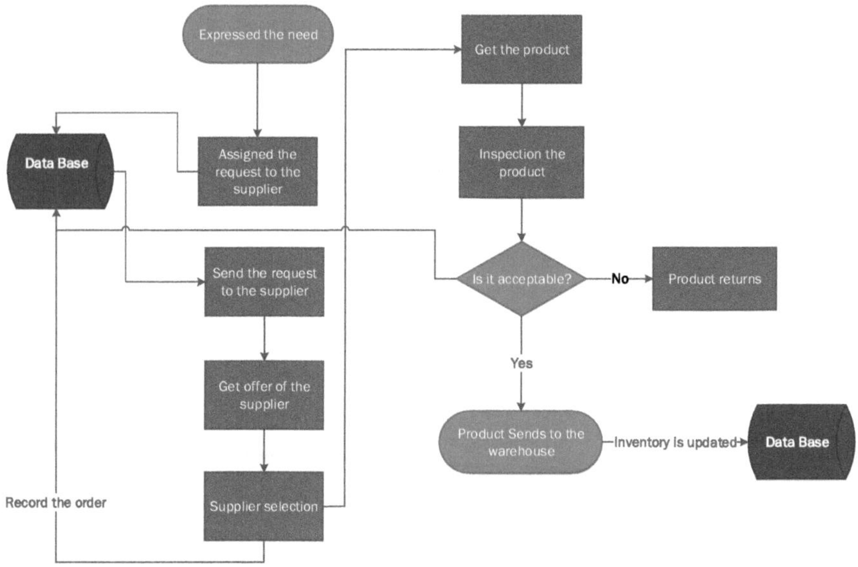

Fig. 9.3 A proposed procurement process

structured processes, organizations can ensure timely delivery and accurate inventory management. The proposed process for procurement is diagrammed in Fig. 9.3.

This study develops a comprehensive if–then rule-based decision model that integrates sales, production, and procurement functions using both real-time customer data and historical transaction records. The system begins by processing customer orders through conditional logic: new customers are automatically assigned unique identifiers while returning customers with verified payment histories receive predetermined discounts. For order fulfillment, the production module evaluates resource availability—including equipment status, workforce capacity, and material inventories—automatically approving feasible orders for manufacturing while rejecting unfulfillable requests with detailed explanations. The material requirements planning component dynamically interacts with procurement, generating automatic purchase orders when raw material stocks fall below threshold levels and activating alternative supply chain contingencies for potential delays. This interconnected framework ensures synchronized operations across departments through: (1) standardized decision protocols that reduce processing latency, (2) data-driven resource allocation that minimizes waste, and (3) automated exception handling that maintains operational continuity. The model's adaptive architecture allows for continuous refinement of decision parameters as additional transaction data accumulates, enabling progressively more accurate forecasting of customer demand patterns, production bottlenecks, and supply chain risks while maintaining flexibility for manual intervention in complex edge cases. By digitizing these critical decision pathways, the system achieves significant improvements in order-to-fulfillment cycle times, working capital efficiency, and customer satisfaction metrics compared to traditional siloed approaches.

Events of Sales Department:

A—If "Customer has ID" Then "Choose form A";

B—If "Customer hasn't ID" Then "Choose form B";

K—If "Production unit accepts the request" Then "Client ID is called";

L—If "Production unit does not accept the request" Then "Order will be canceled";

M—If "The Customer is creditworthy" OR;

N—"The Customer is permanent" Then "Give them discounts or offer";

Events of Production Department:

C—If " Product is in stock" Then "Send the product";

D—If "production is possible" And

E—If "Substances of production is available" Then "Production planning";

F—If "Substance of production isn't available" Then "Buy and check the substance";

J—If "The final product doesn't comply with the standard" Then "The work should be done again";

I—If "The final product comply with the standard" Then "Send the product to the warehouse";

Events of Procurement Department:

G—If "The substance is purchased in accordance with the request" Then "Send the substance to the warehouse";

H—If "The substance is purchased dose not accordance with the request" Then "Return the purchased substance";

Given the uncertainty of customer behavior and the process nature, an indefinite modeling approach is essential to effectively capture and manage these dynamics. Based on the proposed process composition and key sales criteria, the use of possibility theory (Asadi et al., 2015b) provides a robust framework for addressing uncertainty. Below is a detailed explanation:

1. Uncertainty in Customer Behavior and Process Nature

Customer preferences, demands, and decision-making processes are often unpredictable and influenced by external factors. Traditional deterministic models may fail to capture this variability.

Production, procurement, and sales processes are subject to uncertainties such as resource availability, delivery delays, and quality variations. These uncertainties necessitate a flexible modeling approach.

2. Indefinite Modeling Approach

An indefinite modeling approach acknowledges the inherent uncertainties and uses probabilistic or fuzzy methods to represent them. It provides a more realistic representation of real-world scenarios. It enables better decision-making under uncertainty.

3. Benefits of Using Possibility Theory

It allows for the representation of both precise and imprecise information, provides a more comprehensive understanding of uncertain scenarios, and enhances decision-making by considering a range of possible outcomes.

4. Implications for Organizations

Organizations can leverage possibility theory to: Improve forecasting accuracy and resource allocation; Enhance customer satisfaction by better understanding and addressing their needs; Optimize production and procurement processes to minimize risks and delays.

9.5 Possibility Theory

Possibility theory, as proposed by Asadi et al. (2015b), is a mathematical framework for handling uncertainty and imprecision. It extends traditional probability theory by allowing for the representation of possibility distributions rather than strict probabilities.

It captures the range of possible customer actions and preferences, evaluates the feasibility of production and procurement plans under uncertain conditions, and assesses the likelihood of meeting sales targets based on uncertain demand forecasts. The foundational framework of possibility theory was first established by Zadeh (1978) as a theoretical extension of his pioneering work on fuzzy sets and fuzzy logic, providing a mathematical formalism for handling incomplete or imprecise information. This theoretical advancement was subsequently refined and expanded by Dubois and Henri (2001), who contributed significant methodological developments that strengthened its applicability to decision-making under uncertainty. Zadeh's original formulation introduced possibility measures as complementary counterparts to probability distributions, particularly suited for representing epistemic uncertainty where traditional probabilistic approaches may be inadequate. The later contributions by Dubois and colleagues systematized the theoretical underpinnings while demonstrating practical applications in artificial intelligence and operational research domains. Together, these scholarly advancements established possibility theory as a robust alternative to conventional probability theory, especially in scenarios requiring the simultaneous representation of likelihood and plausibility, or when dealing with qualitative uncertainty assessments. The evolution of this theoretical framework has enabled more nuanced modeling of real-world systems characterized by information incompleteness, vagueness, or partial truth, bridging the gap between quantitative precision and qualitative judgment in complex decision environments.

Formulating of Possibility:

The proposed process composition includes steps such as order submission, resource evaluation, production planning, and quality monitoring. Each step is subject to uncertainties that can be modeled using possibility theory. Factors such as delivery timelines, product quality, and customer satisfaction are influenced by uncertainties. Possibility theory helps in evaluating and optimizing these criteria. For simplicity, assume that the universe of discourse Ω is a finite set, and assume that all subsets are measurable. A distribution of possibility is a function *Pos* from 2^{Ω} to [0, 1] such that (Asadi et al., 2015b):

Axiom 1: $pos(\emptyset) = 0$

Axiom 2: $pos(\Omega) = 1$

Axiom 3: $pos(U \cup V) = MAX(POS(U), POS(V))$ for any disjoint subsets U and V.

It follows that, like probability, the possibility measure is determined by its behavior on singletons:

$$POS(U) = MAX_{W \in U} POS(\{W\})$$

Provided that U is finite or count ably infinite.

Axiom 1: can be interpreted as the assumption that Ω is an exhaustive description of future states of the world, because it means that no belief weight is given to elements outside Ω.

Axiom 2: could be interpreted as the assumption that the evidence from which *POS* was constructed is free of any contradiction. Technically, it implies that there is at least one element in Ω with possibility 1.

Axiom 3: corresponds to the additively axiom in probabilities. However there is an important practical difference. Possibility theory is computationally more convenient because Axioms 1–3 imply that

$$pos(U \cup V) = MAX(pos(U), pos(V)) \text{ For } \textit{any} \text{ subsets } U \text{ and } V$$

Because one can know the possibility of the union from the possibility of each component, it can be said that possibility is *compositional* with respect to the union operator. Note however that it is not compositional with respect to the intersection operator (Asadi et al., 2015b). Generally:

$$pos(U \cap V) \leq MIN(pos(U), pos(V))$$

When Ω is not finite, Axiom 3 can be replaced by.

For all index sets I, if the subsets $U_{i, i \in I}$ are pairwise disjoint,

$$pos(\cup_{i \in I} U_i) = Sup_{i \in I} pos(U_i)$$

Necessity:

Whereas probability theory uses a single number, the probability, to describe how likely an event is to occur, possibility theory uses two concepts, the *possibility* and the *necessity* of the event. For any set U, the necessity measure is defined by (Asadi et al., 2015b):

$$Nec(U) = 1 - pos(\overline{U})$$

In the above formula, $\overline{U}$ denotes the complement of U, that is the elements of Ω that do not belong to U. It is straightforward to show that

$$Nec(U) \leq Pos(U) \text{ For } any\ U.$$

And that

$$Nec(U \cap V) = MIN(Nec(U), Nec(V))$$

Note that contrary to probability theory, possibility is not self-dual. That is, for any event U, we only have the inequality:

$$Pos(U) + Pos(\overline{U}) \geq 1$$

However, the following duality rule holds

$$\text{For any event } U\text{, either } Pos(U) = 1\text{, or, } Nec(U) = 0$$

Accordingly, beliefs about an event can be represented by a number and a bit (Asadi et al., 2015b).

Interpretation:

There are four cases that can be interpreted as follows:

$Nec(U) = 1$ Means that U is necessary. U Is certainly true. It implies that $Pos(U) = 1$.

$Pos(U) = 0$ Means that U is impossible. U Is certainly false. It implies that $Nec(U) = 0$.

$Pos(U) = 1$ Means that U is possible. *I* would not be surprised at all if U occurs. It leaves $Nec(U)$ unconstrained.

$Nec(U) = 0$ Means that U is unnecessary. *I* would not be surprised at all if U does not occur. It leaves $Pos(U)$ unconstrained.

The intersection of the last two cases is $Nec(U) = 0$ and $Pos(U) = 1$ meaning that I believe nothing at all about U. Because it allows for indeterminacy like this, possibility theory relates to the graduation of a many-valued logic, such as intuitionistic logic, rather than the classical two-valued logic.

9.6 Numerical Illustrations

MAZIAR Manufacturing Industry is a remarkable example of growth and innovation in the industrial sector. Starting with minimal facilities and small units, the organization has evolved into a prominent industrial group over the past decade. Today, MAZIAR INDUSTRY is known for its high-quality products, advanced machinery, and commitment to meeting customer requirements. Below is a detailed overview of the organization and its products:

MAZIAR INDUSTRY began its operations with limited resources and small-scale units. Over ten years, the organization established a robust framework and expanded into a well-known industrial group. The factory is equipped with state-of-the-art machinery and produces a variety of top-quality products. MAZIAR INDUSTRY continuously works on setting up and expanding new units in compliance with industry standards. The organization is dedicated to fulfilling customer requirements and maintaining high standards of quality and efficiency.

MAZIAR INDUSTRY specializes in manufacturing a range of industrial products, including:

- concrete Beam,
- used in construction projects for structural support and stability,
- lighting Foundation,
- provides a stable base for lighting systems in various applications,
- electrical Cabinet,
- houses electrical components and ensures safety and organization in electrical systems,
- foundation,
- essential for building and infrastructure projects, ensuring structural integrity,
- masts,
- used in telecommunications, lighting, and other applications requiring vertical support,
- boxes Branching,
- facilitates the distribution and organization of electrical or communication systems.

MAZIAR INDUSTRY prioritizes the use of advanced machinery and strict quality control measures to ensure product excellence. The organization is committed to understanding and meeting the specific needs of its customers. MAZIAR INDUSTRY continuously invests in new units and technological advancements to stay ahead in the industry. All products and processes adhere to industry standards, ensuring reliability and safety. MAZIAR INDUSTRY serves as a model for sustainable growth and innovation in the manufacturing sector. The organization's focus on quality, customer satisfaction, and expansion highlights its commitment to excellence. MAZIAR INDUSTRY's products play a vital role in supporting construction, infrastructure, and electrical systems across various industries.

Data Related to the Sales Unit:

A—If “Customer has ID” Then “Choose form A”;

Possibility measure of event A: 0.6
Possibility measure of event $\overline{A}$: 0.4

B—If “Customer hasn’t ID” Then “Choose form B”;

Possibility measure of event B: 0.4
Possibility measure of event $\overline{B}$: 0.6

K—If “Production unit accepts the request” Then “Client ID is called”;

Possibility measure of event C: 0.8
Possibility measure of event $\overline{C}$: 0.3

L—If “Production unit does not accept the request” Then “Order will be canceled”;

Possibility measure of event D: 01
Possibility measure of event $\overline{D}$: 0.9

M—If “The Customer is creditworthy” OR

Possibility measure of event E: 0.7
Possibility measure of event $\overline{E}$: 0.3

N—“The Customer is permanent” Then “Give them discounts or offer”

Possibility measure of event F: 0.8
Possibility measure of event $\overline{F}$: 0.2

Data Related to the Production Unit:

C—If “Product is in stock” Then “Send the product”;

Possibility measure of event C: 0.4
Possibility measure of event $\overline{C}$: 0.5

D—If “production is possible” And

Possibility measure of event D: 0.8
Possibility measure of event $\overline{D}$: 0.3

E—If “Substances of production is available” Then “Production planning”;

Possibility measure of event E: 0.5
Possibility measure of event $\overline{E}$: 0.5

F—If “Substance of production isn’t available” Then “Buy and check the substance”;

Possibility measure of event F: 0.8
Possibility measure of event $\overline{F}$: 0.2

J—If “The final product doesn’t comply with the standard” Then “The work should be done again”;

Possibility measure of event J: 0.2
Possibility measure of event $\overline{J}$: 0.8

I—If "The final product comply with the standard" Then "Send the product to the warehouse";

Possibility measure of event I: 0.9
Possibility measure of event $\overline{I}$: 0.3

Data Related to the Procurement Unit:
G—If "The substance is purchased in accordance with the request" Then "Send the substance to the warehouse";

Possibility measure of event G: 0.7
Possibility measure of event $\overline{G}$: 0.3

H—If "The substance is purchased dose not accordance with the request" Then "Return the purchased substance";

Possibility measure of event H: 0.3
Possibility measure of event $\overline{H}$: 0.8

Computations for Sales Unit:
A—If "Customer has ID" Then "Choose form A";

Possibility measure of event A: 0.6
Possibility measure of event $\overline{A}$: 0.4

$$1 - Pos(A) = 0.6$$
$$Nec(A) = 1 - Pos\left(\overline{A}\right) = 1 - 0.4 = 0.6$$

B—If "Customer hasn't ID" Then "Choose form B";

Possibility measure of event B: 0.4
Possibility measure of event $\overline{B}$: 0.6

$$2 - Pos(B) = 0.4$$
$$Nec(B) = 1 - Pos\left(\overline{B}\right) = 1 - 0.6 = 0.4$$

K—If "Production unit accepts the request" Then "Client ID is called";

Possibility measure of event K: 0.8
Possibility measure of event $\overline{K}$: 0.3

$$3 - Pos(K) = 0.8$$
$$Nec(K) = 1 - Pos\left(\overline{K}\right) = 1 - 0.7 = 0.3$$

L—If "Production unit does not accept the request" Then "Order will be canceled";

Possibility measure of event L: 01.
Possibility measure of event $\overline{L}$: 0.9

$$4 - Pos(L) = 0.1$$
$$Nec(L) = 1 - Pos(\overline{L}) = 1 - 0.9 = 0.1$$

M—If "The Customer is creditworthy" OR

Possibility measure of event M: 0.7
Possibility measure of event $\overline{M}$: 0.3

N—"The Customer is permanent" Then "Give them discounts or offer";

Possibility measure of event N: 0.8
Possibility measure of event $\overline{N}$: 0.2

$$5 - Pos(M \cup N) = MAX[Pos(M), Pos(N)] = MAX[0.8, 0.7] = 0.8$$
$$Nec(M \cup N) = MAX[Nec(M), Nec(N)] = MAX[0.7, 0.8] = 0.8$$
$$Pos(M) = 0.7$$
$$Pos(N) = 0.8$$
$$Nec(M) = 1 - Pos(\overline{M}) = 1 - 0.3 = 0.7$$
$$Nec(N) = 1 - Pos(\overline{N}) = 1 - 0.2 = 0.8$$

Computations for Production Unit:
C—If "Product is in stock" Then "Send the product";

Possibility measure of event C: 0.4
Possibility measure of event $\overline{C}$: 0.5

$$1 - Pos(C) = 0.4$$
$$Nec(C) = 1 - Pos(\overline{C}) = 1 - 0.5 = 0.5$$

D—If "production is possible" And

Possibility measure of event D: 0.8
Possibility measure of event $\overline{D}$: 0.3

$$2 - Pos(D) = 0.8$$
$$Nec(D) = 1 - Pos(\overline{D}) = 1 - 0.3 = 0.7$$

E—If "Substances of production is available" Then "Production planning";

Possibility measure of event E: 0.5
Possibility measure of event $\overline{E}$: 0.5

$$3 - Pos(E) = 0.5$$

$$Nec(E) = 1 - Pos(\overline{E}) = 1 - 0.5 = 0.5$$

F—If "Substance of production isn't available" Then "Buy and check the substance";

Possibility measure of event F: 0.8
Possibility measure of event $\overline{F}$: 0.2

$$4 - Pos(F) = 0.8$$

$$Nec(F) = 1 - Pos(\overline{F}) = 1 - 0.2 = 0.8$$

J—If "The final product doesn't comply with the standard" Then "The work should be done again";

Possibility measure of event J: 0.2
Possibility measure of event $\overline{J}$: 0.8

$$5 - Pos(J) = 0.2$$

$$Nec(J) = 1 - Pos(\overline{J}) = 1 - 0.8 = 0.2$$

C—If "The final product comply with the standard" Then "Send the product to the warehouse";

Possibility measure of event I: 0.9
Possibility measure of event $\overline{I}$: 0.3

$$Pos(I) = 0.9$$

$$Nec(I) = 1 - Pos(\overline{I}) = 1 - 0.3 = 0.7$$

Computations for Procurement Unit:

G—If "The substance is purchased in accordance with the request" Then "Send the substance to the warehouse";

Possibility measure of event G: 0.7
Possibility measure of event $\overline{G}$: 0.3

$$1 - Pos(G) = 0.7$$

$$Nec(G) = 1 - Pos(\overline{G}) = 1 - 0.3 = 0.7$$

H—If "The substance is purchased dose not accordance with the request" Then "Return the purchased substance";

Possibility measure of event H: 0.3
Possibility measure of event $\overline{H}$: 0.8

$$2 - Pos(H) = 0.3$$
$$Nec(H) = 1 - Pos(\overline{H}) = 1 - 0.8 = 0.2$$

The Interpretation of the Events:

To interpret the proposed events and their corresponding measures, we adopt a threshold-based approach using the necessity measure. This choice is grounded in the distinction between possibility and necessity in uncertainty modeling. Possibility represents the required condition for an event to occur. It indicates the potential for an event but does not guarantee its occurrence. Necessity represents both the required and sufficient conditions for an event to occur. It provides a stronger assurance of the event's occurrence compared to possibility.

Since necessity encompasses both required and sufficient conditions, it serves as a more reliable basis for setting thresholds. Using necessity as a threshold ensures that the proposed events are not only possible but also highly likely to occur. The obtained values of necessity are used to define the threshold. Events with necessity measures above the threshold are considered valid and actionable. Events with necessity measures below the threshold are discarded or re-evaluated.

Necessity provides a stronger foundation for decision-making compared to possibility. Ensures that only events with both required and sufficient conditions are considered. Simplifies the interpretation of events by focusing on a single, well-defined measure.

This approach enhances the accuracy and reliability of uncertainty modeling by prioritizing necessity over possibility. It aligns with the goal of making informed decisions based on events that are both possible and necessary.

The interpretations for all the proposed events are given below.

A—If "Customer has ID" Then "Choose form A";

$$1 - Pos(A) = 0.6$$
$$Nec(A) = 1 - Pos(\overline{A}) = 1 - 0.4 = 0.6$$

Interpretation:

1. This event possibility degree is 0.6, this means most customers have ID. But it should be investigated the customer that hasn't customer ID is a new customer or that is a negligence of authorities that have forgotten to give him/her a customer ID.
2. This event necessity degree is 0.6, which means that your customer must have ID is necessary.

Result:

Based on the threshold level that defined for necessity measure equal 0.7, the degree of necessity for this event is 0.6, so we concluded that this event not play an important role in the integrity of our system.

B—If "Customer hasn't ID" Then "Choose form B";

$$2 - Pos(B) = 0.4$$
$$Nec(B) = 1 - Pos(\overline{B}) = 1 - 0.6 = 0.4$$

Interpretation:

1. This possibility degree for this event is 0.4; this means most customers are permanent customer. So range of customers is not high and can it be interpreted as a new customer refer is small and the organization should be more active in advertising and the cause of the problem must be investigated.
2. The necessity degree is 0.4. This means that a customer ID not particular need for the system and the customer can give ID here.

Results:

The degree of necessity for this event is 0.4 and the threshold is not reached, then this event is not necessary for the integrity of the system.

C—If "Product is in stock" Then "Send the product";

$$1 - Pos(C) = 0.4$$
$$Nec(C) = 1 - Pos(\overline{C}) = 1 - 0.5 = 0.5$$

Interpretation:

1. The possibility degree is 0.4; this meaning is that usually the product isn't in the stock and if the product is a commercial, product to be bought and otherwise must be ordered production.
2. The necessity degree is 0.5; this means existence the product in the stock is necessary.

Results:

The necessity degree is 0.5, and this shows the necessity degree of this event isn't reached to the threshold so not play a role in the integrity of the system.

D—If "production is possible" And

$$4 - Pos(D) = 0.8$$
$$Nec(D) = 1 - Pos(\overline{D}) = 1 - 0.3 = 0.7$$

Interpretation:

1. The possibility degree of this event is 0.8 and that is means, the possibility of the production in this organization is high.
2. The necessity degree is 0.7, means this event is necessary for the organization.

Results:

The necessity degree is 0.7, thus this criterion reached to the threshold and this event has an important role in the integrity of the organization.

E—If "Substances of production is available" Then "Production planning";

$$5 - Pos(E) = 0.5$$
$$Nec(E) = 1 - Pos(\overline{E}) = 1 - 0.5 = 0.5$$

Interpretation:

1. The possibility of this happening is 0.5, so normally produced materials are available in the organization.
2. The necessity degree is 0.5, means existence the materials of the production in the organization is essential.

Results:

The necessity degree is 0.5; therefore this event cannot have role in the integrity of the organization.

The Aggregation of Events D and E:

$$Pos(D \cap E) \leq MIN[Pos(D), Pos(E)] = MIN[0.8, 0.5] = 0.5$$
$$Nec(D \cap E) = MIN[Nec(D), Nec(E)] = MIN[0.7, 0.5] = 0.5$$
$$Pos(D) = 0.8$$
$$Pos(E) = 0.5$$
$$Nec(D) = 1 - Pos(\overline{D}) = 1 - 0.3 = 0.7$$
$$Nec(E) = 1 - Pos(\overline{E}) = 1 - 0.5 = 0.5$$

Interpretation:

1. The possibility degree is 0.5, meaning that the possibility of the presence of these two events together with a good degree is possible and usually this event happens.
2. The degree of necessity is 0.5, which means that taken the two events together are necessary and the organization it needs.

Results:

The necessity degree is 0.5, so the presence of these two events together doesn't effective role in the integrity of the organization (system).

F—If "Substance of production isn't available" Then "Buy and check the substance";

$$6 - Pos(F) = 0.8$$
$$Nec(F) = 1 - Pos(\overline{F}) = 1 - 0.2 = 0.8$$

Interpretation:

1. The possibility degree is 0.8, meaning that usually materials needed for production doesn't available in the organization.so should be given the purchase order.
2. The necessity degree is 0.8, which means that in the absence of a required materials purchase order is essential. And it's possible.

Results:

The necessity degree is 0.8 and according to the defined threshold (0.7), this event has effective role in the integrity of the system.

G—If "The substance is purchased in accordance with the request" Then "Send the substance to the warehouse";

$$7 - Pos(G) = 0.7$$

$$Nec(G) = 1 - Pos(\overline{G}) = 1 - 0.3 = 0.7$$

Interpretation:

1. The degree of possibility is 0.7, meaning that normally bought materials is according to the requests.
2. The degree of necessity is 0.7; this means it is necessary for the organization that the purchased materials are in accordance with the request.

Results:

The necessity degree is equal to 0.7, so according to the defined threshold this event has effective role in the integration of the organization.

H—If "The substance is purchased dose not accordance with the request" Then "Return the purchased substance";

$$8 - Pos(H) = 0.3$$

$$Nec(H) = 1 - Pos(\overline{H}) = 1 - 0.8 = 0.2$$

Interpretation:

1. The possibility degree is 0.3, which means it is very low that the purchased materials don't according to the request.
2. The necessity degree is 0.2, means the necessity of this event for the organization is very low.

Results:

The degree of necessity is 0.2, which means this event not plays any role in the integrity of the organization.

I—If "The final product comply with the standard" Then "Send the product to the warehouse";

$$9 - Pos(I) = 0.9$$
$$Nec(I) = 1 - Pos(\overline{I}) = 1 - 0.3 = 0.7$$

Interpretation:

1. The possibility of this event is 0.9, this means the possibility that the product is made according to standard is very high.
2. The necessity degree is 0.7; this means product conformity with defined standards is essential for the organization.

Results:

The necessity degree is 0.7, so this event has role in the integrity of the system.

J—If "The final product doesn't comply with the standard" Then "The work should be done again";

$$10 - Pos(J) = 0.2$$
$$Nec(J) = 1 - Pos(\overline{J}) = 1 - 0.8 = 0.2$$

Interpretation:

1. Possibility degree is 0.2, and it is interpreted, the possibility that the final product isn't according to the standard is very low.
2. The necessity degree is 0.2, so the importance of this event for the organization is minimal.

Results:

The necessity degree is 0.2, so this event has no role in the integration of the organization.

K—If "Production unit accepts the request" Then "Client ID is called";

$$11 - Pos(K) = 0.8$$
$$Nec(K) = 1 - Pos(\overline{K}) = 1 - 0.7 = 0.3$$

Interpretation:

1. Possibility degree of this event is 0.8, thus the probability that product unit accept the demand is very high and event is almost possible.
2. The necessity degree is 0.7, so this event is necessary for the organization.

Results:

The degree of necessity is 0.7 and according to the threshold that we defined for necessity degree this event has effective role for integration of organization.

L—If "Production unit does not accept the request" Then "Order will be canceled";

$$12 - Pos(L) = 0.1$$
$$Nec(L) = 1 - Pos(\overline{L}) = 1 - 0.9 = 0.1$$

Interpretation:

1. The degree of possibility is 0.1, this means that the probability that product unit does not accept the demand is very low and event is almost impossible.
2. The necessity degree is 0.1, so which means it is not necessary event for organization.

Results:

Necessity degree is 0.1, so according to the threshold of this criterion concluded this event hasn't role in the integrity of the organization.

M—If "The Customer is creditworthy" OR

$$13 - Pos(M) = 0.7$$
$$Nec(M) = 1 - Pos(\overline{M}) = 1 - 0.7 = 0.3$$

Interpretation:

1. The degree of possibility is 0.7, means the majority of the customers are creditworthy.
2. The necessity degree is 0.7, which means it is essential for the organization that the customer is creditworthy.

Results:

The degree of necessity is 0.7, according to the threshold that we defined for this criterion this event an effective event for integration of organization.

N—"The Customer is permanent" Then "Give them discounts or offer";

$$14 - Pos(N) = 0.8$$
$$Nec(N) = 1 - Pos(\overline{N}) = 1 - 0.2 = 0.8$$

Interpretation:

1. The degree of possibility is 0.8, means the majority of the customers are permanent.
2. The necessity degree is 0.8, which means it is essential for the organization that the customer is constant and permanent.

The Aggregation of Events M and N:

$$Pos(M \cup N) = MAX[Pos(M), Pos(N)] = MAX[0.8, 0.7] = 0.8$$
$$Nec(M \cup N) = MAX[Nec(M), Nec(N)] = MAX[0.7, 0.8] = 0.8$$

$$Pos(M) = 0.7$$
$$Pos(N) = 0.8$$
$$Nec(M) = 1 - Pos(\overline{M}) = 1 - 0.3 = 0.7$$
$$Nec(N) = 1 - Pos(\overline{N}) = 1 - 0.2 = 0.8$$

Interpretation:

1. The possibility degree is 0.8, so the probability that both events happen and discounts given to the customer is high.
2. The necessity degree is 0.8, meaning that both events occurring together is essential and it is important for organization.

Results:

The necessity of this event is 0.8, thus according to defined threshold, happening this two events together has effective role in integrity of organization.

9.7 Conclusions

In today's competitive business environment, enterprises and small businesses strive to meet customer demands and stakeholder expectations. Effective information sharing and data integration across organizational units play a pivotal role in achieving a competitive advantage. This study focuses on three key departments—Sales, Production, and Procurement—and develops a Decision Support System (DSS) to enhance analysis and decision-making processes.

Right Information Sharing ensure accurate and timely information flow across organizational units. Integrate data from the Sales, Production, and Procurement departments to streamline operations. Develop a DSS to assist in analyzing and making informed decisions in uncertain environments.

The study began by mapping the flow of information within the organization using an organizational flowchart. To analyze decision-making rules in uncertain environments, Possibility Theory was employed. Data was collected from the three departments to estimate the degree of possibility and necessity for each event. For each event, the measure of possibility and degree of necessity were determined. These measures were used to calculate the urgency of each event for the enterprise. The degree of necessity was used as the decision criterion, and a threshold was defined. Events with a necessity degree equal to or higher than the threshold were identified as significant for organizational integration.

Events with a necessity degree equal to or higher than the threshold were found to have a decisive role in organizational integration. Events with a necessity degree below the threshold were deemed ineffective for integration. The study highlighted the importance of focusing on high-necessity events to achieve organizational integrity and competitive advantage.

Enterprises should prioritize events with high necessity degrees to enhance integration and efficiency. The developed DSS can serve as a valuable tool for analyzing and managing uncertainty in decision-making. Effective information sharing and data integration across departments are critical for achieving organizational goals. This study underscores the significance of information sharing, data integration, and decision support systems in achieving organizational integrity. By focusing on high-necessity events, enterprises can streamline operations, enhance decision-making, and maintain a competitive edge in the market.

References and Further Reading

Asadi, M., Fazlollahtabar, H., & Shirazi, B. (2015a). An enterprise management decision making system based on possibility theory. *International Journal of Enterprise Information Systems (IJEIS), 11*(4), 1–27.

Asadi, M., Shirazi, B., & Fazlollahtabar, H. (2015b). A sales management decision making system based on possibility theory. *Journal of Information Technology Software Engineering, 5*(3), 1000159.

Caricato, P., & Grieco, A. (2009). A DSS for production planning focused on customer service and technological aspects. *Robotics and Computer-Integrated Manufacturing, 25*, 871–878.

Chan, S. L., & Ip, W. H. (2011). A dynamic decision support system to predict the value of customer for new product development. *Decision Support System, 52*, 178–188.

Delen, D., & Pratt, D. (2006). An integrated and intelligent DSS for manufacturing systems. *Expert Systems with Applications, 30*, 325–336.

Dubois, D., & Prade, H. (2001). Possibility theory, probability theory and multiple-valued logics: a clarification. *Annals of Mathematics and Artificial Intelligence, 32*, 35–66.

Fazlollahtabar, H. (2022). Intelligent marketing decision model based on customer behavior using integrated possibility theory and K-means algorithm. *Journal of Intelligent Management Decision, 1*(2), 88–96.

Fazlollahtabar, H., Eslami, H., & Salmani, H. (2010). Designing a fuzzy expert system to evaluate alternatives in fuzzy analytic hierarchy process. *Journal of Software Engineering and Applications, 3*(04), 409.

Fazlollahtabar, H., & Gholizadeh, H. (2020). Fuzzy possibility regression integrated with fuzzy adaptive neural network for predicting and optimizing electrical discharge machining parameters. *Computers and Industrial Engineering, 140*, Article 106225.

Hong, Z., & Lee, C. K. M. (2013). A decision support system for procurement risk management in the presence of spot market. *Decision Support Systems, 55*(1), 67–78.

Kulak, O. (2005). A decision support system for fuzzy multi-attribute selection of material handling equipment's. *Expert Systems with Applications, in Press.*

Mir, S. A., & Quadri, S. M. K. (2009). Decision support systems: concepts, progress and issues—A review. In: *Climate change, intercropping, pest control and beneficial microorganisms* (pp. 373–399). Springer Netherlands.

Mojaveri, H. S., & Fazlollahtabar, H. (2012). Designing an integrated AHP based fuzzy expert system and SWOT analysis to prioritize development strategies of Iran agriculture. *Revista De Management Comparat International, 13*(1), 117.

Nagali, V., Hwang, J., Sanghera, D., Gaskins, M., Pridgen, M., Thurston, T., Mackenroth, P., Branvold, D., Scholler, P., & Shoemaker, G. (2008). Procurement risk management (PRM) at Hewlett-Packard company. *Interfaces, 38*(1), 51–60.

Osorio, M.A., Ballinas, A., Jiminez, A., & Sanchez, A. (2008). A decision support system for portfolio optimization on the Mexican market. In *Proceedings of 8th Mexican international conference on computer science* (ENC'08), (pp. 283–294). Baja California.

Ozbayrak, M., & Bell, R. (2003). A knowledge-based support system for the measurement of parts and tools in FSM. *Decision Support Systems, 35*(4), 487–515.

Power, D. J. (2008). Understanding data-driven decision support systems. *Information Systems Management, 25*(2), 149–154.

Wen, W., Wang, W. K., & Wang, T. H. (2005). A hybrid knowledge-based decision support system for enterprise mergers and acquisitions. *Expert Systems with Applications, 28*(3), 569–582.

Zadeh, L. (1978). Fuzzy sets as the basis for a theory of possibility. *Fuzzy Sets and Systems, 1*, 3–28. (Reprinted in (1999) *Fuzzy Sets and Systems, 100*(Supplement), 9–34).

Chapter 10
Bayesian Multi-objective Optimization

Abstract In multi-objective optimization, the utility function plays a critical role in aligning solutions with the priorities and satisfaction of decision-makers (DMs). Traditional approaches often assume a single utility function for each objective function. However, in practical decision-making environments, each objective may have multiple utility functions due to varying priorities, constraints, and uncertainties. This chapter presents a Bayesian-based model for multi-objective optimization, where each objective function is associated with multiple utility functions. By inducing the probability of utilities and handling uncertainty, the model enhances decision-making in complex and practical environments. Illustrative examples demonstrate its effectiveness in both dependent and independent variable scenarios.

Keyword Multi-objective program · Utility function · Bayesian theory

10.1 Introduction

Multi-objective optimization (MOO), also referred to as multi-objective programming, vector optimization, multi-criteria optimization, multi-attribute optimization, or Pareto optimization, is a critical area in multiple criteria decision-making (MCDM). It addresses mathematical optimization problems where multiple objective functions must be optimized simultaneously. MOO involves optimizing more than one objective function at the same time. These objectives often conflict, requiring trade-offs. A solution is Pareto optimal if no objective can be improved without worsening at least one other objective. A human DM may have subjective preferences that guide the selection of a final solution (Komleh & Fazlollahtabar, 2019). Multi-Objective Programming (MOP) models, particularly Goal Programming (GP), are a special case of Distance Function Models (DFM) (Romero, 2001). A classical DFM can be expressed as:

H. Fazlollahtabar, *Uncertain Multi-objective Decision Making: Methods and Industrial Models*, https://doi.org/10.1007/978-981-95-6035-6_10

(DFM)

$$\text{Min}\left\{\sum_{j=1}^{n}\left[w_j\left|f_j^* - f_j(x)\right|\right]^p\right\}^{\frac{1}{p}},$$

$$x \in F$$

where $w_i, f_i^*, f_j(x)$ are the weight factors, aspiration level (goal), achievement function of the jth goal for $j = 1, \ldots, n$ respectively; and $|f_j^* - f_j(x)|$ is the absolute deviation between the achievement function of the j th goal and its aspiration level (Fazlollahtabar et al., 2013).

Solving MOO problems is computationally intensive due to the need to explore multiple solutions. Incorporating DM preferences can be challenging and subjective. Quantifying and communicating trade-offs between objectives requires careful analysis. Multi-objective optimization is a powerful tool for addressing complex decision-making problems with multiple conflicting objectives. By identifying Pareto optimal solutions, quantifying trade-offs, and incorporating DM preferences, MOO provides a robust framework for optimizing real-world systems. Its relationship with Distance Function Models further highlights its versatility and applicability across various domains.

Since the 1970s, significant research has been conducted on solving Multiple-Objective Linear Programming Problems (MOLPPs). These problems involve optimizing multiple conflicting objectives simultaneously and are a key area in Multiple Criteria Decision-Making (MCDM) [see Lai and Hwang (1994)]. There are three primary approaches to solving MOLPPs:

Vector Maximization: Focuses on finding a set of Pareto optimal solutions that represent the best trade-offs between objectives.
Utility Maximization: Incorporates a utility function to reflect the decision-maker's (DM) preferences and priorities.
Aspiration Level Approach: The DM specifies aspiration levels for each objective function, and the goal is to minimize deviations from these levels.

Goal Programming (GP) is the most popular technique for solving Multiple-Objective Decision-Making (MODM) problems. It was first introduced by Charnes et al. (1955) and further developed by Romero (2001). GP aims to find satisfying solutions by minimizing the deviations of achievement functions from their aspiration levels.

Methods in GP:

Lexicographic GP: Prioritizes objectives hierarchically.
Weighted GP (WGP): Assigns weights to objectives based on their importance.
Min–Max (Chebyshev) GP: Minimizes the maximum deviation from the aspiration levels.

The most widely used model is Weighted Goal Programming (WGP), which balances multiple objectives by assigning appropriate weights [see Romero (2001), Arenas-Parra et al. (2010) and Aouni and Kettani (2001)].

Goal Programming (GP) is a powerful and widely used technique for solving Multiple-Objective Linear Programming Problems (MOLPPs). By incorporating aspiration levels and minimizing deviations, GP provides a flexible and practical framework for addressing complex decision-making problems. Its various methods, such as Lexicographic GP, Weighted GP, and Min–Max GP, enable DMs to balance multiple objectives effectively. Despite challenges like subjectivity and complexity, GP remains a cornerstone of Multiple Criteria Decision-Making (MCDM).

Goal programming (GP) serves as a crucial framework for decision-makers (DMs) addressing multi-objective decision-making (MODM) challenges. Since its introduction by Charnes et al. (1955), GP has emerged as a widely adopted approach for tackling multiple objective or criteria decision-making (MODM/MCDM) issues. The core objective of GP is to reduce the absolute deviations from predefined targets. This methodology revolves around achieving specific, predetermined goals. Given resource constraints, DMs aim to minimize the discrepancies between the actual outcomes and these established objectives. The GP framework can be expressed in the following manner:

$$
\begin{aligned}
&\text{(GP)}\\
&\text{Min} \quad \sum_{i=1}^{n} |f_i(x) - g_i|\\
&\text{s.t.} \quad X \in F \ (F \text{ is a feasible set})
\end{aligned}
$$

where $f_i(x)$ is the achievement function of the i th goal. And g_i is the aspiration level of the i th goal. There are various types of methods that can be used to solve the GP model. Some examples are: Lexicographic GP (LGP), Weighted GP (WGP), and Min–Max (Chebyshev) GP. WGP formulation mentioned below:

$$
\begin{aligned}
&\text{(WGP)}\\
&\text{Min} \quad \sum_{i=1}^{n} (\alpha_i d_i^+ + \beta_i d_i^-)\\
&\text{s.t.} \quad f_i(x) - d_i^+ + d_i^- = g_i, \quad i = 1, 2, \ldots, n,\\
&\qquad\ \ d_i^+, d_i^- \geq 0, \quad i = 1, 2, \ldots, n,\\
&\qquad\ \ x \in F,
\end{aligned}
$$

The positive and negative deviations from the objective function (goal) are represented by overachievement and underachievement, respectively. These deviations are assigned specific weights (α_i, β_i), which reflect their relative importance. In the context of GP, the weights w_i correspond to the i-th objective function (goal), while $f_i(X)$ and g_i denote the linear functions and target levels associated with the goals, respectively. The terms d_i^+ and d_i^- represent the positive and negative deviations from the target level of the i-th goal.

In scenarios where resource limitations or insufficient information exist, it may be challenging to define precise target levels. As a result, the actual outcomes might exceed the initially set target levels, indicating that the decision-maker (DM) underestimated the initial target setting. This suggests that higher target levels could have been achieved given the available resources and information. Typically, GP solution methods focus on a single target level for the right-hand side of a constraint, which may not account for such variations.

The achievement function of MCGP is defined to be:

$$\begin{aligned} &\text{Min} \quad \sum_{i=1}^{n} w_i |f_i(X) - g_{i1} \; or \; g_{i2} \text{ or} \cdots \text{or } g_{im}|, \\ &\text{s.t.} \quad X \in F \quad (F, \text{ a feasible set}), \end{aligned}$$

where g_{ij} is the jth aspiration level (for $j = 1, 2, \ldots, m$) corresponding to the ith goal (for $i = 1, 2, \ldots, n$). Other variables are as defined in WGP.

The corresponding MCGP model is shown as follows [see Aouni and kettani (2001)]:

$$\begin{aligned} &\text{Min} \sum_{i=1}^{n} w_i(d_i^+ + d_i^-) \\ &\text{s.t.} \;\; f_i(x) - d_i^+ + d_i^- = \sum_{j=1}^{m} g_{ij} s_{ij}(B), \quad i = 1, \ldots, n, \\ &\qquad d_i^+, d_i^- \geq 0, \quad i = 1, \ldots, n, \\ &\qquad s_{ij}(B) \in R_i(x), \quad i = 1, \ldots, n, \quad j = 1, \ldots, m, \\ &\qquad X \in F \quad (F, \text{ a feasible set}), \end{aligned}$$

where $S_{ij}(B)$ is the jth function of binary serial numbers corresponding to the ith goal, $R_i(x)$ is a function of resource limitations, and the other notations are as defined in WGP.

A utility function reflects the decision-maker's (DM) preferences and provides greater flexibility regarding objectives or attributes (refer to Al-Nowaihi et al., 2008; Podinovski, 2010; Yu et al., 2009). As noted in previous studies (Licalzi & Sorato, 2006), four types of utility functions are typically considered: concave, convex, S-shaped, and reverse S-shaped. In this work, we introduce a novel model for addressing MODM problems where each goal is associated with multiple utility functions. To the best of our knowledge, such problems have not been explored in prior research. We leverage Bayesian theory to determine the probabilities of different utility functions. Abbasian et al. (2015) proposed a method for handling unconstrained multi-objective problems with multiple utility functions by employing Bayesian theory to derive probability values as aspiration levels for objective functions.

A utility function represents the level of satisfaction a decision-maker (DM) derives from a specific objective. Traditionally, a single utility function is assigned to each objective function. However, in real-world applications, multiple utility functions may be required to capture varying DM preferences and constraints. Bayesian theory is utilized to estimate the probability distribution of utilities for each objective function, allowing DMs to assess utility probabilities in uncertain conditions. The proposed model addresses both constrained and unconstrained multi-objective problems, associating each objective function with multiple utility functions. Conditional probability is employed to determine the likelihood of different utilities. Bayesian theory is applied to compute these probabilities, effectively managing uncertainty. To illustrate the model's effectiveness, various examples are provided, covering cases with both dependent and independent variables. By incorporating multiple utility functions and their associated probabilities, DMs can make more informed choices. This framework offers a structured approach for optimizing objectives under uncertain conditions.

10.2 Modeling Unconstrained and Constrained MOLP Problems

In real-world scenarios, decision-making processes across various industries and trade sectors operate under diverse conditions, leading to different utility values for each objective. This variation arises due to environmental factors influencing the achievement of optimal solutions. Traditionally, existing literature assigns a single utility function to each objective. However, to better align with real-life conditions and improve solution quality, it is necessary to determine the probability of each utility function.

To achieve this, we employ Bayesian theory to estimate the likelihood of utility functions corresponding to a given objective function (refer to Olshausen, 2004). Bayesian theory serves as a probabilistic framework for classifying events based on their likelihood of occurrence or non-occurrence. Since directly computing event probabilities is often complex, Bayesian theory provides a means to evaluate conditional probabilities instead (Fazlollahtabar & Ashena, 2013). Our approach applies Bayes' rule as follows:

Bayes' rule:

If $B_1, \ldots, B_k$ partition a space S, and A is an ideal event, then

$$p(B_r|A) = \frac{p(B_r \cap A)}{p(A)} = \frac{p(B_r) \cdot p(A|B_r)}{\sum\limits_i p(B_i) \cdot p(A|B_i)}, \quad r = 1, \ldots, k,$$

where $P(B_r|A)$ denotes the conditional probability for the occurrence of event B_r, given that event A has occurred [see Olshausen (2004)].

Multi-objective without restrictions model can be expressed as follows:

$$\begin{aligned} &(\max \text{ or } \min)\ F(X) = \{f_1(x), f_2(x), \ldots, f_n(x)\} \\ &\text{s.t.} \qquad X \in F \quad (F \text{ is a feasible set}) \end{aligned}$$

A constrained multi-objective problem can be expressed as follows:

$$\begin{aligned} &\text{Min } F(X) = \{f_1(x), \ldots, f_n(x)\} \\ &\text{s.t.} \quad g_j(x) \begin{bmatrix} \leq \\ \geq \\ = \end{bmatrix} 0, \quad j = 1, \ldots, m \\ &\qquad x \in F. \end{aligned}$$

Here, we consider utilities $u_{i1}, \ldots, u_{i,p_i}$ for each objective function f_i.

Assuming that the utility functions associated with the ith objective function are mutually exclusive $u_{11}, \ldots, u_{1,p_i}$, we analyze their probabilities separately under conditions of variable dependence and independence. The total probability and summation rules are then applied to derive overall probabilities. In this study, we first examine how variable independence influences utility functions and later incorporate both dependent and independent cases through illustrative examples (Fazlollahtabar & Ashena, 2013).

The probability of the utilities using Bayesian theory is calculated as follows:

$$\begin{aligned} p(u_{ir}|x_j) &= \frac{p(u_{ir} \cap x_j)}{p(x_j)} \\ &= \frac{p(x_j|u_{ir})p(u_{ir})}{\sum_{k=1}^{p_i} p(x_j|u_{ik})p(u_{ik})}, \quad \forall i, j, r, \ i = 1, \ldots, n, \ j = 1, \ldots, n, \ r = 1, \ldots, p_i. \end{aligned} \tag{10.1}$$

In the context of Bayesian probability, a decision-maker (DM) can assign prior probabilities to utility functions to evaluate the associated conditional probabilities. These prior probabilities represent initial beliefs or assumptions about the likelihood of utility functions before considering new evidence. Once updated with additional information or data, the resulting conditional probability of a utility function u_{ir} is referred to as the posterior probability. This posterior probability reflects the revised likelihood of the utility function, incorporating both the prior knowledge and the new evidence.

From the definitions mentioned above now we have:

$$\begin{aligned} &(\text{max or min}) \ F(X) = \{f_1(x), f_2(x), \ldots, f_n(x)\} \\ &\text{s.t.} \qquad\qquad X \in F \quad (F \text{ is a feasible set}) \end{aligned}$$

For f_1, $p(u_{11}), p(u_{12}), \ldots, p(u_{1,p_1})$ are the aspiration levels achieved. For f_2, $p(u_{21}), p(u_{22}), \ldots, p(u_{2,p_2})$ are the aspiration levels achieved. Until f_n we have $p(u_{n1}), p(u_{n2}), \ldots, p(u_{n,p_n})$.

We can use the definition of the sum of the probabilities as combining all of these final utilities for each of the objectives separately.

This way, we can use the total probability rule for the objective function f_i to reach the final probabilities as follows:

$$p(u_{ir}) = \sum_{j=1}^{n} p(u_{ir}|x_j)p(x_j), \quad \forall i, r, \quad i = 1, \ldots, n, \quad r = 1, \ldots, p_i. \tag{10.2}$$

For each objective f_i, let $p(u_{i1}), \ldots, p(u_{i,p_i})$ represent the achieved aspiration levels. To combine the final probabilities for each objective, we apply the principle of summing probabilities, as outlined in Eq. (10.2). Since each objective is associated with a single probability, we can treat this probability as the aspiration level for the corresponding objective function. This allows us to frame the problem using a Goal Programming (GP) formulation, where the objective is to minimize the deviations between the achieved goals and their respective aspiration levels. This approach can be further refined using Weighted Goal Programming (WGP), which assigns weights to the deviations based on their importance. The problem can then be expressed as the following optimization program:

$$\begin{aligned}
\text{Min } & \sum_{i=1}^{n} w_i(d_i^+ + d_i^-) \\
\text{s.t. } & f_i(x) - d_i^+ + d_i^- = g_i, \; i = 1, \ldots, n, \\
& d_i^+, d_i^-, x \geq 0, \\
& x \in F,
\end{aligned}$$

where g_i is sum of probabilities of the utility functions corresponding to the ith goal.

When the sum of probabilities is not applied, each objective function may have multiple probability values. These values can be interpreted as aspiration levels for the respective objective function. However, to the best of our knowledge, traditional Goal Programming (GP) approaches are not equipped to handle problems with multiple aspiration levels. To address this limitation, we propose using a Multi-Choice Goal Programming (MCGP) approach. MCGP is specifically designed to manage multiple aspiration levels for each objective, providing a more flexible and comprehensive solution. The problem can be formulated as follows:

$$\begin{aligned}
\text{Min } & \sum_{i=1}^{n} w_i(d_i^+ + d_i^-) \\
\text{s.t. } & f_i(x) - d_i^+ + d_i^- = \sum_{j=1}^{m} g_{ij} s_{ij}(B), \quad i = 1, \ldots, n, \\
& d_i^+, d_i^- \geq 0, \quad i = 1, \ldots, n, \\
& s_{ij}(B) \in R_i(x), \quad i = 1, \ldots, n, \; j = 1, \ldots, m, \\
& X \in F; \quad (F, \text{ a feasible set}),
\end{aligned}$$

where g_{ij} is the jth aspiration level corresponding to the ith goal.

10.3 Illustrative Examples

10.3.1 Unconstrained Multi-objective Model

Let us consider a problem with the following goal and utility functions, this problem can be solved by current GP approaches if we consider the sum of the utility function probabilities as its aspiration level for each goal.

$$\text{Goal1: max } f_1(x) = 3x_1 + 2x_2 + x_3$$

Utility functions of $f_1(x)$:

$$\begin{cases} u_{11} = 2x_1 + 3x_2 + 2x_3 \\ u_{12} = x_1 + 2x_2 \\ u_{13} = x_1 + x_2 + x_3 \end{cases}$$

Our inputs are: $p(u_{11}) = 0.25, p(u_{12}) = 0.5, p(u_{13}) = 0.25.p(u_{11}) = 0.25, p(u_{12}) = 0.5, p(u_{13}) = 0.25$.

$$\begin{aligned} p(x_1|u_{11}) &= 0.35 \\ p(x_2|u_{11}) &= 0.3 \\ p(x_3|u_{11}) &= 0.35 \\ p(x_1|u_{12}) &= 0.5 \\ p(x_2|u_{12}) &= 0.5 \\ p(x_1|u_{13}) &= 0.23 \\ p(x_2|u_{13}) &= 0.74 \\ p(x_3|u_{13}) &= 0.03 \end{aligned}$$

If x_1, x_2, x_3 independently affect on utility functions, then we can have the following probabilities:

$$\begin{aligned} p(x_1) &= p(x_1|u_{11})p(u_{11}) + p(x_1|u_{12})p(u_{12}) + p(x_1|u_{13})p(u_{13}) \\ &= (0.35)(0.25) + (0.5)(0.5) + (0.23)(0.25) \cong 0.39 = 0.4 \\ p(x_2) &= p(x_2|u_{11})p(u_{11}) + p(x_2|u_{12})p(u_{12}) + p(x_2|u_{13})p(u_{13}) \\ &= (0.3)(0.25) + (0.5)(0.5) + (0.74)(0.25) = 0.51 \\ p(x_3) &= p(x_3|u_{11})p(u_{11}) + p(x_3|u_{13})p(u_{13}) \\ &= (0.35)(0.25) + (0.03)(0.25) \cong 0.09 \end{aligned}$$

$$p(u_{11}|x_1) = \frac{p(x_1|u_{11})p(u_{11})}{p(x_1)} = \frac{(0.35)(0.25)}{0.4} \cong 0.22$$
$$p(u_{11}|x_2) = \frac{p(x_2|u_{11})p(u_{11})}{p(x_2)} = \frac{(0.3)(0.25)}{0.51} \cong 0.15$$
$$p(u_{11}|x_3) = \frac{p(x_3|u_{11})p(u_{11})}{p(x_3)} = \frac{(0.35)(0.25)}{0.09} \cong 0.97$$

By using the sum of the probability law, we will have:

$$p(u_{11}) = p(u_{11}|x_1)p(x_1) + p(u_{11}|x_2)p(x_2) + p(u_{11}|x_3)p(x_3),$$
$$p(u_{11}) = (0.22)(0.4) + (0.15)(0.51) + (0.97)(0.09) \cong 0.25.$$

If x_1, x_2, x_3 dependently affect on utility functions, then we can have these probabilities:

Suppose that:

$$p(x_2|x_1) = 0.1$$
$$p(x_3|x_1) = 0.75$$
$$p(x_3|x_2) = 0.35$$
$$p(x_2|u_{11}x_1) = 0.1$$
$$p(x_3|x_1u_{11}) = 0.5$$
$$p(x_3|x_2u_{11}) = 0.45$$
$$p(x_3|u_{11}x_1x_2) = 0.15$$
$$p(x_3|x_1x_2) = 0.64$$

Then we have:

$$p(u_{11}|x_1x_2) = \frac{p(x_1|u_{11})p(u_{11})p(x_2|u_{11}x_1)}{p(x_1)p(x_2|x_1)} = \frac{(0.35)(0.25)(0.1)}{(0.4)(0.1)} \cong 0.22$$
$$p(u_{11}|x_1x_3) = \frac{p(x_1|u_{11})p(u_{11})p(x_3|x_1u_{11})}{p(x_1)p(x_3|x_1)} = \frac{(0.35)(0.25)(0.5)}{(0.4)(0.75)} = 0.15$$
$$p(u_{11}|x_2x_3) = \frac{p(x_2|u_{11})p(u_{11})p(x_3|x_2u_{11})}{p(x_2)p(x_3|x_2)} = \frac{(0.3)(0.25)(0.45)}{(0.51)(0.35)} \cong 0.19$$

$$p(u_{11}|x_1x_2x_3) = \frac{p(x_3|u_{11}x_1x_2)p(x_2|u_{11}x_1)p(x_1|u_{11})p(u_{11})}{p(x_3|x_1x_2)p(x_2|x_1)p(x_1)}$$
$$= \frac{(0.15)(0.1)(0.35)(0.25)}{(0.64)(0.1)(0.4)} \cong 0.05.$$

By using the sum of the probability law, we will have:

$$p(x_1x_2) = p(x_2|x_1)p(x_1) = (0.1)(0.4) = 0.04$$
$$p(x_1x_3) = p(x_3|x_1)p(x_1) = (0.75)(0.4) = 0.3$$
$$p(x_2x_3) = p(x_3|x_2)p(x_2) = (0.35)(0.51) \cong 0.18$$
$$p(x_1x_2x_3) = p(x_3|x_1x_2)p(x_1x_2) = (0.64)(0.04) \cong 0.03$$

$$p(u_{11}) = p(u_{11}|x_1x_2)p(x_1x_2) + p(u_{11}|x_1x_3)p(x_1x_3)$$
$$+ p(u_{11}|x_2x_3)p(x_2x_3) + p(u_{11}|x_1x_2x_3)p(x_1x_2x_3)$$
$$= p(u_{11}) = (0.22)(0.04) + (0.15)(0.3)$$
$$+ (0.19)(0.18) + (0.05)(0.03) \cong 0.09.$$

Other utility functions have the same steps as the steps taken in the first utility function. Inputs that are necessary for our computations are:

$$p(x_2|u_{12}x_1) = 0.1$$
$$p(x_2|x_1u_{13}) = 0.3$$
$$p(x_3|x_1u_{13}) = 0.35$$
$$p(x_3|x_2u_{13}) = 0.21$$
$$p(x_3|u_{13}x_1x_2) = 0.73$$
$$p(x_3|x_1x_2) = 0.64.$$

Then:

$$p(u_{12}) = p(u_{12}|x_1)p(x_1) + p(u_{12}|x_2)p(x_2) = (0.62)(0.4) + (0.5)(0.51) \cong 0.5\ .$$

$$p(u_{12}|x_1x_2) = \frac{p(x_1|u_{12})p(u_{12})p(x_2|u_{12}x_1)}{p(x_1)p(x_2|x_1)} = \frac{(0.5)(0.5)(0.1)}{(0.4)(0.1)} = 0.62.$$

$$p(u_{13}) = p(u_{13}|x_1)p(x_1) + p(u_{13}|x_2)p(x_2) + p(u_{13}|x_3)p(x_3)$$
$$= (0.14)(0.4) + (0.36)(0.51) + (0.08)(0.09) = 0.25.$$

$$p(u_{13}) = p(u_{13}|x_1x_2)p(x_1x_2) + p(u_{13}|x_1x_3)p(x_1x_3) + p(u_{13}|x_2x_3)p(x_2x_3)$$
$$+ p(u_{13}|x_1x_2x_3)p(x_1x_2x_3)$$
$$= (0.43)(0.04) + (0.07)(0.3) + (0.22)(0.18) + (0.49)(0.03) \cong 0.09.$$

For the sake of simplicity, but without the loss of generality, we considered two other goals with their utility functions.

$$\begin{aligned}&\text{Goal 2:} \quad \max\ f_2(x) = 3x_2 + 2x_3\\&\text{Goal 3:} \quad \max\ f_3(x) = 3.5x_1 + 5x_2 + 3x_3.\end{aligned}$$

Utility functions of $f_2(x), f_3(x)$:

$$\begin{cases} u_{21} = x_2 + x_3 \\ u_{22} = 2x_2 + 2x_3 \end{cases}$$

$$\begin{cases} u_{31} = 2x_1 + 3x_3 \\ u_{32} = x_1 + 5x_2 + 3x_3. \end{cases}$$

Some operations like the above have been done on these utilities; we just considered the final probabilities as expressed below:

$$p(u_{21}) = p(u_{21}|x_2)p(x_2) + p(u_{21}|x_3)p(x_3) = (0.4)(0.19) + (0.15)(0.06) \cong 0.08.$$

$$p(u_{21}|x_2x_3) = \frac{p(x_2|u_{21})p(u_{21})p(x_3|u_{21}x_2)}{p(x_2)p(x_3|x_2)} = \frac{(0.17)(0.45)(0.29)}{(0.19)(0.4)} = 0.29.$$

$$p(u_{22}) = p(u_{22}|x_2)p(x_2) + p(u_{22}|x_3)p(x_3) = (0.62)(0.19) + (0.83)(0.06) \cong 0.17.$$

$$p(u_{22}|x_2x_3) = \frac{p(x_2|u_{22})p(u_{22})p(x_3|u_{22}x_2)}{p(x_2)p(x_3|x_2)} = \frac{(0.47)(0.25)(0.49)}{(0.19)(0.4)} = 0.76.$$

$$p(u_{31}) = p(u_{31}|x_1)p(x_1) + p(u_{31}|x_3)p(x_3) = (0.44)(0.18) + (0.61)(0.28) \cong 0.25.$$

$$p(u_{31}|x_1x_3) = \frac{p(x_1|u_{31})p(u_{31})p(x_3|u_{31}x_1)}{p(x_1)p(x_3|x_1)} = \frac{(0.21)(0.38)(0.33)}{(0.18)(0.5)} = 0.29.$$

$$\begin{aligned}p(u_{32}) &= p(u_{32}|x_1)p(x_1) + p(u_{32}|x_2)p(x_2) + p(u_{32}|x_3)p(x_3)\\&= (0.55)(0.18) + (1)(0.03) + (0.38)(0.28) \cong 0.24.\end{aligned}$$

$$\begin{aligned}p(u_{32}) &= p(u_{32}|x_1x_2)p(x_1x_2) + p(u_{32}|x_1x_3)p(x_1x_3) + p(u_{32}|x_2x_3)p(x_2x_3)\\&\quad + p(u_{32}|x_1x_2x_3)p(x_1x_2x_3)\\&= (0.22)(0.04) + (0.15)(0.09) + (0.12)(0.02) + (0.1)(0.03) = 0.03.\end{aligned}$$

Now we solve this problem in 2 cases:

Case I: The Independence of Variables

In this case, we can have two forms of problem to be solved; first, we solve the GP model as mentioned below:

$$\begin{aligned}\text{Min } & (d_1^+ + d_1^-) + (d_2^+ + d_2^-) + (d_3^+ + d_3^-)\\ \text{s.t. } & 3x_1 + 2x_2 + x_3 - d_1^+ + d_1^- = 100\\ & 3x_2 + 2x_3 - d_2^+ + d_2^- = 25\\ & 3.5x_1 + 5x_2 + 3x_3 - d_3^+ + d_3^- = 49.\end{aligned}$$

We solve the program using LINGO to obtain the optimal solutions $(x_1, x_2, x_3) = (2.09, 8.3\overline{3}, 0)$ as shown in Table 10.1.

Second, based on the proposed MCGP method [8], this problem can be formulated as the following program:

$$\begin{aligned}\text{Min } & d_1^+ + d_1^- + d_2^+ + d_2^- + d_3^+ + d_3^-\\ \text{s.t. } & 3x_1 + 2x_2 + x_3 - d_1^+ + d_1^- = 25z_1z_2 + 50z_1(1 - z_2) + 25(1 - z_1)z_2,\\ & 3x_2 + 2x_3 - d_2^+ + d_2^- = 8z_3 + 17(1 - z_3)\\ & 3.5x_1 + 5x_2 + 3x_3 - d_3^+ + d_3^- = 25z_4 + 24(1 - z_4)\\ & d_i^+, d_i^- \geq 0, \quad i = 1, 2, 3,\\ & X \in F \quad (f \text{ is a feasible set})\end{aligned},$$

where z_1, z_2, z_3, z_4 are binary variables, d_i^+, d_i^- are the positive and negative variables, respectively.

Solve this problem using LINGO to obtain the optimal solutions $(x_1, x_2, x_3, z_1, z_2, z_3, z_4) = (0, 0, 8, 0, 0, 0, 1)$ as shown in Table 10.1.

Table 10.1 Solutions of Case I and Case II

	Case 1		Case 2	
	GP	MCGP	GP	MCGP
x_1	2.09	0	0	0
x_2	$8.3\overline{3}$	0	35	0
x_3	0	8	0	$9.6\overline{6}$
z_1	–	0	–	0
z_2	–	0	–	1
z_3	–	0	–	1
z_4	–	1	–	1

Case II: The Dependence of Variables

Formulate this problem using WGP as follows: Suppose that the weight attached to deviations has value one, and then solve it by LINGO to obtain the optimal solutions $(x_1, x_2, x_3) = (0, 35, 0)$ as shown in Table 10.1.

$$\begin{aligned}&\text{Min } w_1(d_1^+ + d_1^-) + w_2(d_2^+ + d_2^-) + w_3(d_3^+ + d_3^-)\\&\text{s.t. } 3x_1 + 2x_2 + x_3 - d_1^+ + d_1^- = 80\\&3x_2 + 2x_3 - d_2^+ + d_2^- = 105\\&3.5x_1 + 5x_2 + 3x_3 - d_3^+ + d_3^- = 32\end{aligned}$$

Now solve this problem using an MCGP method as shown below:

$$\begin{aligned}&\text{Min } d_1^+ + d_1^- + d_2^+ + d_2^- + d_3^+ + d_3^-\\&\text{s.t. } 3x_1 + 2x_2 + x_3 - d_1^+ + d_1^- = 9z_1z_2 + 62z_1(1 - z_2) + 9(1 - z_1)z_2,\\&3x_2 + 2x_3 - d_2^+ + d_2^- = 29z_3 + 76(1 - z_3)\\&3.5x_1 + 5x_2 + 3x_3 - d_3^+ + d_3^- = 29z_4 + 3(1 - z_4)\\&d_i^+, d_i^- \geq 0, \ i = 1, 2, 3,\\&x \in F\end{aligned}$$

Similarly, we solve the program using LINGO to obtain the optimal solutions $(x_1, x_2, x_3, z_1, z_2, z_3, z_4) = (0, 0, 9.6\overline{6}, 0, 1, 1, 1)$ as shown in Table 10.1.

The outcomes from Case I and Case II are summarized in Tables 10.2 and 10.3, which provide a detailed comparison between the Goal Programming (GP) and Multi-Choice Goal Programming (MCGP) models. As illustrated in Table 10.2, the MCGP model achieves a total deviation value of 8.66, significantly outperforming the GP model, which records a total deviation of 77.05. This indicates that MCGP offers a more effective solution.

In the GP model, the results show that one goal achieves 0.23, reaching its aspiration level of 100, while another goal approximates its aspiration level of 25, and a third goal approaches its aspiration level of 49. However, Table 10.2 reveals that these values differ when the MCGP model is applied. The MCGP solutions are superior to those of GP, as they achieve a perfect balance across all three goals. This demonstrates that MCGP's ability to incorporate multiple aspiration levels leads to more refined and optimal solutions. In essence, MCGP's methodology allows for

Table 10.2 Comparisons of GP model and MCGP model in Case I

	GP	Percentage of goal achievement (%)	MCGP	Percentage of goal achievement (%)
Goal g_1	$22.93 \cong 23$	$22.93 \cong 23$	8	Not achieved (released)
Goal g_2	$24.99 \cong 25$	$99.96 \cong 100$	16	94.12
Goal g_3	$48.96 \cong 49$	$99.92 \cong 100$	24	96
Total deviation	77.05		$8.6\overline{6}$	

Table 10.3 Comparisons of GP model and MCGP model in Case II

	GP	Percentage of goal achievement (%)	MCGP	Percentage of goal achievement (%)
Goal g_1	70	87.5	9.66	93.17
Goal g_2	105	100	19.32	66.62
Goal g_3	175	18.29	28.98	$99.93 \cong 100$
Total deviation	153		$10.3\overline{3}$	

a more comprehensive consideration of aspiration levels, resulting in better overall outcomes.

From Table 10.3, it is evident that the goal g_1 demonstrates a higher percentage of goal achievement in both the GP and MCGP models compared to the results in Table 10.2. However, when comparing the GP model results between the two tables, the goal g_2 maintains an equal percentage of goal achievement. In contrast, the MCGP model shows a lower percentage of goal achievement in Table 10.3 relative to Table 10.2.

Interestingly, while the goal g_3 exhibits a reduced percentage of goal achievement in the GP model when compared to Table 10.2, it achieves a higher percentage of goal achievement in the MCGP model relative to the same table. This highlights the variability in performance between the two models across different scenarios and underscores the adaptability of MCGP in optimizing goal achievement under varying conditions.

10.3.2 Constrained Multi-objective Model

In this section, we examine a constrained Multi-Objective Linear Programming (MOLP) problem where each goal is associated with multiple utility functions. We then analyze how the independence and dependence of variables influence these utility functions (Abbasian et al., 2018). The goals and constraints for the problem are as follows:

$$
\begin{aligned}
&\text{Goal 1}: \quad \max f_1(x) = 3x_1 + 2x_2 + x_3 \\
&\text{Goal 2}: \quad \max f_1(x) = 3x_2 + 2x_3 \\
&\text{Goal 3}: \quad \max f_3(x) = 3.5x_1 + 5x_2 + 3x_3. \\
&\qquad \text{s.t.} \\
&\qquad x_2 + x_3 \geq 10 \\
&\qquad x_2 \geq 4 \\
&\qquad x_1 + x_2 + x_3 \geq 15.
\end{aligned}
$$

The utility functions corresponding to the goals are:

$$\begin{cases} u_{11}(x) = 2x_1 + 3x_2 + 2x_3 \\ u_{12}(x) = x_1 + 2x_2 \\ u_{13}(x) = x_1 + x_2 + x_3 \end{cases}$$

$$\begin{cases} u_{21} = x_2 + x_3 \\ u_{22} = 2x_2 + 2x_3 \end{cases}$$

$$\begin{cases} u_{31} = 2x_1 + 3x_3 \\ u_{32} = x_1 + 5x_2 + 3x_3. \end{cases}$$

Starting with Goal 1, the inputs are: $p(u_{11}) = 0.7$, $p(u_{12}) = 0.02$, $p(u_{13}) = 0.28$. The necessary computations are:

$$\begin{aligned} p(x_1|u_{11}) &= 0.2 \\ p(x_2|u_{11}) &= 0.1 \\ p(x_3|u_{11}) &= 0.3 \\ p(x_1|u_{12}) &= 0.4 \\ p(x_2|u_{12}) &= 0.6 \\ p(x_1|u_{13}) &= 0.1 \\ p(x_2|u_{13}) &= 0.6 \\ p(x_3|u_{13}) &= 0.3. \end{aligned}$$

If x_1, x_2, and x_3 are assumed to independently affect the utility functions, then we arrive at the following probabilities:

$$\begin{aligned} p(x_1) &= p(x_1|u_{11})p(u_{11}) + p(x_1|u_{12})p(u_{12}) + p(x_1|u_{13})p(u_{13}) \\ &= (0.2)(0.7) + (0.4)(0.02) + (0.1)(0.28) \cong 0.18 \\ p(x_2) &= p(x_2|u_{11})p(u_{11}) + p(x_2|u_{12})p(u_{12}) + p(x_2|u_{13})p(u_{13}) \\ &= (0.1)(0.7) + (0.6)(0.02) + (0.6)(0.28) = 0.25 \\ p(x_3) &= p(x_3|u_{11})p(u_{11}) + p(x_3|u_{13})p(u_{13}) \\ &= (0.3)(0.7) + (0.3)(0.28) = 0.29 \end{aligned}$$

$$\begin{aligned} p(u_{11}|x_1) &= \frac{p(x_1|u_{11})p(u_{11})}{p(x_1)} = \frac{(0.2)(0.7)}{0.18} \cong 0.77 \\ p(u_{11}|x_2) &= \frac{p(x_2|u_{11})p(u_{11})}{p(x_2)} = \frac{(0.1)(0.7)}{0.25} \cong 0.28 \\ p(u_{11}|x_3) &= \frac{p(x_3|u_{11})p(u_{11})}{p(x_3)} = \frac{(0.3)(0.7)}{0.29} \cong 0.72. \end{aligned}$$

Using the total probability rule, we obtain:

$$
\begin{aligned}
p(u_{11}) &= p(u_{11}|x_1)p(x_1) + p(u_{11}|x_2)p(x_2) + p(u_{11}|x_3)p(x_3) \\
p(u_{11}) &= (0.77)(0.18) + (0.28)(0.25) + (0.72)(0.29) \cong 0.42.
\end{aligned}
$$

On the other hand, if x_1, x_2, x_3 are assumed to dependently affect the utility functions, then with the following data:

$$
\begin{aligned}
p(x_2|x_1) &= 0.4 \\
p(x_3|x_1) &= 0.4 \\
p(x_3|x_2) &= 0.2 \\
p(x_2|u_{11}x_1) &= 0.5 \\
p(x_3|x_1u_{11}) &= 0.45 \\
p(x_3|x_2u_{11}) &= 0.3 \\
p(x_3|u_{11}x_1x_2) &= 0.1 \\
p(x_3|x_1x_2) &= 0.8,
\end{aligned}
$$

we get the following probabilities:

$$
\begin{aligned}
p(u_{11}|x_1x_2) &= \frac{p(x_1|u_{11})p(u_{11})p(x_2|u_{11}x_1)}{p(x_1)p(x_2|x_1)} = \frac{(0.2)(0.7)(0.5)}{(0.18)(0.4)} \cong 0.97 \\
p(u_{11}|x_1x_3) &= \frac{p(x_1|u_{11})p(u_{11})p(x_3|x_1u_{11})}{p(x_1)p(x_3|x_1)} = \frac{(0.2)(0.7)(0.45)}{(0.18)(0.4)} = 0.87 \\
p(u_{11}|x_2x_3) &= \frac{p(x_2|u_{11})p(u_{11})p(x_3|x_2u_{11})}{p(x_2)p(x_3|x_2)} = \frac{(0.1)(0.7)(0.3)}{(0.25)(0.2)} \cong 0.42 \\
p(u_{11}|x_1x_2x_3) &= \frac{p(x_3|u_{11}x_1x_2)p(x_2|u_{11}x_1)p(x_1|u_{11})p(u_{11})}{p(x_3|x_1x_2)p(x_2|x_1)p(x_1)} \\
&= \frac{(0.1)(0.5)(0.2)(0.7)}{(0.8)(0.4)(0.18)} \cong 0.12.
\end{aligned}
$$

We then have:

$$
\begin{aligned}
p(x_1x_2) &= p(x_2|x_1)p(x_1) = (0.4)(0.18) = 0.07 \\
p(x_1x_3) &= p(x_3|x_1)p(x_1) = (0.4)(0.5) = 0.07 \\
p(x_2x_3) &= p(x_3|x_2)p(x_2) = (0.2)(0.25) \cong 0.05 \\
p(x_1x_2x_3) &= p(x_3|x_1x_2)p(x_1x_2) = (0.8)(0.07) \cong 0.06.
\end{aligned}
$$

By using the total probability rule, we obtain:

$$
\begin{aligned}
p(u_{11}) = {} & p(u_{11}|x_1x_2)p(x_1x_2) + p(u_{11}|x_1x_3)p(x_1x_3) + p(u_{11}|x_2x_3)p(x_2x_3) \\
& + p(u_{11}|x_1x_2x_3)p(x_1x_2x_3)
\end{aligned}
$$

$$p(u_{11}) = (0.97)(0.07) + (0.87)(0.07) + (0.42)(0.05) + (0.12)(0.06) \cong 0.16.$$

The same calculations can be carried out for the other two utility functions. Using the given inputs,

$$\begin{aligned} p(x_2|u_{12}x_1) &= 0.7 \\ p(x_2|x_1u_{13}) &= 0.35 \\ p(x_3|x_1u_{13}) &= 0.24 \\ p(x_3|x_2u_{13}) &= 0.02 \\ p(x_3|u_{13}x_1x_2) &= 0.77 \\ p(x_3|x_1x_2) &= 0.8, \end{aligned}$$

we get

$$p(u_{12}) = p(u_{12}|x_1)p(x_1) + p(u_{12}|x_2)p(x_2) = (0.02)(0.18) + (0.06)(0.25) \cong 0.02$$

$$p(u_{12}|x_1x_2) = \frac{p(x_1|u_{12})p(u_{12})p(x_2|u_{12}x_1)}{p(x_1)p(x_2|x_1)} = \frac{(0.4)(0.02)(0.7)}{(0.18)(0.4)} = 0.08$$

$$p(u_{12}) = p(u_{12}|x_1x_2)p(x_1x_2) = (0.08)(0.07) = 0.01$$

$$\begin{aligned} p(u_{13}) &= p(u_{13}|x_1)p(x_1) + p(u_{13}|x_2)p(x_2) + p(u_{13}|x_3)p(x_3) \\ &= (0.16)(0.18) + (0.67)(0.25) + (0.29)(0.29) = 0.28 \end{aligned}$$

$$\begin{aligned} p(u_{13}) &= p(u_{13}|x_1x_2)p(x_1x_2) + p(u_{13}|x_1x_3)p(x_1x_3) + p(u_{13}|x_2x_3)p(x_2x_3) \\ &\quad + p(u_{13}|x_1x_2x_3)p(x_1x_2x_3) \\ &= (0.14)(0.07) + (0.09)(0.07) + (0.07)(0.05) + (0.13)(0.06) \cong 0.03. \end{aligned}$$

We continue with Goal 2 and Goal 3.

Next, assume the following data for Goal 2:

$$\begin{aligned} p(u_{21}) &= 0.5 \\ p(u_{22}) &= 0.5 \end{aligned}$$

$$\begin{aligned} p(x_2|u_{21}) &= 0.6 \\ p(x_3|u_{21}) &= 0.2 \\ p(x_2|u_{22}) &= 0.3 \\ p(x_3|u_{22}) &= 0.6 \\ p(x_3|x_2) &= 0.8 \\ p(x_3|u_{21}x_2) &= 0.6. \end{aligned}$$

Doing similar computations as before, we arrive at the final probabilities as follows:

$$p(u_{21}) = p(u_{21}|x_2)p(x_2) + p(u_{21}|x_3)p(x_3) = (0.66)(0.45) + (0.25)(0.4) \cong 0.4$$

$$p(u_{21}|x_2x_3) = \frac{p(x_2|u_{21})p(u_{21})p(x_3|u_{21}x_2)}{p(x_2)p(x_3|x_2)} = \frac{(0.6)(0.5)(0.6)}{(0.45)(0.8)} = 0.5$$

$$p(u_{21}) = p(u_{21}|x_2x_3)p(x_2x_3) = (0.5)(0.36) = 0.18$$

$$p(u_{22}) = p(u_{22}|x_2)p(x_2) + p(u_{22}|x_3)p(x_3) = (0.33)(0.45) + (0.75)(0.4) \cong 0.45$$

$$p(u_{22}|x_2x_3) = \frac{p(x_2|u_{22})p(u_{22})p(x_3|u_{22}x_2)}{p(x_2)p(x_3|x_2)} = \frac{(0.3)(0.5)(0.6)}{(0.39)(0.8)} = 0.25$$

$$p(u_{22}) = p(u_{22}|x_2x_3)p(x_2x_3) = (0.25)(0.36) = 0.09.$$

Using the following inputs for Goal 3,

$$\begin{aligned}
p(u_{31}) &= 0.4 \\
p(u_{32}) &= 0.2 \\
p(x_1|u_{31}) &= 0.4 \\
p(x_3|u_{31}) &= 0.2 \\
p(x_1|u_{32}) &= 0.2 \\
p(x_2|u_{32}) &= 0.5 \\
p(x_3|u_{32}) &= 0.1 \\
p(x_3|x_1) &= 0.9 \\
p(x_3|u_{31}x_1) &= 0.8
\end{aligned}$$

$$\begin{aligned}
p(x_2|u_{32}x_1) &= 0.43 \\
p(x_2|x_1) &= 0.6 \\
p(x_3|u_{32}x_1) &= 0.3
\end{aligned}$$

$$\begin{aligned}
p(x_3|x_2) &= 0.4 \\
p(x_3|u_{32}x_2) &= 0.32 \\
p(x_3|u_{32}x_1x_2) &= 0.65 \\
p(x_3|x_1x_2) &= 0.75,
\end{aligned}$$

we get

$$p(u_{31}) = p(u_{31}|x_1)p(x_1) + p(u_{31}|x_3)p(x_3) = (0.57)(0.28) + (0.57)(0.14) \cong 0.24$$

$$p(u_{31}|x_1x_3) = \frac{p(x_1|u_{31})p(u_{31})p(x_3|u_{31}x_1)}{p(x_1)p(x_3|x_1)} = \frac{(0.4)(0.4)(0.8)}{(0.28)(0.9)} \cong 0.51$$

$$p(u_{31}) = p(u_{31}|x_1x_3)p(x_1x_3) = (0.51)(0.25) \cong 0.13$$

$$\begin{aligned} p(u_{32}) &= p(u_{32}|x_1)p(x_1) + p(u_{32}|x_2)p(x_2) + p(u_{32}|x_3)p(x_3) \\ &= (0.43)(0.28) + (1)(0.3) + (0.43)(0.14) \cong 0.48 \end{aligned}$$

$$\begin{aligned} p(u_{32}) &= p(u_{32}|x_1x_2)p(x_1x_2) + p(u_{32}|x_1x_3)p(x_1x_3) + p(u_{32}|x_2x_3)p(x_2x_3) \\ &\quad + p(u_{32}|x_1x_2x_3)p(x_1x_2x_3) \\ &= (0.31)(0.17) + (0.14)(0.25) + (0.8)(0.12) + (0.27)(0.13) \cong 0.22. \end{aligned}$$

Next, we are able to solve the problem in the two considered cases.

Case I: Independence of Variables

Here, we need to consider two models. First, we solve the following GP model:

$$\begin{aligned} \text{Min } & (d_1^+ + d_1^-) + (d_2^+ + d_2^-) + (d_3^+ + d_3^-) \\ \text{s.t. } & 3x_1 + 2x_2 + x_3 - d_1^+ + d_1^- = 72 \\ & 3x_2 + 2x_3 - d_2^+ + d_2^- = 85 \\ & 3.5x_1 + 5x_2 + 3x_3 - d_3^+ + d_3^- = 72 \\ & x_2 + x_3 \geq 10 \\ & x_2 \geq 4 \\ & x_1 + x_2 + x_3 \geq 15. \end{aligned}$$

We solved the program using LINGO software [see Scharge (2008)] and obtained the optimal solutions $(x_1, x_2, x_3) = (0, 28.33, 0)$ as shown in Table 10.1, under the column entitled GP corresponding to case I.

Second, based on the proposed MCGP approach, the problem is formulated as follows:

$$\begin{aligned} \text{Min } & d_1^+ + d_1^- + d_2^+ + d_2^- + d_3^+ + d_3^- \\ \text{s.t. } & 3x_1 + 2x_2 + x_3 - d_1^+ + d_1^- = 42z_1z_2 + 2z_1(1 - z_2) + 28(1 - z_1)z_2, \\ & 3x_2 + 2x_3 - d_2^+ + d_2^- = 40z_3 + 45(1 - z_3) \\ & 3.5x_1 + 5x_2 + 3x_3 - d_3^+ + d_3^- = 24z_4 + 48(1 - z_4) \\ & d_i^+, d_i^- \geq 0, \quad i = 1, 2, 3, \\ & x_2 + x_3 \geq 10 \\ & x_2 \geq 4 \\ & x_1 + x_2 + x_3 \geq 15, \end{aligned}$$

where z_1, z_2, z_3, and z_4 are binary variables, d_i^+ and d_i^- are the positive and negative variables, respectively.

We solved this problem using LINGO [see Scharge (2008)] again to obtain the optimal solution $(x_1, x_2, x_3, z_1, z_2, z_3, z_4) = (0, 10, 5, 0, 1, 1, 0)$ as shown in Table 10.1, under the column entitled MCGP corresponding to case I.

Case II: Dependence of Variables

Here, we formulated the problem using the WGP model. Suppose that all the weights attached to the deviations have the value one. We then have the following problem:

$$\begin{aligned}
\text{Min } & w_1(d_1^+ + d_1^-) + w_2(d_2^+ + d_2^-) + w_3(d_3^+ + d_3^-) \\
\text{s.t. } & 3x_1 + 2x_2 + x_3 - d_1^+ + d_1^- = 20 \\
& 3x_2 + 2x_3 - d_2^+ + d_2^- = 27 \\
& 3.5x_1 + 5x_2 + 3x_3 - d_3^+ + d_3^- = 35 \\
& x_2 + x_3 \geq 10 \\
& x_2 \geq 4 \\
& x_1 + x_2 + x_3 \geq 15.
\end{aligned}$$

We solved the program using LINGO [see, Scharge (2008)] to obtain the optimal solution $(x_1, x_2, x_3) = (0.5, 4, 10.5)$ as shown in Table 10.1, under the column entitled WGP corresponding to case II.

Next, the problem is formulated as an MCGP problem as follows:

$$\begin{aligned}
\text{Min } & d_1^+ + d_1^- + d_2^+ + d_2^- + d_3^+ + d_3^- \\
\text{s.t. } & 3x_1 + 2x_2 + x_3 - d_1^+ + d_1^- = 16z_1z_2 + 1z_1(1 - z_2) + 3(1 - z_1)z_2, \\
& 3x_2 + 2x_3 - d_2^+ + d_2^- = 18z_3 + 9(1 - z_3) \\
& 3.5x_1 + 5x_2 + 3x_3 - d_3^+ + d_3^- = 13z_4 + 22(1 - z_4) \\
& d_i^+, d_i^- \geq 0, \quad i = 1, 2, 3, \\
& x_2 + x_3 \geq 10 \\
& x_2 \geq 4 \\
& x_1 + x_2 + x_3 \geq 15.
\end{aligned}$$

The program was solved using LINGO [see Scharge (2008)] to obtain the optimal solution $(x_1, x_2, x_3, z_1, z_2, z_3, z_4) = (0, 4, 11, 1, 1, 1, 0)$ as shown in Table 10.4, under the column entitled MCGP corresponding to case II.

Tables 10.5 and 10.6 provide a summary of the results derived from the Goal Programming (GP) and Multi-Choice Goal Programming (MCGP) models, respectively. As shown in Table 10.5, the MCGP model achieves a total deviation value of 20, which is significantly lower than the total deviation value of 85 recorded by the GP model. This indicates that MCGP offers a more effective solution.

In the GP model for Case I, the results reveal that Goal 1 attains 78.7% of its aspiration level of 72, Goal 2 precisely meets its aspiration level of 85, and Goal 3 achieves 51% of its aspiration level of 72. However, Table 10.5 demonstrates that these values

Table 10.4 Solutions corresponding to cases I and II

	Case I		Case II	
	GP	MCGP	WGP	MCGP
x_1	0	0	0.5	0
x_2	$28.3\overline{3}$	10	4	4
x_3	0	5	10.5	11
z_1	–	0	–	1
z_2	–	1	–	1
z_3	–	1	–	1
z_4	–	0	–	0

Table 10.5 Comparison of GP and MCGP models in case I

	GP	Percentage of goal achievement (%)	MCGP	Percentage of goal achievement (%)
Goal 1	56.66	78.7	25	89.3
Goal 2	85	100	40	100
Goal 3	141.65	51	65	74
Total deviation	85		20	

Table 10.6 Comparison of GP and MCGP models in case II

	GP	Percentage of goal achievement (%)	MCGP	Percentage of goal achievement (%)
Goal 1	20	100	19	84.2
Goal 2	33	81.8	34	53
Goal 3	53.25	65.7	53	41.5
Total deviation	24.25		50	

differ when the MCGP model is applied. The MCGP solutions outperform those of the GP model, as the percentages of goal achievement achieved by MCGP are consistently higher than those obtained by GP. This highlights the superior performance of MCGP in optimizing goal attainment across multiple objectives.

From Table 10.6, it is observed that the MCGP model records a total deviation value of 50, which is 9higher than the total deviation value of 24.25 achieved by the GP model in Case II. Specifically, the GP model in Case II shows that Goal 1 fully meets its aspiration level of 20 (100%), Goal 2 achieves 81.8% of its aspiration level of 27, and Goal 3 attains 65.7% of its aspiration level of 35. However, the percentages of goal achievement in the MCGP model are lower than those in the GP model for this case.

Interestingly, when the GP model is solved by individually selecting the aspiration levels from the MCGP model for each goal, the GP model results in larger deviation values compared to the MCGP model. This demonstrates that the MCGP model outperforms the GP model in such scenarios, as it provides more balanced and optimized solutions across multiple goals.

10.4 Conclusions

This study presents an advanced methodology for solving multi-objective linear programming (MOLP) problems through the integration of multiple utility functions per objective. Departing from conventional approaches that employ a single utility function for each goal, our technique assigns probabilistic weights to multiple utility representations of each objective. These weighted utilities serve as dynamic aspiration levels within both Goal Programming (GP) and Multi-Choice Goal Programming (MCGP) formulations.

The proposed framework was rigorously evaluated through comparative analysis of two distinct scenarios:

- systems with interdependent variables;
- systems with independent variables.

Key findings demonstrate that decision-makers can attain optimal solutions using either GP or MCGP implementations. However, the MCGP approach consistently outperformed standard GP in solution quality, particularly in complex optimization environments requiring flexible target achievement. This performance advantage stems from MCGP's inherent capacity to:

- Accommodate variable aspiration levels.
- Manage conflicting objective hierarchies.
- Resolve solution space discontinuities.

The results validate the efficacy of probabilistic multi-utility assignment as a paradigm for enhancing multi-criteria decision analysis, offering practitioners a more nuanced tool for navigating real-world optimization challenges where objectives frequently exhibit nonlinear relationships and competing priorities.

References and Further Reading

Abbasian, P., Mahdavi-Amiri, N., & Fazllolahtabar, H. (2015). Modeling and optimizing of a non-constrained multi-objective problem having multiple utility functions using the bayesian theory. *Journal of Information Sciences and Computing Technologies, 4*(3), 332–342.

Abbasian, P., Mahdavi-Amiri, N., & Fazlollahtabar, H. (2018). Multiple utility constrained multi-objective programs using Bayesian theory. *Journal of Industrial Engineering International, 14*, 111–118.

Al-Nowaihi, A., Bradley, I., & Dhami, S. (2008). A note on the utility function under prospect theory. *Economics Letters, 99*, 337–339.

Aouni, B., & Kettani, O. (2001). Goal programming model: A glorious history and a promising future. *European Journal of Operational Research, 133*, 225–231.

Arenas-Parra, M., Bilbao-Terol, A., Perez-Gladish, B., & Rodrıguez-Urıa, M. V. (2010). A new approach of Romero's extended lexicographic goal programming: Fuzzy extended lexicographic goal programming. *Soft Computing, 14*, 1217–1226.

Charnes, A., Cooper, W. W., & Ferguson, R. O. (1955). Optimal estimation of executive compensation by linear Programming. *Management Science, 1*, 138–151.

Fazlollahtabar, H., & Ashena, M. (2013). A Bayesian network decision support system for order management in new product development. *Computer Science and Information Technology, 1*, 82–89.

Fazlollahtabar, H., Akbari, F., & Mahdavi, I. (2013). A fuzzy goal programming for optimizing service industry market using virtual intelligent agent. *Journal of Industrial and Production Engineering, 30*(1), 20–29.

Komleh, A. A., & Fazlollahtabar, H. (2019). Stochastic multi-criteria acceptability analysis for technology transfer evaluation: A case study in construction digging. *International Journal of Mathematical, Engineering and Management Sciences, 4*(4), 1031.

Lai, Y.-J., & Hwang, C.-L. (1994). *Fuzzy multi objective decision making: methods and applications*. Springer-Verlag.

LiCalzi, M., & Sorato, A. (2006). The Pearson system of utility functions. *European Journal of Operational Research, 172*, 560–573.

Olshausen, B. (2004). *Bayesian probability theory*. University of California.

Podinovski, V. V. (2010). Set choice problems with incomplete information about the preferences of the decision maker. *European Journal of Operational Research, 207*, 371–379.

Romero, C. (2001). Extended lexicographic goal programming: A unifying approach. *Omega, 29*, 63–71.

Scharge, L. (2008). LINGO Release 11.0. LINDO System, Inc.

Yu, B.W.-T., Pang, W. K., Troutt, M. D., & Hou, S. H. (2009). Objective comparisons of the optimal portfolios corresponding to different utility functions. *European Journal of Operational Research, 199*, 604–610.

Chapter 11
Stochastic Multi-objective Optimization

Abstract This study develops a novel mathematical formulation for cellular manufacturing systems under parameter uncertainty, where processing times and product demands are modeled as stochastic variables characterized by their expected values and standard deviations derived from sampling data. The optimization framework simultaneously addresses three critical manufacturing objectives: (1) minimizing inter-cell and intra-cell material movements through strategic machine-cell assignment, (2) eliminating production bottlenecks via balanced workload distribution, and (3) maximizing overall profitability through systematic production cost reduction. The model's effectiveness is demonstrated through numerical case studies, with computational analysis revealing significant solution quality improvements. To address the combinatorial complexity of large-scale implementations, we propose an innovative heuristic solution approach featuring a proprietary neighborhood search algorithm for efficient solution space exploration, and an adaptive resource allocation methodology for dynamic cell configuration.

Keywords Cellular manufacturing system (CMS) · Stochastic parameters · Mathematical programming · Pricing · Multi-objective

11.1 Introduction

The current global manufacturing landscape demands unprecedented levels of efficiency and adaptability to maintain competitiveness. Facing pressures from shortened product life cycles, accelerated time-to-market expectations, and diversified customer requirements, modern production systems must deliver high-quality outputs rapidly while minimizing costs (Fazlollahtabar & Aghasi, 2014). Traditional manufacturing approaches like job shops and flow lines often prove inadequate in meeting these challenges, particularly regarding their limited capacity to accommodate design changes or demand fluctuations without substantial capital investment.

Cellular Manufacturing Systems (CMS) have emerged as a strategic solution, applying group technology principles to organize production around dedicated

H. Fazlollahtabar, *Uncertain Multi-objective Decision Making: Methods and Industrial Models*, https://doi.org/10.1007/978-981-95-6035-6_11

machine clusters for processing part families. This configuration offers distinct advantages over conventional systems, including:

- significant reductions in setup times and work-in-process inventory;
- enhanced material handling efficiency through optimized inter-cell and intra-cell movement strategies;
- improved production quality and faster market responsiveness;
- greater flexibility for mid-volume, mid-variety production mixes.

The effectiveness of CMS implementations heavily depends on strategic planning and accurate modeling of key operational factors such as machine reliability (affecting defect rates and bottlenecks) and material flow dynamics (Chattopadhyay et al., 2013; Fazlollahtabar et al., 2010).

This research develops an advanced CMS model incorporating stochastic elements to better reflect real-world manufacturing conditions. The framework addresses two critical uncertainty factors of variable part demand patterns and fluctuating processing times.

Building on established mathematical programming approaches (Majazi Dalfard, 2013; Paydar & Saidi-Mehrabad, 2013), the model integrates production planning advantages identified by Chen and Cao (2004), including strategic inventory management for demand fulfillment.

For uncertainty quantification, the study employs continuous probability distributions to characterize random parameters and queuing theory applications for system performance analysis (Egilmez et al., 2012).

This methodology provides a robust alternative to conventional uncertainty modeling techniques like stochastic programming or robust optimization (Gholizadeh et al., 2020), particularly when complete probability information is unavailable, and parameters must be bounded within established intervals (Ghezavati et al., 2008; Snyder, 2006).

The proposed model enables manufacturers to optimize cellular layouts for dynamic production requirements, balance efficiency and flexibility objectives, make informed strategic decisions about CMS implementation, and maintain profitability in volatile market conditions.

By addressing both structural design and operational planning within a unified stochastic framework, this approach offers a comprehensive solution for modern manufacturing challenges.

11.2 Literature Review

Research on cellular manufacturing (CM) has led to the development of diverse methodologies, which can be broadly classified into cluster analysis, graph partitioning, descriptive procedures (including part family and machine group identification), mathematical programming (such as linear, integer, quadratic, goal, and dynamic programming), and artificial intelligence techniques (Fazlollahtabar et al.,

2015). Recent studies emphasize the need for more integrated models and solution approaches to CM design (Paydar & Saidi-Mehrabad, 2013). While many cell formation techniques focus on minimizing intercellular movements, these objectives do not always ensure optimal cell performance or meet application-specific requirements.

This research builds on foundational work by Nsakanda et al. (2006) and Chen and Cao (2004, 2005), who address critical aspects of CMS design, including intercellular and intracellular costs, outsourcing considerations, part demands, machine capacity constraints, multiple process plans, alternative part routings, and processing sequences. Solimanpur et al. (2004) further contribute by integrating production planning with CMS, enhancing operational efficiency.

Recent advances in CM research explore various dimensions, including:

- Cell Formation and Group Scheduling: Addressing exceptional elements and applying meta-heuristics for large-scale problems (Wu et al., 2006).
- Probabilistic Demand Modeling: Tavakkoli-Moghaddam et al. (2007) incorporate probabilistic demand into CMS design, assigning probabilities to product mixes and part-machine matrices.
- Uncertainty Handling: Studies by Rafiei and Ghodsi (2013) and Chattopadhyay et al. (2013) examine CMS under uncertain conditions.
- Flexible Cell Structures: Tavana et al. (2014) propose a cascading cell system, where child cells inherit machines from parent cells while adding supplementary capacity. This design allows dynamic rerouting of parts when parent cells face demand fluctuations, improving system adaptability.

These developments highlight the shift toward more robust and flexible CMS frameworks capable of accommodating real-world manufacturing complexities, such as variable demand, machine redundancy, and production scalability. Future research directions may explore hybrid approaches combining AI-driven optimization with stochastic modeling to further enhance CM performance in dynamic environments.

11.3 Proposed Stochastic Multi-objective Modeling

This study presents a comprehensive cell formation model incorporating stochastic processing parameters and advanced logistical considerations. The framework builds upon established work in stochastic CMS design (Salarian et al., 2014) while introducing novel operational constraints and performance metrics.

Key Model Assumptions:

1. Processing Time Variability:
    - Part processing times (λ_{ij}) follow normal distributions with known expectation $E(\lambda_{ij})$ and variance var(λ_{ij}) within a (1–α) confidence interval.
    - Part demand (D_i) similarly follows normal distribution patterns.

2. Production Flow Characteristics:
   - unconstrained material availability with guaranteed demand fulfillment;
   - single rework allowance for defective parts;
   - uniform movement cost application for both inter-cell and intra-cell transfers;
   - distance calculations based on fixed facility coordinates.
3. Machine State Modeling:
   - Three operational statuses:
     - optimal production (failure rate = 0),
     - rework-enabled production,
     - scrap-generating production.
   - Material handling vehicle selection impacts bottleneck formation.
4. Spatial Configuration:
   - multiple candidate locations for machine placement;
   - reference source point for coordinate system establishment;
   - reverse production rate computation from finished goods.

Process Plan Requirements:

The model incorporates binary movement indicators ($M_{ijj'}$) where:
$M_{ijj'} = 1$ when part i requires sequential processing from machine j to j′;
$M_{ijj'} = 0$ otherwise.
Material flow quantification follows:
Transport volume for part $i = \left\lceil \frac{Q_i}{CM_{ib}} \right\rceil$.

Layout Constraint:

Number of candidate locations $\geq$ Σ(machine count) (Salarian & Fazlollahtabar, 2015).

Implementation Notes:

1. The normal distribution assumption enables probabilistic capacity planning.
2. Machine state differentiation allows precise quality cost accounting.
3. Coordinate-based distance computation facilitates realistic layout optimization.
4. Reverse production rate calculation supports pull-system implementations.

Model Advancements:

- Integrates stochastic processing with deterministic layout optimization.
- Captures both quality costs (rework/scrap) and logistical costs (material handling).
- Accommodates flexible facility configurations through candidate locations.

This framework provides manufacturers with a robust tool for:

- cellular system design under uncertainty,
- bottleneck prediction and mitigation,
- trade-off analysis between quality and throughput.

Indices

$i \in \{1 \ldots I\}$	Index for parts.
$j, j' \in \{1 \ldots M\}$	Index for machines.
$k, k' \in \{1 \ldots C\}$	Index for cells.
$f, f' \in \{1 \ldots \theta\}$	Index for candidate location.
$b \in \{1 \cdots B\}$	Index for material handling device.
$\phi \in \{1, 2, 3\}$	Index for status of machines in encountering with parts.
$n \in \{1, \ldots, N_i\}$	Index for stages of processing plan for each part.

Parameters

$M_{ijj'}$	1 if part i is moved from machine j to machine j' directly according to process plan and 0 otherwise.
$r_{ijj'}$	Number of repetitions of $M_{ijj'}$.
$\eta_{ifkf'k'}$	The movement cost for part i between location f, f' in corresponding cells k, k'.
a_{ij}	1 if part i requires to be processed on machine j and 0 otherwise.
H_i	Number of transmission of part i.
$z_{1-\alpha/2}$	Standard normal value for $1 - \alpha/2$ level in two-sided confidence interval.
λ_{ij}	Service rate of machine j to part i in unit of time.
$E(\cdot)$	The expected value of a parameter.
$\mathrm{var}(\cdot)$	The variance value of a parameter.
N_{fk}	1 if candidate location f is fixed in cell k and 0 otherwise.
G_k	Number of candidate location in cell k.
θ	Summation of candidate location in all cells.
CM_{ib}	Capacity of material handling device b to convey part i.
$d_{ff'}$	Distance between candidate location f and f' that computed by rectilinear method.
A_{ib}	Cost of unit measure of movement for part i that is conveyed via material handling device b.
C_b	Purchase cost of material handling device b.
C'_b	The annual cost of maintenance and repair for material handling device b.
X_f, X'_f	The abscissa of candidate location f and f'.
Y_f, Y'_f	The ordinate of candidate location f and f'.
$\beta_{kk'}$	Penalty cost of movement of part i from cell k to k'.
SC_{ij}	The set up and processing cost for part i that processed on machine j.
PC_i	The purchase cost of each unit of part i.
rc_{ij}	The rework cost of part i on machine j.
$\mathrm{lp}_i, \mathrm{up}_i$	The lower bound and upper bound of selling price for part i.

$u_{ij}^{n\emptyset}$ 1 if part i exist on machine j with status Ø in stage n of the process plan; 0 otherwise.
Q_i The amount of input material for part i.
D_i Demand for part i.
$q_{i,n}$ Input for part i in stage n.
$\hat{p}_{ij}^{n\emptyset}$ The probability of completeness in status Ø for part i on machine j in stage n.
$\tilde{P}_{ij}^{n\emptyset}$ The probability of failure in status Ø for part i on machine j in stage n.
γ_{ij} The percent of scrap for part i on machine j in rework status.
$1-\gamma_{ij}$ The percent of complete output part i from rework on machine j in rework status.

Remarks

1. The movement cost is 0 if $f = f'$ but if $N = N'$ only the intra-cell will exist.
2. Number of repetitions of parameter $M_{ijj'}$, $r_{ijj'}$ is the number of direct movements from machine j to machine j' for part i.
3. The service rate of machine j to part i in unit of time is computed by: $\lambda_{ij} = E(\lambda_{ij}) + \sqrt{\mathrm{var}(\lambda_{ij})} \cdot z_\alpha$.

Decision Variables

L_{jf} 1 if machine j is in candidate location f and 0 otherwise.
δ_j Productivity coefficient for machine j.
V_{ib} 1 if part i is carried by material handling device b and 0 otherwise.
pr_i Selling price of product i.

The nonlinear mathematical model:

$$\max(w^1) = \left(\begin{array}{l} \sum_{i=1}^{I} \mathrm{pr}_i \cdot D_i - \left[\sum_{i=1}^{I} \sum_{b=1}^{B} \sum_{j=1}^{m} \sum_{\substack{j'=1 \\ j' \neq j}}^{m} \sum_{f=1}^{\theta} \sum_{\substack{f'=1 \\ f' \neq f}}^{\theta} \sum_{k=1}^{c} \sum_{k'=1}^{c} (a_{ij} \cdot L_{jf} \cdot N_{fk} \right. \\ \cdot a_{ij'} \cdot L_{j'f'} \cdot N_{f'k'} \cdot M_{ijj'} \cdot r_{ijj'} \cdot \eta_{ifkf'k'b} \cdot v_{ib} \cdot H_i) + \sum_{i=1}^{I} \sum_{b=1}^{B} (c_b + c_b') \cdot v_{ib} \\ \left. + \sum_{n=1}^{N_i} \mathrm{sc}_{ij} \cdot q_{i,n} \cdot u_{ij}^{n\varphi} + \sum_{i=1}^{I} pc_i \cdot Q_i + \sum_{n=1}^{N_i} \sum_{j=1}^{m} \hat{p}_{ij}^{n3} \cdot Q_i \cdot \mathrm{rc}_{ij} \cdot u_{ij}^{n3} \right] \end{array} \right)$$

$$\max(w^2) = \sum_{j=1}^{m} \delta_j$$

s.t.

$$\sum_{j=1}^{m} L_{jf} \leq 1 \quad \forall f \in \{1 \cdots \theta\} \tag{11.1}$$

$$\sum_{f=1}^{\theta} L_{jf} = 1 \quad \forall j \in \{1 \cdots \theta\} \tag{11.2}$$

$$\theta = \sum_{k=1}^{c} G_k \tag{11.3}$$

$$\sum_{f=1}^{\theta} \sum_{j=1}^{m} L_{jf} \leq \theta \tag{11.4}$$

$$H_i = \left\lceil \frac{Q_i}{CM_{ib}} \right\rceil \tag{11.5}$$

$$\eta_{ifkf'k'b} = (d_{ff'} \cdot A_{ib}) + \beta_{kk'} \tag{11.6}$$

$$\sum_{b=1}^{B} v_{ib} = 1 \tag{11.7}$$

$$d_{ff'} = \left| x_f - x'_{f'} \right| + \left| y_f - y'_{f'} \right| \tag{11.8}$$

$$\delta_j = \sum_{k=1}^{C} \sum_{i=1}^{I} \frac{\mathrm{CM}_{ib} \cdot v_{ib} \cdot a_{ij} \cdot L_{jf} \cdot N_{fk}}{\lambda_{ij}} \tag{11.9}$$

$$\lambda_{ij} = E(\lambda_{ij}) + \sqrt{\mathrm{var}(\lambda_{ij})} \cdot z_{(1-\alpha/2)} \tag{11.10}$$

$$D_j = E(D_i) + \sqrt{\mathrm{var}(D_i)} \cdot z_{(1-\alpha/2)} \tag{11.11}$$

$$\delta_j \leq 1 \tag{11.12}$$

$$\mathrm{lp}_i \leq \mathrm{pr}_i \leq \mathrm{up}_i \tag{11.13}$$

$$q_{i,n} = D_i \quad \forall n = N_{i+1} \tag{11.14}$$

$$q_{i,n} = \frac{q_{i,n+1}}{\hat{p}_{ij}^{n\varphi} \cdot u_{ij}^{n\varphi}} \quad \forall n \in \{2, \ldots, N_i\} \tag{11.15}$$

$$Q_i = q_{i,n} \quad \forall n = 1 \tag{11.16}$$

$$\hat{p}_{ij}^{n\varphi} = \begin{cases} 1 - \tilde{p}_{ij}^{n\varphi} & \forall \varphi = 1 \text{ or } 2 \\ \tilde{p}_{ij}^{n\varphi} \cdot (1 - \gamma_{ij}) \cdot (1 - \tilde{p}_{ij}^{n\varphi}) & \forall \varphi = 3 \end{cases} \tag{11.17}$$

$$\delta_j, pr_i \geq 0 \tag{11.18}$$

$$L_{jf} \in \{0, 1\} \tag{11.19}$$

$$v_{ib} \in \{0, 1\} \tag{11.20}$$

This study presents a novel mathematical model with dual optimization objectives for cellular manufacturing systems, integrating both cost efficiency and productivity enhancement considerations:

Objective Functions

1. Cost Minimization Objective:
   - Reduces inter-cell and intra-cell material movement expenses.
   - Optimizes machine location assignment costs.
   - Minimizes material handling equipment expenditures.
2. Productivity Maximization Objective:
   - Enhances overall system throughput by maximizing $\Sigma(\mu_j)$.
   - Prevents bottleneck formation through machine productivity coefficients.
   - Balances workload across manufacturing cells.

Core Model Constraints

1. Machine-Location Allocation:

Equation (11.1): Ensures single-machine occupancy per candidate location.

Equation (11.2): Mandates complete machine assignment to locations.

Equation (11.3–11.4): Maintains feasible location-to-machine ratios.

2. Material Flow Management:

Equation (11.5): Quantifies part transmission iterations.

Equation (11.6): Computes total material movement distances.

Equation (11.7): Enforces material handling device utilization.

3. Spatial Configuration:

Equation (11.8): Implements rectilinear distance measurement between locations.

4. Production Performance Metrics:

Equation (11.9): Calculates machine productivity via queuing theory (input/output ratios).

Equation (11.10–11.11): Models stochastic processing times and demand variability.

Equation (11.12): Explicitly prevents bottleneck conditions.

5. Production Planning Constraints:

Equation (11.13): Bounds product selling price within market limits.

Equation (11.14): Aligns final production output with demand requirements.

Equation (11.15–11.16): Manages inter-stage production quantities accounting for quality factors.

Equation (11.17): Computes perfect part probability considering three machine states (normal, rework, scrap), and conditional rework probabilities.

6. Variable Specifications:

Equations (11.18–11.20): Define decision variable domains and signatures.

Model Innovations:

- dual optimization of both economic and operational performance;
- integration of queuing theory with facility layout optimization;
- comprehensive quality-aware production planning;
- bottleneck prediction and prevention mechanisms;
- stochastic demand and processing time incorporation.

Linearization

Because the mathematical model contains certain nonlinear relationships, a linearization approach is applied to simplify the solution procedure and enhance the model's effectiveness (Salarian & Fazlollahtabar, 2015).

Consider.

Linearization

Since some nonlinear relations are included in the mathematical model, linearization is performed to facilitate the solution process and increase the capability of the model (Salarian & Fazlollahtabar, 2015).

Consider,

$$L_{jf} * V_{ib} = T_{ibjf}$$

as a substitution for the product of two binary variables. Also, the following constraints are added,

$$\left(T_{ibjf} - 1\right) \geq (v_{ib} - 1) + \left(L_{jf} - 1\right)$$

$$2\left(T_{ibjf} - 1\right) \leq (v_{ib} - 1) + \left(L_{jf} - 1\right)$$

Again, a nonlinearity due to product of two binary variables ($L_{j'f'} * T_{ibjf}$) is fulfilled as below,

$$L_{j'f'} * T_{ibjf} = \xi_{ibjfj'f'}.$$

Considering two constraints given below,

$$\left(\xi_{ibjfj'f'} - 1\right) \geq \left(T_{ibjf} - 1\right) + \left(L_{j'f'} - 1\right)$$

$$2(\xi_{ibjfj'f'} - 1) \leq \left(T_{ibjf} - 1\right) + \left(L_{j'f'} - 1\right).$$

Therefore the linear model is in hand.

The linear mathematical model

$$\max(w^1) = \left(\begin{array}{l} \sum_{i=1}^{I} \mathrm{pr}_i \cdot D_i - \left[\sum_{i=1}^{I} \sum_{b=1}^{B} \sum_{j=1}^{m} \sum_{\substack{j'=1 \\ j'\neq j}}^{m} \sum_{f=1}^{\theta} \sum_{\substack{f'=1 \\ f'\neq f}}^{\theta} \sum_{k=1}^{c} \sum_{k'=1}^{c} (a_{ij} \cdot N_{fk} \cdot a_{ij'} \cdot N_{f'k'} \right. \\ \cdot M_{ijj'} \cdot r_{ijj'} \cdot \eta_{ifkf'k'b} \cdot H_i \cdot \xi_{ibjfj'f'}) + \sum_{i=1}^{I} \sum_{b=1}^{B} (c_b + c'_b) \cdot v_{ib} \\ \left. + \sum_{n=1}^{N_i} \mathrm{sc}_{ij} \cdot q_{i,n} \cdot u_{ij}^{n\varphi} + \sum_{i=1}^{I} \mathrm{pc}_i \cdot Q_i + \sum_{n=1}^{N_i} \sum_{j=1}^{m} \hat{p}_{ij}^{n3} \cdot Q_i \cdot \mathrm{rc}_{ij} \cdot u_{ij}^{n3} \right] \end{array} \right)$$

$$\max(w^2) = \sum_{j=1}^{m} \delta_j$$

s.t.

Equations (11.1)–(11.20)

$$\left(T_{ibjf} - 1\right) \geq (v_{ib} - 1) + \left(L_{jf} - 1\right) \tag{11.21}$$

$$2\left(T_{ibjf} - 1\right) \leq (v_{ib} - 1) + \left(L_{jf} - 1\right) \tag{11.22}$$

$$\left(\xi_{ibjfj'f'} - 1\right) \geq \left(T_{ibjf} - 1\right) + \left(L_{j'f'} - 1\right) \tag{11.23}$$

$$2(\xi_{ibjfj'f'} - 1) \leq \left(T_{ibjf} - 1\right) + \left(L_{j'f'} - 1\right) \tag{11.24}$$

$$\xi_{ibjfj'f'}, T_{ibjf} \in \{0, 1\} \tag{11.25}$$

11.4 Solution Approach

The solution methodology is developed based on two key components: a search algorithm and a solution architecture. The following section examines these elements in detail (Salarian & Fazlollahtabar, 2015).

11.4.1 Search Algorithm

Step 0: Initial Basic Feasible Solution (BFS) Setup

To establish the initial Basic Feasible Solution (BFS), an artificial cell (denoted as "0") and artificial routes are introduced into the manufacturing system. Cell "0" serves as the system's source node, with artificial routes linking it to other cells. The primary routes of the system are contained in set B, while set A remains empty at this stage.

Step 1: Machine/Part Scheme Selection and Optimality Check

A critical decision in this step involves selecting a machine/part assignment scheme. Concurrently, the waiting times of the material handling device on non-basic routes are recomputed. If any route violates its optimality condition, it becomes a candidate for inclusion in the basis

Upon selecting an entering route (k, k')—where a violation exists—its introduction into the system creates a unique cycle W with the existing basis routes.

Step 2: Cycle Elimination and Flow Augmentation

To resolve cycle W, one of its constituent routes must be removed from the basis. The cycle is eliminated by augmenting the flow by the minimum required machine/part allocation, redirecting the flow to adjacent cells. Augmenting flow along an initially empty route enhances the solution's objective value.

The process involves:

- Identifying all routes within cycle W.
- Selecting the leaving route based on the cycle's structure.

Step 3: Pivoting and Optimality Verification

Replacing the leaving route with the entering route and restructuring the system constitutes a pivot operation. After pivoting, the machine/part allocation for each route is recalculated.

- If all routes satisfy the optimality conditions, the current BFS is optimal.
- Otherwise, a new violating route (k, k') is selected, and the algorithm repeats from Step 1.

11.4.2 Architecture of Solution Approach

Solution Approach: Architecture and Operations

This section outlines the structure and workflow of the proposed solution methodology. Initially, the Cell Generator creates a small set of machine cells, which are then integrated with the remaining (initially empty) cells. These remaining cells serve as the workspace for the heuristic search algorithm, which ultimately assigns each machine to a dedicated individual cell.

During runtime, the system dynamically updates the status of material handling devices and machines for each part. When a part is allocated to a machine, it is removed from the handling device's assignment list and no longer considered among the remaining parts. However, if a part is en route to a machine, it cannot be removed until the inter-arrival time between preceding and subsequent parts elapses. Additionally, the Cell Generator produces new cells whenever idle handling devices are detected.

Algorithmic Efficiency and Applicability

The heuristic search algorithm, combined with the proposed solution architecture, constitutes a comprehensive and scalable method capable of addressing problems of unrestricted size. To demonstrate its practicality, the solution framework has been systematically designed and implemented.

This algorithm is particularly advantageous for:

- Large-scale problems exceeding the limitations of conventional methods.
- Time-constrained scenarios requiring rapid computational solutions.

Due to its simplicity and adaptability, the proposed heuristic search method offers both ease of use and operational efficiency.

11.5 Numerical Example

To demonstrate the validity and practical applicability of our proposed mathematical model, we present a comprehensive numerical case study. The problem instance consists of:

- Production elements: 2 distinct parts,
- Manufacturing resources: 6 machines distributed across 3 cells,
- Layout considerations: 10 potential machine locations,
- Material handling: 3 dedicated transport devices.

The complete parameter set for this validation study includes:

1. Machine service rates for all parts,
2. Part-specific machine requirements,

3. Machine operational costs (setup and processing),
4. Quality control metrics (rework costs and scrap percentages),
5. Machine availability statuses.

These critical parameters are systematically presented in Table 11.1, providing a complete foundation for model evaluation and solution analysis.

The process plan of these parts is depicted below in Fig. 11.1 (Each circle shows a machine, and the digits in the circles are the number of machines).

According to Fig. 11.1, $\tilde{u}_{ij}^{n\varphi}$ only the items given below gets value of 1 and otherwise is zero:

$$\tilde{u}_{1,5}^{1,2} = 1, \tilde{u}_{1,2}^{2,2} = 1, \tilde{u}_{1,6}^{3,3} = 1, \tilde{u}_{1,1}^{4,1} = 1, \tilde{u}_{1,4}^{5,3} = 1, \tilde{u}_{1,3}^{6,1} = 1, \tilde{u}_{1,5}^{7,2} = 1, \tilde{u}_{1,2}^{8,2} = 1, \tilde{u}_{1,4}^{9,3} = 1$$

$$\tilde{u}_{2,1}^{1,1} = 1, \tilde{u}_{2,3}^{2,1} = 1, \tilde{u}_{2,1}^{3,1} = 1, \tilde{u}_{2,6}^{4,3} = 1, \tilde{u}_{2,4}^{5,3} = 1, \tilde{u}_{2,2}^{6,2} = 1, \tilde{u}_{2,3}^{7,1} = 1, \tilde{u}_{2,1}^{8,1} = 1, \tilde{u}_{2,5}^{9,2} = 1$$

The processing time and demand are obtained using a mean and standard deviation extracted from a sampling process. The purchase cost of parts, the lower bound and upper bound of selling price for parts, and the demand for parts are shown in Table 11.2.

The movement of parts and the repetition of a certain movement occurrence are shown in Tables 11.3 and 11.4.

The purchase cost and the annual cost of maintenance and repair for the material handling device b are shown in Table 11.5. The coordinates of candidate locations f and f' are computed in Table 11.6.

The operational parameters for material handling are presented across three reference tables. Table 11.7 specifies unit transportation cost for each part type when

Table 11.1 Input parameters of parts and machines

Parameters	Part number	Machine number					
		1	2	3	4	5	6
λ_{ij}	1	130	160	145	200	100	170
	2	120	135	160	135	210	220
a_{ij}	1	1	1	1	1	1	1
	2	1	1	1	1	1	1
SC_{ij}	1	150	120	180	90	45	120
	2	100	130	140	40	80	60
rc_{ij}	1	–	–	–	80	–	55
	2	–	–	–	65	–	75
	1	–	–	–	0.035	–	0.045
	2	–	–	–	0.05	–	0.025
ϕ	–	1	2	1	3	2	3

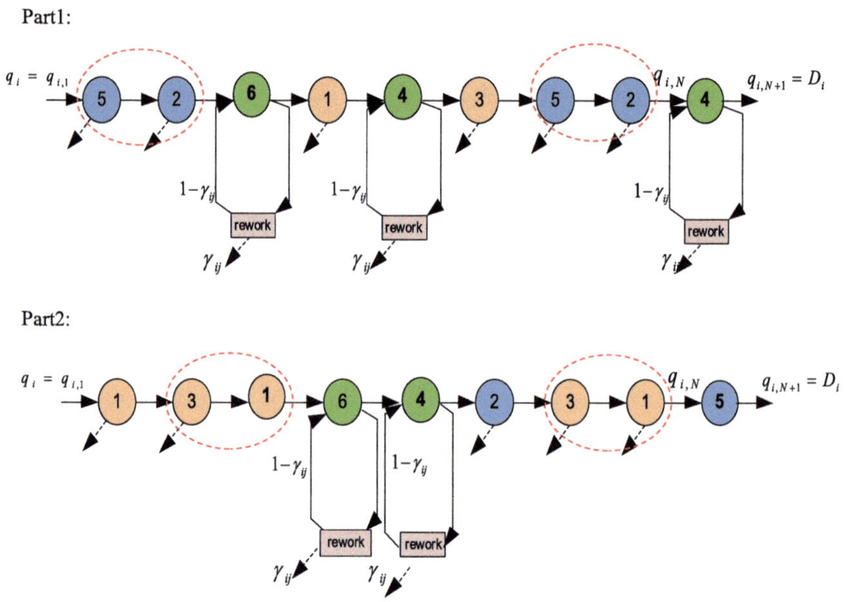

Fig. 11.1 Processing route for parts 1 and 2

Table 11.2 Input parameters of each part

Part number	PC_i	lp_i	up_i	D_i
1	1800	4500	5800	96,540
2	2100	3200	6400	10,345

moved by material handling device, and maximum transfer capacity of each material handling device for every part type. Table 11.8 details penalty costs associated with intercellular movement of parts between cell and quantity of available candidate locations within cell. Table 11.9 documents the presence and distribution of candidate locations across all manufacturing cells.

The probabilities of failure for a machine in various statuses are given in Table 11.10.

Using the input parameters specified in the previous section and applying our proposed solution methodology, we implemented the mathematical model in MATLAB 7.1. The computational results yield the following optimal allocations for the decision variables. Machine-to-location assignments were determined as follows:

$L_{14} = 1, L_{27} = 1, L_{39} = 1, L_{42} = 1, L_{510} = 1, L_{62} = 1$ and others are 0.

The allocations of parts to material handling devices are:

$V_{13} = 1, V_{22} = 1.$

The machine's coefficients are:

$\delta_1 = 0.18, \delta_2 = 0.43, \delta_3 = 0.73, \delta_4 = 0.09, \delta_5 = 0.66, \delta_6 = 0.81.$

Table 11.3 Direct movement of parts between any two machines

	1	2	3	4	5	6
$M_{1jj'}$ *(part 1)*						
1	0	0	0	1	0	0
2	0	0	0	1	0	1
3	0	0	0	0	1	0
4	0	0	1	0	0	0
5	0	1	0	0	0	0
6	1	0	0	0	0	0
$M_{2jj'}$ *(part 2)*						
1	0	0	1	0	1	1
2	0	0	1	0	0	0
3	1	0	0	0	0	0
4	0	1	0	0	0	0
5	0	0	0	0	0	0
6	0	0	0	1	0	0

Table 11.4 Number of repetitions of $M_{ijj'}$ for parts

	1	2	3	4	5	6
$r_{1jj'}$ *(part 1)*						
1	0	0	0	1	0	0
2	0	0	0	1	0	1
3	0	0	0	0	1	0
4	0	0	1	0	0	0
5	0	**2**	0	0	0	0
6	1	0	0	0	0	0
$r_{2jj'}$ (part 2)	1	2	3	4	5	6
1	0	0	1	0	1	1
2	0	0	1	0	0	0
3	**2**	0	0	0	0	0
4	0	1	0	0	0	0
5	0	0	0	0	0	0
6	0	0	0	1	0	0

Table 11.5 Purchase and annual cost of maintenance for the material handling device

Parameters	Material handling device number		
	1	2	3
C_b	5000	4300	3200
$C_{b'}$	300	310	290

Table 11.6 The coordinates of locations

Location	1	2	3	4	5	6	7	8	9	10
X_f	60	120	90	225	320	240	320	120	180	180
Y_f	285	285	225	240	255	195	195	90	120	60

Table 11.7 Transmission cost and capacity for each device

Parameters	Part number	Material handling device number		
		1	2	3
A_{ib}	1	20	18	16
	2	22	22	18
CM_{ib}	1	50	30	20
	2	60	45	30

Table 11.8 Input parameters for cells

	Cell number		1	2	3
Inter cell movement Penalty cost	$\beta_{kk'}$	1	0	500	400
		2	500	0	200
		3	400	200	0
	Number of location (G_k)		3	4	3

Table 11.9 Location in each cells

N_{fk}	Cell number		
Location number	1	2	3
1	1	0	0
2	1	0	0
3	1	0	0
4	0	1	0
5	0	1	0
6	0	1	0
7	0	1	0
8	0	0	1
9	0	0	1
10	0	0	1

Table 11.10 The probability of failure of the machine in its status Ø

	φ		
$\forall i, j, n$	1	2	3
$\tilde{p}_{ij}^{n\varphi}$	0	0.08	0.05

The selling prices for products are:
$pr_1 = 5300$, $pr_2 = 6100$.
Thus, the optimal values were found in negligible computational time and effort.

11.6 Conclusions

This study develops a stochastic mathematical model for cellular manufacturing systems, where critical parameters such as processing times and product demands are treated as random variables, estimated through statistical sampling (mean and standard deviation). The model aims to optimize machine layout configurations by minimizing inter-cell and intra-cell material movements, eliminating production bottlenecks, and ultimately maximizing profit through reduced production costs. To demonstrate its applicability, a numerical example was implemented, while a novel heuristic methodology—featuring an original search algorithm and allocation framework—was proposed to address computational complexity in large-scale instances. The combined approach ensures both theoretical rigor and practical scalability for real-world manufacturing systems.

References and Further Reading

Chattopadhyay, M., Sengupta, S., Ghosh, T., Dan, P. K., & Mazumdar, S. (2013). Neuro-genetic impact on cell formation methods of Cellular Manufacturing System design: A quantitative review and analysis. *Computers & Industrial Engineering, 64*(1), 256–272.

Chen, M., & Cao, D. (2004). Coordinating production planning in cellular manufacturing environment using Tabu search. *Computers & Industrial Engineering, 46*, 571–588.

Chen, M., & Cao, D. (2005). A robust cell formation approach for varying product demands. *International Journal of Production Research, 49*(8), 1587–1605.

Egilmez, G., Süer, G. A., & Huang, J. (2012). Stochastic cellular manufacturing system design subject to maximum acceptable risk level. *Computers & Industrial Engineering, 63*(4), 842–854.

Fazlollahtabar, H., & Aghasi, E. (2014). An integrated stochastic multi-criteria acceptability analysis and mathematical optimisation for service marketing. *International Journal of Services and Operations Management, 17*(1), 38–49.

Fazlollahtabar, H., Mahdavi-Amiri, N., & Muhammadzadeh, A. (2015). A genetic optimization algorithm for nonlinear stochastic programs in an automated manufacturing system. *Journal of Intelligent & Fuzzy Systems, 28*(3), 1461–1475.

Fazlollahtabar, H., Rezaie, B., & Kalantari, H. (2010). Mathematical programming approach to optimize material flow in an AGV-based flexible jobshop manufacturing system with performance analysis. *The International Journal of Advanced Manufacturing Technology, 51*(9–12), 1149–1158.

Ghezavati, V. R., Jabal-Ameli, M. S., & Makui, A. (2009). A new heuristic method for distribution networks considering service constraint and coverage radius. *Expert Systems with Applications,36*(3), 5620–5629.

Gholizadeh, H., Fazlollahtabar, H., & Khalilzadeh, M. (2020). A robust fuzzy stochastic programming for sustainable procurement and logistics under hybrid uncertainty using big data. *Journal of Cleaner Production, 258*, Article 120640.

Majazi Dalfard, V. (2013). New mathematical model for problem of dynamic cell formation based on number and average length of intra and intercellular movements. *Applied Mathematical Modelling, 37*(4), 1884–1896.

Nsakanda, A. L., Diaby, M., & Price, W. L. (2006). Hybrid genetic approach for solving large-scale capacitated cell formation problems with multiple routings. *European Journal of Operational Research, 171*, 1051–1070.

Paydar, M. M., & Saidi-Mehrabad, M. (2013). A hybrid genetic-variable neighborhood search algorithm for the cell formation problem based on grouping efficacy. *Computers and Operations Research, 40*(4), 980–990.

Rafiei, H., & Ghodsi, R. (2013). A bi-objective mathematical model toward dynamic cell formation considering labor utilization. *Applied Mathematical Modelling, 37*(4), 2308–2316.

Salarian, R., & Fazlollahtabar, H. (2015). Multioptimization in a cellular manufacturing system having stochastic parameters considering pricing. *Journal for Manufacturing Science and Production, 15*(3), 257–265.

Salarian, R., Fazlollahtabar, H., & Mahdavi, I. (2014). Inter-cell movement minimisation in a cellular manufacturing system having stochastic parameters. *International Journal of Services and Operations Management, 17*(1), 67–87.

Snyder, L. V. (2006). Facility location under uncertainty: A review. *IIE Transactions, 38*, 537–554.

Solimanpur, M., Prem, V., & Ravi, S. (2004). A heuristic to minimize makespan of cell scheduling problem. *International Journal of Production Economics, 88*, 231–241.

Tavakkoli-Moghaddam, R., Javadian, N., Javadi, B., & Safaei, N. (2007). Design of a facility layout problem in cellular manufacturing systems with stochastic demands. *Applied Mathematics and Computation, 184*, 721–728.

Tavana, M., Fazlollahtabar, H., & Hassanzadeh, R. (2014). A bi-objective stochastic programming model for optimising automated material handling systems with reliability considerations. *International Journal of Production Research, 52*(19), 5597–5610.

Wu, X. D., Chu, C. H., Wang, Y. F., & Yan, W. L. (2006). Concurrent design of cellular manufacturing systems: A genetic algorithm approach. *International Journal of Production Research, 44*(6), 1217–1241.

Chapter 12
Artificial Intelligence Application for Multi-objective Optimization

Abstract The convergent product concept is realized through the systematic integration of functions and sub-functions to form a final product. This integration is modeled as a web-based network, where nodes represent base functions and sub-functions, and arcs denote connections between them. The objective is to identify an optimal functional tree within this network that maximizes product value in a web-based environment. The modeling approach begins with an algorithmic framework to establish relationships between base functions and sub-functions. A mathematical formulation is then applied to assign quantitative measures, where arcs are characterized by benefit values and nodes by implementation costs, with customer value being explicitly incorporated through the benefit structure. To solve this optimization problem, the conventional Steiner tree methodology is extended into a multi-objective framework and solved using an ant colony optimization algorithm, demonstrated through a numerical example. Further enhancing this approach, a particle swarm optimization technique is implemented within a fuzzy Steiner tree framework for the convergent product network. This advanced formulation incorporates fuzzy mathematical principles to represent uncertain arc benefits and node costs. The fuzzy Steiner tree methodology is subsequently adapted into a bi-objective optimization model to determine the optimal network configuration, with its practical application illustrated through a detailed case study.

Keywords Convergent product · Web-based (digital) network · Ant colony optimization · Fuzzy bi-objective programming · Steiner tree · Particle Swarm Optimization (PSO)

12.1 Introduction

The ongoing convergence in electronics and communications has enabled the integration of novel functionalities into existing base products, exemplified by mobile television in smartphones or internet capabilities in PDAs. This technological evolution

H. Fazlollahtabar, *Uncertain Multi-objective Decision Making: Methods and Industrial Models*, https://doi.org/10.1007/978-981-95-6035-6_12

presents manufacturers with the critical challenge of determining which new functionalities will optimally enhance product value. As Fazlollahtabar and Mohseninejad (2015) demonstrate, this requires evaluating both congruent additions (e.g., electronic Yellow Pages for PDAs) and incongruent features (e.g., satellite radio integration). The decision-making process is further complicated by the base product's fundamental nature—whether utilitarian (e.g., PDAs) or hedonic (e.g., MP3 players)—as established by Fazlollahtabar and Aghasi (2012).

Conceptually, convergent product development parallels product assembly processes, where discrete components combine to form a complete product. Effective assembly modeling, as discussed by Mahdavi et al. (2008, 2010), must identify critical design features that influence manufacturing optimization. Key capabilities include:

- detection of part interferences and clearances prior to physical assembly;
- calculation of mass properties affecting product stability;
- generation of 3D exploded views to identify DFA violations (Fazlollahtabar & Tehranchian, 2015).

Advanced modeling enables comprehensive analyses—from basic kinematic evaluations to complex tolerance assessments—either through native tools or third-party software. Tolerance analysis proves particularly vital, as it preempts costly tolerance stack issues that typically manifest only during physical assembly (Fazlollahtabar & Ehsani, 2010; Shirazi et al., 2010). While tolerance optimization can mitigate these risks, designers must balance precision requirements with manufacturing costs, guided by process-specific tolerance charts (Fazlollahtabar & Gholizadeh, 2020).

For systematic optimization, tree-based methodologies—particularly the Steiner tree approach (Gholizadeh & Fazlollahtabar, 2019)—offer robust frameworks for decision-making in both assembly processes and convergent product development, enabling manufacturers to navigate the complexities of feature integration while maintaining production efficiency and product quality.

The Steiner Tree Problem (STP) represents a fundamental challenge in combinatorial optimization and graph theory with broad theoretical significance and practical applications in network design. Formally defined, given:

- An undirected, connected graph $G = (V, E)$, where
  - V represents the vertex set;
  - E denotes the edge set with associated weights $\{c_e | e \in E\}$.
- A designated subset of terminal nodes $N \subseteq V$.

The STP seeks to identify the minimum-weight subtree $T \subseteq G$ that spans all terminals in N, potentially incorporating additional non-terminal vertices (Steiner nodes) from $N = V - N$ to achieve optimal connectivity.

Key computational aspects include:

1. NP-Hardness: The problem is provably NP-hard for most graph classes of practical interest (Johnson, 1985), establishing inherent computational complexity.
2. Algorithmic Challenge: The combinatorial explosion of possible Steiner node configurations creates significant solution space complexity.
3. Dual Nature: The op tional inclusion of Steiner nodes differentiates STP from simpler spanning tree problems.

This canonical formulation serves as the foundation for numerous variants addressing real-world network design challenges, where optimal interconnection of critical nodes often requires judicious use of intermediate connection points.

The Steiner tree problem in graphs was formally introduced by Hwang and Richards (1992), sparking extensive research into both exact and heuristic solution methods. Their foundational work demonstrated that a Steiner minimal tree (SMT) for terminal set X in network $G = (V, \ E)$ could be derived by enumerating minimum spanning trees (MSTs) of subgraphs induced by supersets of X. Lawler (1976) later refined this approach by proving that the number of potential Steiner points is bounded by $|X| - 2$, thereby significantly reducing the search space. While such restrictions can sometimes enable polynomial-time solutions for special graph classes, the general problem remains NP-hard for arbitrary graphs (Johnson, 1985).

Polynomial-Time Special Cases: Two tractable instances of the STP are well-established:

Terminal set $N = V$: The SMT reduces to an MST, solvable via Prim's or Kruskal's algorithm.
Terminal set $|N| = 2$: The SMT is equivalent to the shortest path between the two nodes, computable using Dijkstra's algorithm.

Exact and Approximation Methods: Hwang and Richards (1992) surveyed exact approaches, including:

- dynamic programming (Dreyfuss & Wagner, 1971),
- lagrangian relaxation,
- branch-and-cut methods.

For the generalized Steiner problem (GSP), Duin and Volgenant (1989) pioneered graph reduction techniques to improve scalability. Approximation algorithms were subsequently developed to balance solution quality and computational efficiency.

Heuristic and Metaheuristic Advances: Notable heuristics include Shortest Path Heuristic (SPH) (Takahashi & Matsuyama, 1980), Path-Distance Heuristic (PDH) (Winter & MacGregor Smith, 1992), and Improved performance-ratio algorithms (Robins & Zelikovsky, 2000, 2005).

Recent work has focused on meta-heuristics to handle large-scale instances, Genetic Algorithms (GA), GRASP, and Tabu search, Parallel implementations (e.g., hybrid local search with parallel GRASP, parallel GA) to mitigate computational costs (Fazlollahtabar et al., 2012; Abbasian et al., 2015; Golneshini & Fazlollahtabar, 2019).

12.2 The Proposed Model

Our proposed product digital network framework incorporates a systematic approach for functionality classification based on customer preferences. The model operates through the following key steps:

1. Functionality Definition and Presentation
    - A comprehensive set of product functionalities is predefined within the system.
    - Upon customer selection of a specific product from the manufacturer's portfolio, all associated functionalities are dynamically displayed via a web interface.
2. Interactive Customer Classification Process
    - Customers actively categorize each functionality as either:
        - base functions (core capabilities),
        - sub-functions (supporting features).
    - The classification interface features an intuitive "Classify" submission button.
3. Verification and Feedback
    - The system generates a dedicated review webpage displaying the customer's classification choices.
    - This output serves as both confirmation and documentation of preferences (Hassanzadeh et al., 2012).

The complete workflow is visually represented in Fig. 12.1, illustrating the seamless transition from product selection through final classification.

Within our analytical framework, we conduct a comprehensive assessment of all product functionalities—encompassing both core (primary) and supplementary (additional) features. This evaluation incorporates multiple critical attributes that collectively determine the product's overall value proposition. To formalize this assessment, we employ the following mathematical representations:

Mathematical Notations

i and j	Index for functions and sub-functions;	i and $j = 1, \ldots, n + m$
K	Index for attributes;	$k = 1, \ldots, p$
F_{ijk}	The score of triplet comparison of functions (or sub-functions) with functions (or sub-functions) considering different attributes.	

The three-dimensional comparison matrix F is shown in Fig. 12.2. Note that customers fill in this matrix using numerical values $F_{ijk} \in [0, 1]$.

This matrix is normalized to remove the scales. The normalized values are shown by $\overset{norm}{F_{ijk}}$. A threshold value of θ is considered in a way that the $f_{ij} = \frac{\sum_{k=1}^{p} \overset{norm}{F_{ijk}}}{p} \geq \theta$ are

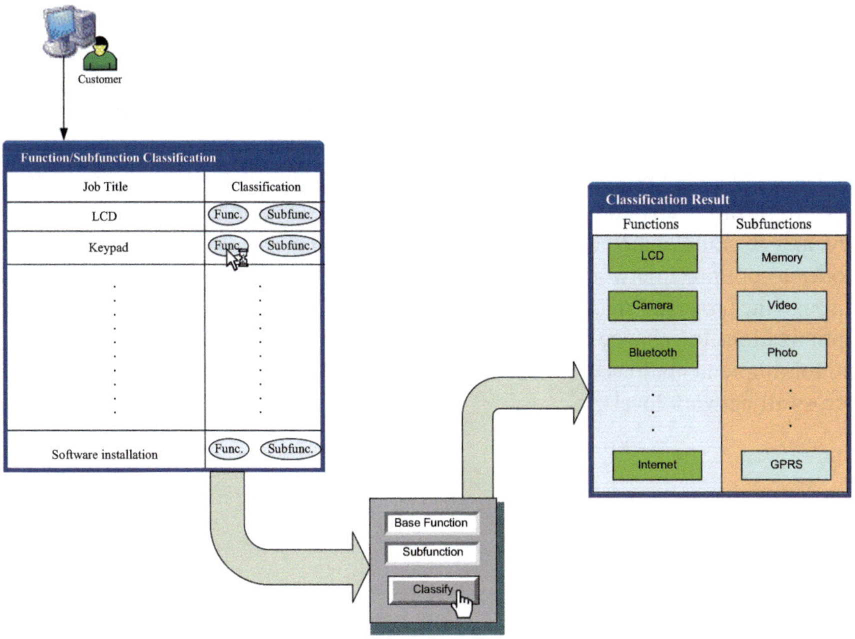

Fig. 12.1 The classification process

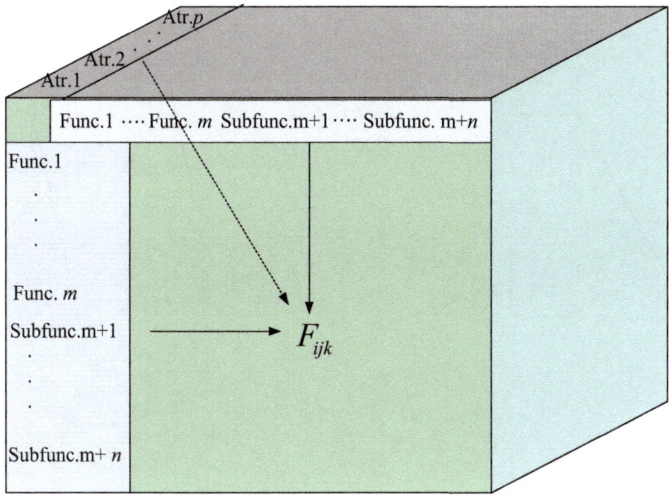

Fig. 12.2 The three-dimensional comparison matrix

chosen to be assigned as links. The f_{ij} is a value that customer considers for arc (i, j). These links configure a network called the purified network, as shown in Fig. 12.3.

The network refinement process involves defining arcs through two systematic procedures: leveling and clustering. In the leveling phase, a hierarchical structure is established by assigning base functions to level zero, while sub-functions connecting to the immediately preceding level are sequentially allocated to higher levels (level one, etc.), ultimately forming an l-level network (Hassanzadeh et al., 2012). This leveling is complemented by a clustering process that groups functionally related nodes to optimize network connectivity. The complete methodology is formalized through the proposed algorithm (Algorithm 1), which operationalizes these structural optimizations to enhance the product value framework introduced earlier. The algorithm's implementation ensures both topological efficiency and functional coherence across all network levels.

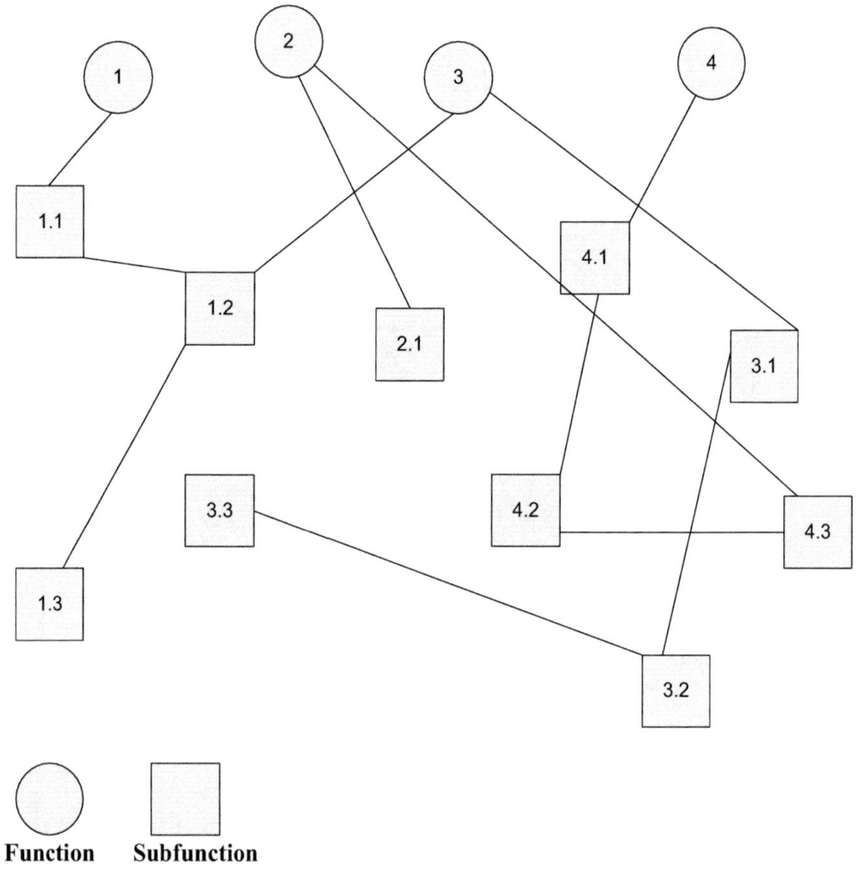

Fig. 12.3 A purified network

Algorithm 1 Leveling to configure a leveled network.

Within the leveled network architecture, each node is assigned a specific cost parameter while the value contribution of individual links is quantitatively assessed. This valuation is achieved through an optimized clustering methodology where sub-functions sharing a common parent node are systematically grouped at each hierarchical level. The clustering process generates distinct nodal aggregates that are subsequently reconfigured into an enhanced network topology. As visually demonstrated in Fig. 12.4, this dual-phase approach (leveling followed by clustering) establishes the structural foundation for network optimization. The refined network then undergoes further improvement through application of the Steiner tree algorithm (Hassanzadeh et al., 2012), which identifies the most efficient connectivity patterns. The complete clustering procedure is formalized in the following algorithm: (Algorithm 2).

Algorithm 2 Clustering of levels in a leveled network.

Step 0: Set each node at level 0 to be a cluster.

Step 1: **For** i=1 till l **do**

{Form clusters at level i}

Cluster all sub-functions at level i linked to a single parent at level i–1;

Solve a zero/one mathematical program for level i (we will discuss the corresponding mathematical program later on);

Perform purification of benefits and costs at level i (as discussed later on);

End for.

Step 2: **Stop**.

Building upon the clustered network architecture, we employ an enhanced Steiner tree formulation that preserves all base functions while simultaneously optimizing three critical objectives: (1) cost minimization, (2) profit maximization, and (3) total customer value maximization across the convergent product's value-adding arcs. While traditional Steiner tree applications focus solely on arc cost minimization, our extended approach incorporates a multi-objective framework that maintains the problem's structural alignment with Steiner tree theory while addressing broader business requirements.

The network model incorporates dual cost parameters for each node i:

c_{i1}: software cost,
c_{i2}: hardware cost,

with each arc (i,j) possessing a composite benefit value $p_{ii'}$ derived from its constituent nodes i and i'. Recognizing the NP-hard complexity inherent to Steiner tree problems in such networks, we implemented preprocessing through leveling and clustering phases to reduce computational complexity. The clustering phase enforces

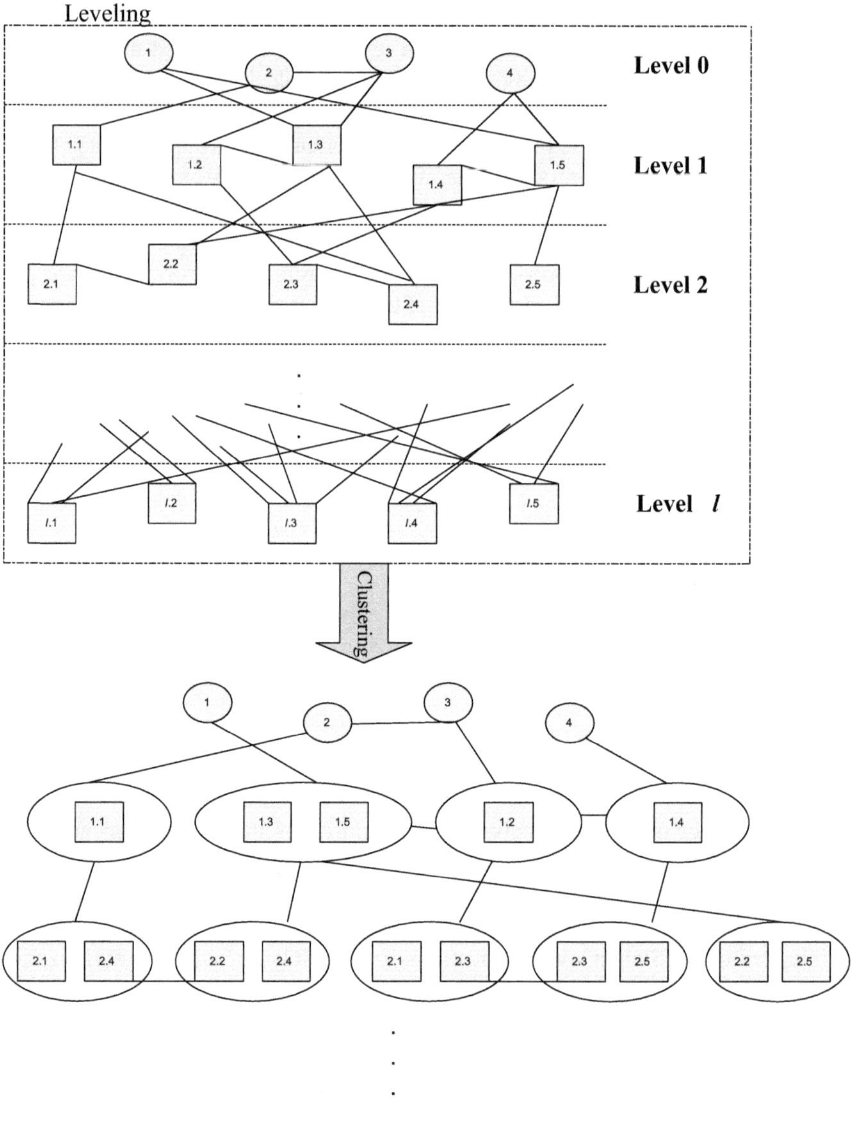

Fig. 12.4 Leveling and clustering processes

strict single-cluster membership per node at each level l through a zero-one integer programming formulation that minimizes total allocation costs (Hassanzadeh et al., 2012).

The complete methodology comprises:

- the multi-objective Steiner tree optimization framework,
- level-specific zero-one cluster assignment programs,
- a purification procedure for cluster refinement.

12.3 Deterministic Model

The Zero/One Mathematical Program for Level l

$$\min \quad T = \sum_{i \in n_l} \sum_{j \in m_{li}} (\alpha_{ij} c_{ij1} + \beta_{ij} c_{ij2}) z_{ij}$$

$$\sum_{j \in m_{li}} z_{ij} = 1, \qquad i = 1, \ldots, n_l$$

$$z_{ij} = \begin{cases} 1, & \text{if node } i \text{ is in cluster } j \\ 0, & \text{otherwise,} \end{cases}$$

where $\beta_{ij}, \alpha_{ij} \in [0, 1], \ \forall i, j$, and

m_{li}: set of indices of clusters at level l where node i is included,

n_l: set of indices of different nodes at level l,

c_{ij2}: hardware cost of node i in cluster j ($c_{ij2} = c_{i2}$, for all i and j),

c_{ij1}: software cost of node i in cluster j ($c_{ij1} = c_{i1}$, for all i and j),

α_{ij}: the software reduction cost coefficient of node i in cluster j,

β_{ij}: the hardware reduction cost coefficient of node i in cluster j.

Purifying Benefits, Costs, and Customers Total Value at Level l

To determine the cost for each cluster at level l, we use

$$c_{jl} = \sum_{i \in n_l} (\alpha_{ij} c_{ij1} + \beta_{ij} c_{ij2}) - \sum_{\forall i, i'} p_{ii'},$$

with c_{ij1}, c_{ij2}, and n_l as defined above, $\beta_{ij}, \ \alpha_{ij} \in [0, 1], \ \forall i, j$, and

c_{jl}: cost of cluster j at level l,

α_{ij}: the software reduction cost coefficient of node i in cluster j,

β_{ij}: the hardware reduction cost coefficient of node i in cluster j,

$p_{ii'}$: the benefit of an arc connecting node i in cluster j to node i' in cluster j.

Also, to adjust the combined arc benefits in clusters, the following equation is used:

$$p_{jj'} = (1 + \gamma_{jj'}) \sum_{\forall i, i'} p_{ii'},$$

where $\forall i, i', j\ , j'$, and

$p_{jj'}$: the adjusted arc benefit connecting cluster j to cluster j',

$p_{ii'}$: the benefit of an arc connecting node i in cluster j to node i' in cluster j',

$\gamma_{jj'}$: the added value configured from nodes in clusters j and j'.

Also, to adjust the combined arc customers total value in clusters, the following equation is used:

$$f_{jj'} = (1 + \gamma_{jj'}) \sum_{\forall i\,,\, i'} f_{ii'},$$

where $\forall i, i', j\ , j'$, and

$f_{jj'}$: the adjusted arc customers total value connecting cluster j to cluster j',

$f_{ii'}$: the customers total value of an arc connecting node i in cluster j to node i' in cluster j',

$\gamma_{jj'}$: the added value configured from nodes in clusters j and j'.

Through systematic refinement, Algorithms 1 and 2 are consolidated into Algorithm 3 by incorporating the previously established framework. This transformation introduces two critical enhancements to the clustering process:

1. Unique Cluster Membership Enforcement:
   - Each node is strictly assigned to exactly one cluster at level l.
   - Selection prioritizes nodes with minimal implementation costs ($c_{i1} + c_{i2}$).
2. Computational Efficiency Improvement:
   - Step 3 of Algorithm 3 replaces the zero-one integer programming formulation.
   - Eliminates the computational overhead associated with solving discrete optimization problems at each level.
   - Maintains solution quality while significantly reducing processing requirements.

The resulting methodology (Algorithm 3) preserves all theoretical guarantees of the original formulation while achieving superior computational performance. This optimization is particularly valuable given the NP-hard nature of the underlying Steiner tree problem, as it reduces the solution space without compromising the multi-objective optimization goals (cost minimization, profit maximization, and customer value enhancement).

Algorithm 3 Leveling and clustering in the network.

Step 0: Set the base functions at level 0. Let $l = 0$.

Step 1: **While** sub-functions exist for processing **do**

Find sub-functions with a link to a function (or sub-function) at level l and place them at level $l + 1$; Let $l = l + 1$;

End while.
{l is the number of levels}
Step 2: Set each node at level 0 to be a cluster.
Step 3: **For** $i = 1$ till l **do**
{Form clusters at level i}
Cluster all sub-functions at level i linked to a single parent at level i – 1;
While $|n_i| > 0$ **do**
Select $k \in n_i$ such that $\alpha_{kp}c_{kp1} + \beta_{kp}c_{kp2} = \min_{j \in m_{ik}}\left\{\alpha_{kj}c_{kj1} + \beta_{kj}c_{kj2}\right\}$. Set $z_{kp} = 1$
and $z_{kj} = 0 \ , \ \forall j \in m_{ik} \ , \ j \neq p;$
$n_i \leftarrow n_i - \{k\}.;$
End while;
For $j = 1$ till q_i **do** {q_i is the number of clusters in the level i}

$$c_{ji} = \sum_{i \in n_l} (\alpha_{ij}c_{ij1} + \beta_{ij}c_{ij2}) - \sum_{\forall i,i'} p_{ii'};$$

End for;
For $j = 1$ till q_i **do**
For $j' = 1$ till q_i **do**

$$p_{jj'} = (1 + \gamma_{jj'}) \sum_{\forall i\,,\,i'} p_{ii'};$$

$$f_{jj'} = (1 + \gamma_{jj'}) \sum_{\forall i\,,\,i'} f_{ii'};$$

End for;
End for;
End for.
Step 4: **Stop**.

12.3.1 Mathematical Formulation

We derive our optimization framework from the classical undirected Dantzig–Fulkerson–Johnson model (Bellman & Zadeh, 1970), adapted for the Convergent Product Steiner Tree Problem (CPSTP). The model employs two key binary decision variables:

Structural Variables:

$$y_i = \{1, \text{ if cluster } i \text{ is selected; } 0, \text{ otherwise}$$

$$x_{ij} = \{1, \text{ if arc}(i, j) \text{ is activated; } 0, \text{ otherwise}$$

Network Characterization:

Let $E(S)$ denote the edge set where both endpoints belong to subset S

$N \subset V$ represents the terminal nodes (base functions)

The mathematical formulation can then be expressed as follows:

$$\text{Maximize} \sum_{(i,j)\in E} p_{ij}.x_{ij}, \tag{12.1}$$

$$\text{Maximize} \sum_{(i,j)\in E} f_{ij}.x_{ij}, \tag{12.2}$$

$$\text{Minimize} \sum_{i\in V} c_i.y_i, \tag{12.3}$$

Such that,

$$\sum_{(i,j)\in E} x_{ij} = \sum_{i\in V} y_i - 1, \tag{12.4}$$

$$\sum_{(i,j)\in E(S)} x_{ij} \leq \sum_{i\in S-\{k\}} y_i, \quad \forall k \in S \subseteq V, \quad \forall S : |S| \geq 2, \tag{12.5}$$

$$y_h = 1, \qquad \forall h \in N, \tag{12.6}$$

$$x_{ij} \in \{0, 1\}, \quad \forall i, j \in E, \tag{12.7}$$

$$y_i \in \{0, 1\}, \quad \forall i \in V. \tag{12.8}$$

The optimization model pursues three primary objectives: (1) maximizing total accrued benefits, (2) minimizing aggregate costs, and (3) improving overall customer value. The constraint system enforces critical solution properties through four key components:

Constraint (12.4) establishes the fundamental tree structure requirement by equating the total number of selected clusters to one less than the number of activated links. Constraints (12.5) implement connectivity conditions, with their total count growing combinatorially based on the problem size $2^{|V|} - |V| - 1$. Constraints (12.6) mandate the inclusion of all terminal clusters in the solution. Finally, relations (12.7) and (12.8) formally define the binary nature of all decision variables.

The model's computational complexity manifests in two dimensions:

The variable set expands proportionally with cluster count.

The constraint system grows exponentially in relation to the number of clusters.

This dual expansion characteristic underscores the NP-hard nature of the problem formulation while precisely capturing the trade-offs between solution quality and computational tractability in the convergent product network optimization context.

12.3.2 Ant Colony Optimization

Our enhanced ant colony algorithm incorporates several fundamental components: pheromone matrices, initialization procedures, heuristic parameters, transition rules, non-dominated solution archiving, and update mechanisms (both local and global). The approach extends the classical Ant Colony System (ACS) framework by integrating maximum selection in the transition rule alongside dual update strategies.

Pheromone Management System: The implementation utilizes three specialized pheromone matrices, each dedicated to tracking a specific optimization objective. These matrices undergo independent updates after every iteration, incorporating information from all solutions generated during that cycle.

Initialization and Parameter Configuration: Initial parameter values and heuristic settings are established through empirical testing. The algorithm executes multiple experimental trials, with each trial producing a potential Steiner tree solution. This methodology parallels traditional ACO approaches where:

- Standard ACO: Single ant constructs complete cyclic tours (TSP).
- Our adaptation: Multiple ants collaboratively build tree structures.

Cooperative Tree Construction Process: The Steiner tree generation employs a distributed ant system where:

1. Initial placement: One ant occupies each mandatory terminal vertex.
2. Movement mechanics: Ants traverse edges probabilistically, favoring pheromone-rich paths.
3. Path management: Individual tabu lists prevent cyclic revisitation.
4. Convergence mechanism: Path merging occurs when ants intersect, gradually forming unified subtrees.

The process terminates when all active ants merge into a single coherent structure, yielding the candidate Steiner tree for that trial. This collaborative approach effectively transforms multiple independent paths into an optimized tree configuration through pheromone-mediated cooperation.

For ant m positioned at vertex i, the transition to vertex $j \notin T^{(m)}$ occurs only when $(i, j) \in E$. Each vertex $j \in V$ possesses two ant-specific potentials:

$$\psi_j^{(m)} = \max_k (P_{jk}),$$

and

$$\delta_j^{(m)} = \max_k (F_{jk}),$$

where

$$k \in \bigcup_{m' \neq m} T^{(m')}.$$

The above equations denote the maximum path weight between j and k (sum of edge weights along longest path). Benefit P_{jk} and customer value F_{jk} metrics are aggregated across all connecting paths.

Desirability Function:

The vertex selection probability incorporates:

$$\eta_{ij}^{(m)} = \frac{p_{ij} + \gamma \psi_j^{(m)}}{H},$$
$$\theta_{ij}^{(m)} = \frac{f_{ij} + \kappa \delta_j^{(m)}}{K}$$
$$\phi_j^{(m)} = \frac{1}{c(j)},$$

where $\gamma, \kappa > 0$ are tuning parameters that control exploration–exploitation balance $H = \sum_{i=1}^{n} \sum_{j=1}^{n} p_{ij}$, $K = \sum_{i=1}^{n} \sum_{j=1}^{n} f_{ij}$. The algorithm preferentially selects edges with higher η values.

Pheromone Update Rules:

Dual update mechanisms maintain solution diversity:

Local Update (After Each Move):

$$\tau_{ij} = (1 - \rho)\tau_{ij} + \rho \Delta \tau_{ij}$$
$$\tau'_{ij} = (1 - \rho)\tau'_{ij} + \rho \Delta \tau'_{ij}$$
$$\tau_j = (1 - \rho)\tau_j + \rho \Delta \tau_j,$$

where $\rho \in (0, 1)$ is the evaporation rate.

Global Update (After Trial Completion):

Applies only to edges/nodes in non-dominated solutions
Uses incremental updates:

$$\Delta \tau_{ij} = \begin{cases} Q \cdot w(P), \text{ if } & (i,j) \in E_S \\ 0, & \text{otherwise} \end{cases} \tag{12.9}$$

where $w(P) = \frac{\sum p_{ij}}{H}$ such that $(i,j) \in E_S$,

$$\Delta\tau'_{ij} = \begin{cases} Q \cdot w(F), \text{ if } \ (i,j) \in E_S \\ 0, \qquad\qquad \text{otherwise} \end{cases} \tag{12.10}$$

where $w(F) = \frac{\sum f_{ij}}{K}$ such that $(i,j) \in E_S$, and

$$\Delta\tau_j = \begin{cases} \dfrac{Q}{w(C)}, & \text{if } \ j \in V_S \\ 0, & \text{otherwise,} \end{cases} \tag{12.11}$$

where $w(C) = \sum c_j$ such that $j \in V_S$.

where Q is constant and denominators represent:
(12.9): Path cost inverse
(12.10): Benefit aggregate
(12.11): Customer value measure

Transition Rule: The following probability distribution shows how an ant moves from node i to its subsequent node:
If $q \le q_0$ then

$$\text{let } p_{ij} = \begin{cases} 1, & \text{if } j = \arg \max\limits_{u \in w(i)} \left\{ (\tau_{ij})^{\alpha} (\eta_{ij}^{m})^{\beta} \cdot (\tau'_{ij})^{\alpha} (\theta_{ij}^{m})^{\beta} \cdot (\tau_i)^{\alpha} (\phi_j^{m})^{\beta} \right\} \\ 0, & \text{otherwise,} \end{cases}$$

Else

$$p_{ij} = \begin{cases} \dfrac{(\tau_{ij})^{\alpha} (\eta_{ij}^{m})^{\beta} \cdot (\tau'_{ij})^{\alpha} (\theta_{ij}^{m})^{\beta} \cdot (\tau_i)^{\alpha} (\phi_j^{m})^{\beta}}{\sum\limits_{u \in w(i)} (\tau_{ij})^{\alpha} (\eta_{ij}^{m})^{\beta} \cdot (\tau'_{ij})^{\alpha} (\theta_{ij}^{m})^{\beta} \cdot (\tau_i)^{\alpha} (\phi_j^{m})^{\beta}}, & \text{if } \ j \in w(i) \\ 0, & \text{otherwise,} \end{cases}$$

where $w(i) = \bigcup\limits_{\forall m} T^{(m)}$ is a set of outgoing nodes from node i not previously visited by the ants.

Keeping Non-dominated Solutions Procedure: The proposed ant colony optimization algorithm operates iteratively, with each ant generating a candidate solution in every iteration. Upon creation, each solution undergoes a dominance check against the current Pareto optimal set to determine if it qualifies as a non-dominated solution. If the new solution either (1) dominates existing solutions or (2) introduces incomparable improvements across the multiple objectives (cost, benefit, and customer value), the Pareto archive is dynamically updated by removing dominated entries and incorporating the new solution. This ensures the preservation of only high-quality, non-dominated solutions throughout the optimization process. The complete workflow—encompassing solution generation, dominance evaluation, Pareto set maintenance, and termination checks—is systematically illustrated in Fig. 12.5. By continuously refining the Pareto front, the algorithm efficiently converges toward a diverse and optimal set of solutions while balancing exploration and exploitation through its pheromone-based mechanisms.

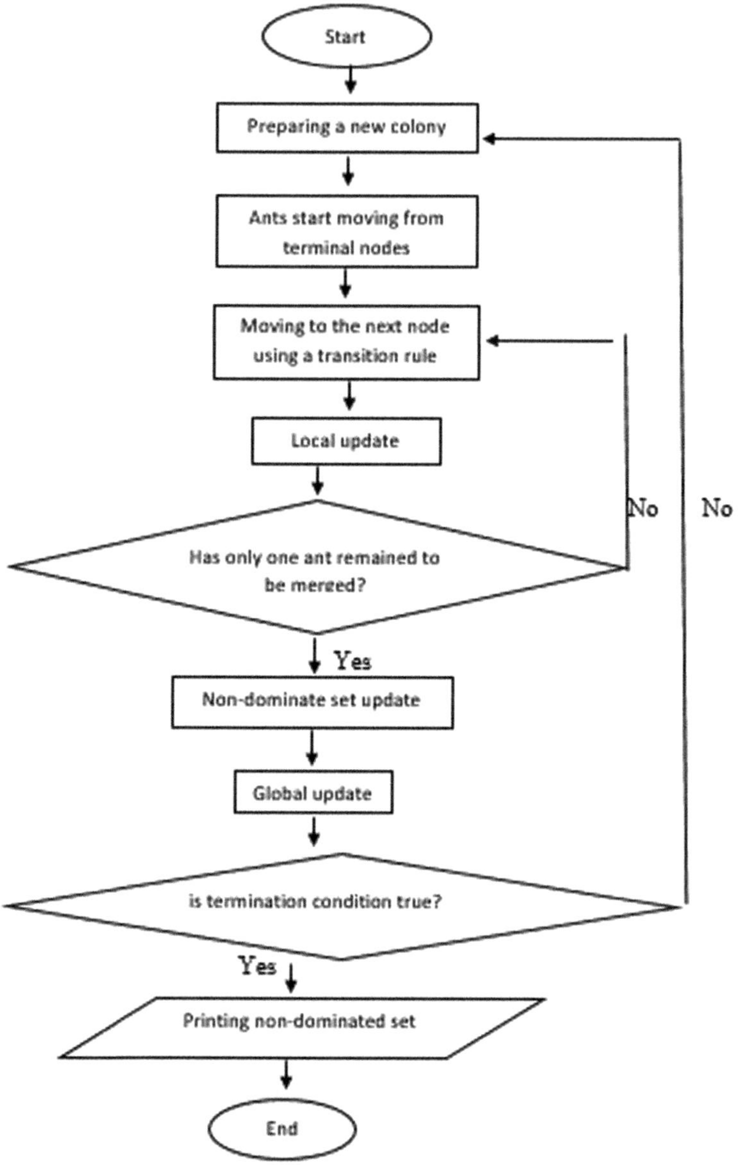

Fig. 12.5 Flow chart of the multi-objective ant colony algorithm

12.3.3 Experimental Study

To validate the proposed multi-objective Steiner tree approach, we conducted an experimental study on an undirected graph $G = (i, j)$, where V represents clusters with associated costs $c_i \geq 0$ and E denotes links with profits $p_e \geq 0$. The goal is to identify a tree structure that maximizes total profit ($\sum p_e$) from selected links while minimizing cluster activation costs ($\sum c_i$), creating a trade-off between network coverage and expense. The three-dimensional relationship between functions, sub-functions, and attributes is formalized in Table 12.1, with arithmetic means computed to derive a consolidated comparison matrix. Using a threshold value of 0.561 (the mean of the dataset in Table 12.1), we generated a thresholded matrix (Table 12.2) and visualized the resulting network in Fig. 12.6, illustrating the optimal balance between profit retention and cost efficiency. This structured approach ensures that the solution avoids excessive cluster costs while maintaining high-revenue connectivity, demonstrating the method's applicability in real-world convergent product optimization.

Subsequently, the leveling process (comprising the zeroth and first steps of Algorithm 3) is executed, resulting in the leveled network illustrated in Fig. 12.7. The clustered network, which corresponds to the second and third steps of Algorithm 3, is depicted in Fig. 12.8.

The cost vectors, the benefit matrix, the customers total value matrix, and the matrices $\alpha = [\alpha_{ij}],\ \beta = [\beta_{ij}]$ are obtained to be:

$$C_1 = (150, 210, 180, 20, 30, 30, 50, 10, 20, 20),$$
$$C_2 = (300, 450, 600, 70, 80, 50, 40, 20, 20, 20).$$

Table 12.1 The three-dimensional comparison matrix for all the attributes for the deterministic network

Attributes	B_1	B_2	B_3	S_1	S_2	S_3	S_4	S_5	S_6	S_7
B_1	0	0.59	0.58	0.6	0.53	0.66	0.46	0.52	0.48	0.46
B_2	–	0	0.46	0.36	0.46	0.63	0.86	0.58	0.43	0.43
B_3	–	–	0	0.59	0.56	0.53	0.9	0.63	0.33	0.53
S_1	–	–	–	0	0.58	0.59	0.6	0.53	0.6	0.6
S_2	–	–	–	–	0	0.36	0.43	0.43	0.7	0.83
S_3	–	–	–	–	–	0	0.63	0.6	0.63	0.58
S_4	–	–	–	–	–	–	0	0.46	0.6	0.43
S_5	–	–	–	–	–	–	–	0	0.65	0.63
S_6	–	–	–	–	–	–	–	–	0	0.63
S_7	–	–	–	–	–	–	–	–	–	0

Table 12.2 The thresholded comparison matrix for all the attributes

Attributes	B_1	B_2	B_3	S_1	S_2	S_3	S_4	S_5	S_6	S_7
B_1	0	1	1	1	0	1	0	0	0	0
B_2	–	0	0	0	0	1	1	1	0	0
B_3	–	–	0	1	0	0	1	1	0	0
S_1	–	–	–	0	1	1	1	0	1	1
S_2	–	–	–	–	0	0	0	0	1	1
S_3	–	–	–	–	–	0	1	1	1	1
S_4	–	–	–	–	–	–	0	0	1	0
S_5	–	–	–	–	–	–	–	0	1	1
S_6	–	–	–	–	–	–	–	–	0	1
S_7	–	–	–	–	–	–	–	–	–	0

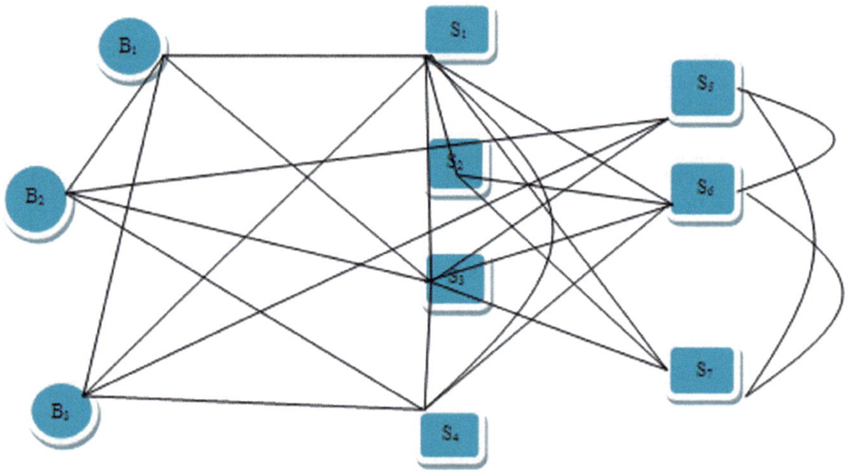

Fig. 12.6 The configured thresholded network for deterministic model

$$\alpha = \begin{bmatrix} - & - & - \\ - & - & - \\ - & - & - \\ 0.65 & - & 0.7 \\ - & - & - \\ 1 & 0.8 & - \\ - & 0.6 & 0.9 \\ - & 0.7 & 0.65 \\ - & - & - \\ - & - & - \end{bmatrix} \quad \beta = \begin{bmatrix} - & - & - \\ - & - & - \\ - & - & - \\ 0.9 & - & 0.9 \\ - & - & - \\ 0.9 & 0.7 & - \\ - & 0.6 & 0.9 \\ - & 0.7 & 0.6 \\ - & - & - \\ - & - & - \end{bmatrix}$$

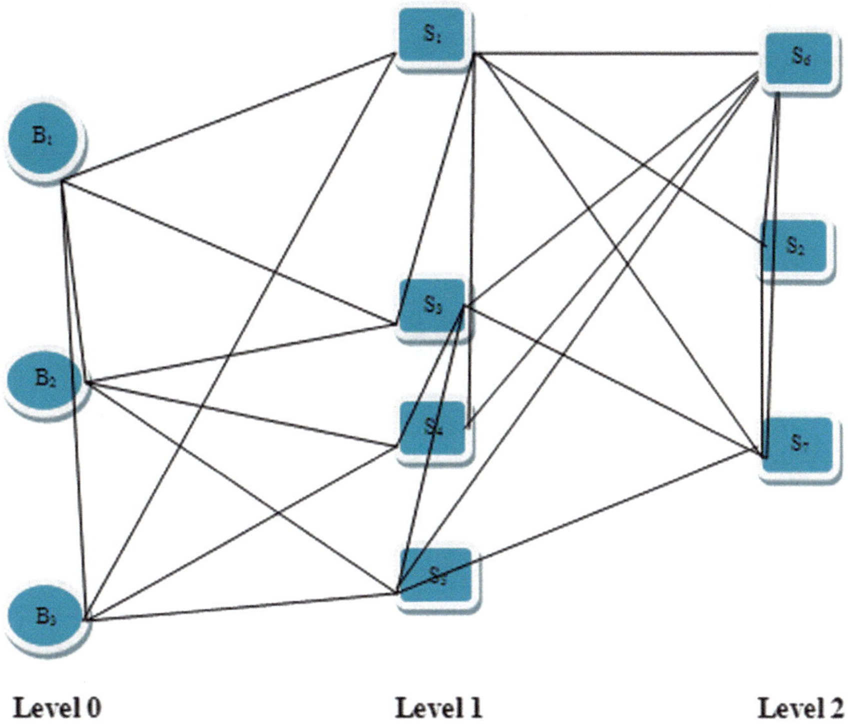

Fig. 12.7 The configured leveled network for deterministic model

$$p = \begin{bmatrix} - & 1500 & 1300 & 100 & - & 80 & - & - & - & - \\ - & - & - & - & - & 70 & 90 & 120 & - & - \\ - & - & - & 60 & - & - & 100 & 110 & - & - \\ - & - & - & - & 50 & 90 & 110 & - & 60 & 40 \\ - & - & - & - & - & - & - & - & 70 & 30 \\ - & - & - & - & - & - & 80 & 70 & 40 & 20 \\ - & - & - & - & - & - & - & - & 30 & - \\ - & - & - & - & - & - & - & - & 20 & 10 \\ - & - & - & - & - & - & - & - & - & 20 \\ - & - & - & - & - & - & - & - & - & - \end{bmatrix}.$$

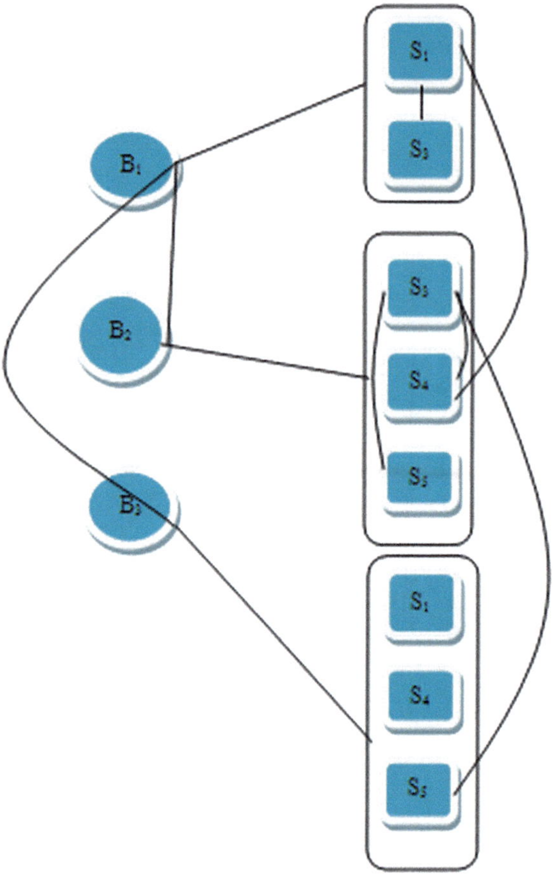

Fig. 12.8 The configured clustered network (the second and the third steps of Algorithm 3)

$$
f = \begin{bmatrix}
- & 0.59 & 0.58 & 0.6 & - & 0.66 & - & - & - & - \\
- & - & - & - & - & 0.63 & 0.86 & 0.58 & - & - \\
- & - & - & 0.59 & - & - & 0.9 & 0.63 & - & - \\
- & - & - & - & 0.58 & 0.59 & 0.6 & - & 0.6 & 0.6 \\
- & - & - & - & - & - & - & - & 0.7 & 0.83 \\
- & - & - & - & - & - & 0.63 & 0.6 & 0.63 & 0.58 \\
- & - & - & - & - & - & - & - & 0.6 & - \\
- & - & - & - & - & - & - & - & 0.65 & 0.63 \\
- & - & - & - & - & - & - & - & - & 0.63 \\
- & - & - & - & - & - & - & - & - & -
\end{bmatrix}
$$

For level 1, using iteration 1 of the while loop in step 3 of Algorithm 3, we obtain:

$$(\alpha_{41} c_{411} + \beta_{41} c_{412}) = 76$$

$$(\alpha_{43}c_{431} + \beta_{43}c_{432}) = 77$$

$$(\alpha_{61}c_{611} + \beta_{61}c_{612}) = 75$$
$$(\alpha_{62}c_{621} + \beta_{62}c_{622}) = 59$$

$$(\alpha_{72}c_{721} + \beta_{72}c_{722}) = 54$$
$$(\alpha_{73}c_{731} + \beta_{73}c_{732}) = 81$$

$$(\alpha_{82}c_{821} + \beta_{82}c_{822}) = 21$$
$$(\alpha_{83}c_{831} + \beta_{83}c_{832}) = 18.5$$

Therefore, $z_{41} = 1$, $z_{62} = 1$, $z_{72} = 1$ and $z_{83} = 1$ with other variables equal to zero. The configured network up to level 1 is shown in Fig. 12.9.

In Fig. 12.10, the next iteration of Algorithm 3 for clustering is performed, and the purified network is obtained as shown in Fig. 12.11.

After purifying the benefits, costs, and customers total value for level 2, the final cost, benefit, and customers total value matrices are formed as follows:

Fig. 12.9 The configured clustered network for level 1

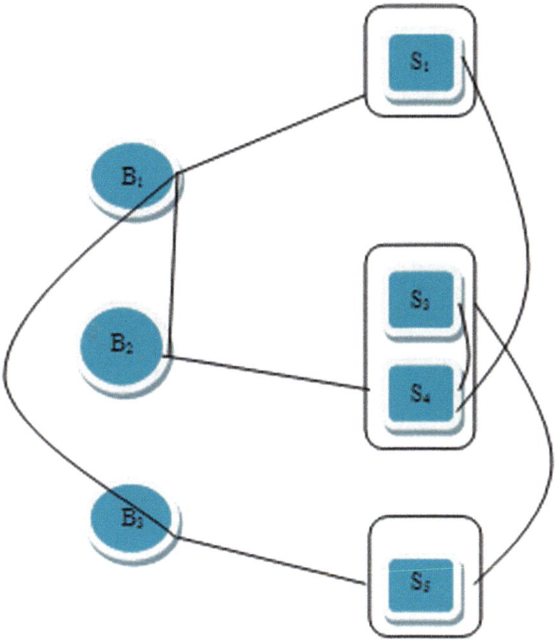

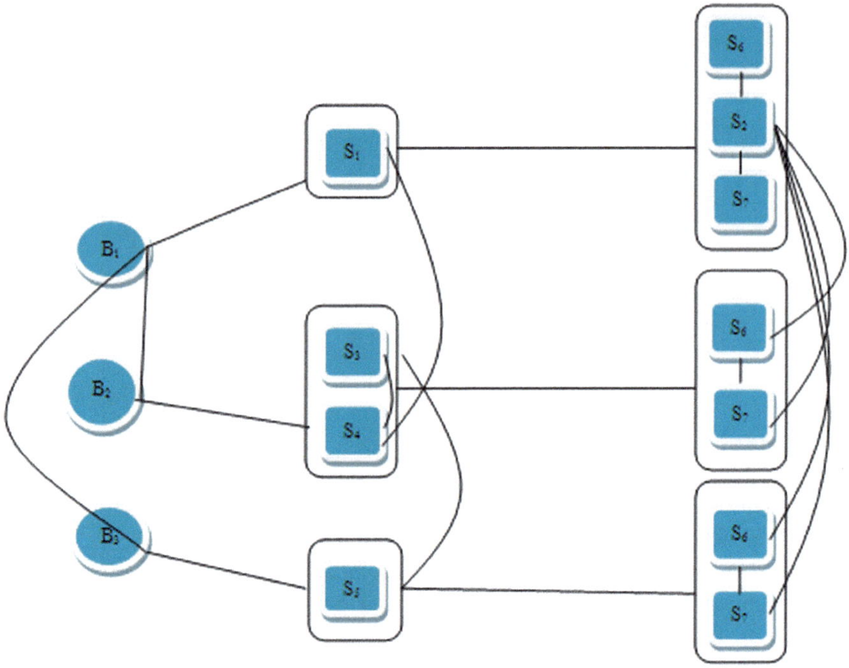

Fig. 12.10 The configured clustered network

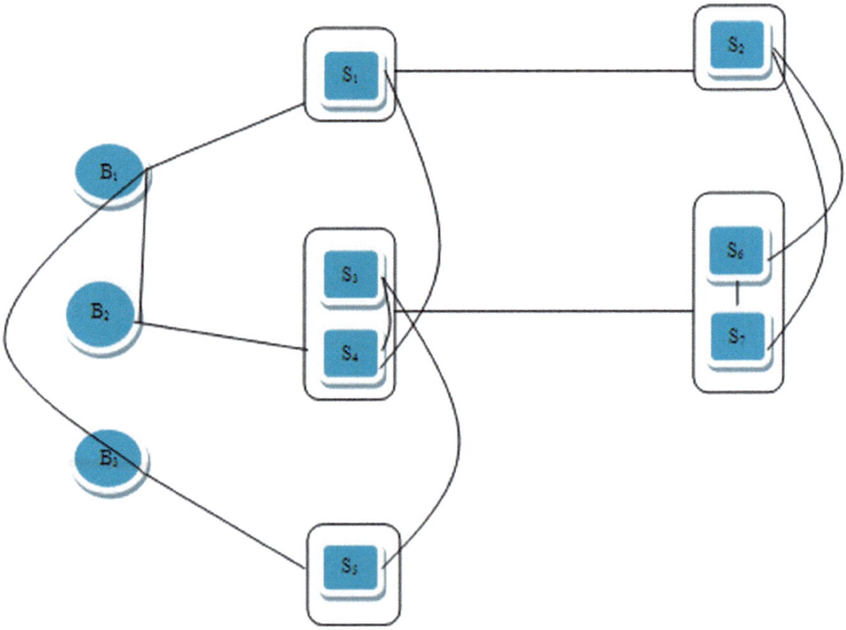

Fig. 12.11 The configured purified network

$$\tilde{p} = \begin{bmatrix} - & 1500 & 1300 & 100 & - & - & - & - \\ - & - & - & - & - & 208 & - & - \\ - & - & - & - & - & - & 110 & - \\ - & - & - & - & 50 & 132 & - & - \\ - & - & - & - & - & - & - & 130 \\ - & - & - & - & - & - & 84 & 135 \\ - & - & - & - & - & - & - & - \\ - & - & - & - & - & - & - & - \end{bmatrix}$$

$$\tilde{c} = (450\ 660\ 780\ 90\ 110\ 33\ 30\ 34)$$

$$\tilde{f} = \begin{bmatrix} - & 0.59 & 0.58 & 0.6 & - & - & - & - \\ - & - & - & - & - & 1.94 & - & - \\ - & - & - & - & - & - & 0.63 & - \\ - & - & - & - & 0.58 & 0.72 & - & - \\ - & - & - & - & - & - & - & 1.99 \\ - & - & - & - & - & - & 0.72 & 2.72 \\ - & - & - & - & - & - & - & - \\ - & - & - & - & - & - & - & - \end{bmatrix}$$

With respect to these matrices, the Steiner tree model is:

$$\begin{aligned} \max \quad X &= 1500x_{12} + 1300x_{13} + 100x_{14} + 208\,x_{26} + 110x_{37} \\ &\quad + 50x_{45} + 132\,x_{46} + 130\,x_{58} + 84\,x_{67} + 135\,x_{68} \\ \max \quad X' &= 0.59x_{12} + 0.58x_{13} + 0.6x_{14} + 1.94\,x_{26} + 0.63x_{37} \\ &\quad + 0.58x_{45} + 0.72x_{46} + 1.99x_{58} + 0.72\,x_{67} + 2.72x_{68} \\ \min \quad Y &= 450y_1 + 660y_2 + 780y_3 + 90y_4 + 110y_5 + 33y_6 \\ &\quad + 30y_7 + 34y_8 \end{aligned}$$

s.t.

$$x_{12} + x_{13} + x_{14} + x_{26} + x_{37} + x_{45} + x_{46} + x_{58} + x_{67} + x_{68}$$
$$= y_1 + y_2 + y_3 + y_4 + y_5 + y_6 + y_7 + y_8 - 1$$

$x_{14} \le y_1$

$x_{14} \le y_4$

$x_{26} \le y_2$

$x_{26} \le y_6$

$x_{37} \le y_3$

$x_{37} \le y_7$

$x_{46} \le y_4$

$x_{46} \le y_6$

$$x_{58} \leq y_5$$
$$x_{58} \leq y_8$$
$$x_{12} \leq y_1$$
$$x_{12} \leq y_2$$
$$x_{67} \leq y_6$$
$$x_{67} \leq y_7$$
$$x_{13} \leq y_1$$
$$x_{13} \leq y_3$$
$$x_{45} \leq y_5$$
$$x_{45} \leq y_4$$
$$x_{68} \leq y_6$$
$$x_{68} \leq y_8$$
$$x_{14} + x_{26} + x_{12} + x_{46} \leq y_1 + y_2 + y_4$$
$$x_{14} + x_{26} + x_{12} + x_{46} \leq y_1 + y_2 + y_6$$
$$x_{14} + x_{26} + x_{12} + x_{46} \leq y_1 + y_6 + y_4$$
$$x_{14} + x_{26} + x_{12} + x_{46} \leq y_6 + y_2 + y_4$$
$$x_{45} + x_{58} + x_{68} + x_{46} \leq y_4 + y_5 + y_6$$
$$x_{45} + x_{58} + x_{68} + x_{46} \leq y_4 + y_5 + y_8$$
$$x_{45} + x_{58} + x_{68} + x_{46} \leq y_4 + y_8 + y_6$$
$$x_{45} + x_{58} + x_{68} + x_{46} \leq y_8 + y_5 + y_6$$
$$x_{12} + x_{26} + x_{67} + x_{37} + x_{13} \leq y_1 + y_2 + y_3 + y_6$$
$$x_{12} + x_{26} + x_{67} + x_{37} + x_{13} \leq y_1 + y_2 + y_3 + y_7$$
$$x_{12} + x_{26} + x_{67} + x_{37} + x_{13} \leq y_1 + y_2 + y_7 + y_6$$
$$x_{12} + x_{26} + x_{67} + x_{37} + x_{13} \leq y_1 + y_7 + y_3 + y_6$$
$$x_{12} + x_{26} + x_{67} + x_{37} + x_{13} \leq y_7 + y_2 + y_3 + y_6$$
$$x_{14} + x_{46} + x_{67} + x_{37} + x_{13} \leq y_1 + y_4 + y_3 + y_6$$
$$x_{14} + x_{46} + x_{67} + x_{37} + x_{13} \leq y_1 + y_4 + y_3 + y_7$$
$$x_{14} + x_{46} + x_{67} + x_{37} + x_{13} \leq y_1 + y_4 + y_7 + y_6$$
$$x_{14} + x_{46} + x_{67} + x_{37} + x_{13} \leq y_1 + y_7 + y_3 + y_6$$
$$x_{14} + x_{46} + x_{67} + x_{37} + x_{13} \leq y_7 + y_4 + y_3 + y_6$$
$$x_{14} + x_{45} + x_{58} + x_{26} + x_{12} + x_{68} \leq y_1 + y_2 + y_4 + y_5 + y_6$$
$$x_{14} + x_{45} + x_{58} + x_{26} + x_{12} + x_{68} \leq y_1 + y_2 + y_4 + y_5 + y_8$$
$$x_{14} + x_{45} + x_{58} + x_{26} + x_{12} + x_{68} \leq y_1 + y_2 + y_4 + y_8 + y_6$$
$$x_{14} + x_{45} + x_{58} + x_{26} + x_{12} + x_{68} \leq y_1 + y_2 + y_8 + y_5 + y_6$$
$$x_{14} + x_{45} + x_{58} + x_{26} + x_{12} + x_{68} \leq y_1 + y_8 + y_4 + y_5 + y_6$$
$$x_{14} + x_{45} + x_{58} + x_{26} + x_{12} + x_{68} \leq y_8 + y_2 + y_4 + y_5 + y_6$$

$$x_{14} + x_{45} + x_{58} + x_{68} + x_{37} + x_{67} + x_{13} \le y_1 + y_3 + y_4 + y_5 + y_6 + y_7$$
$$x_{14} + x_{45} + x_{58} + x_{68} + x_{37} + x_{67} + x_{13} \le y_1 + y_3 + y_4 + y_5 + y_6 + y_8$$
$$x_{14} + x_{45} + x_{58} + x_{68} + x_{37} + x_{67} + x_{13} \le y_1 + y_3 + y_4 + y_5 + y_8 + y_7$$
$$x_{14} + x_{45} + x_{58} + x_{68} + x_{37} + x_{67} + x_{13} \le y_1 + y_3 + y_4 + y_8 + y_6 + y_7$$
$$x_{14} + x_{45} + x_{58} + x_{68} + x_{37} + x_{67} + x_{13} \le y_1 + y_3 + y_8 + y_5 + y_6 + y_7$$
$$x_{14} + x_{45} + x_{58} + x_{68} + x_{37} + x_{67} + x_{13} \le y_1 + y_8 + y_4 + y_5 + y_6 + y_7$$
$$x_{14} + x_{45} + x_{58} + x_{68} + x_{37} + x_{67} + x_{13} \le y_8 + y_3 + y_4 + y_5 + y_6 + y_7$$
$$y_h = 1, \qquad \forall h \in \{1, 2, 3\}$$
$$x_{ij} \in \{0, 1\}, \qquad \forall i, j$$
$$y_i \in \{0, 1\}, \qquad \forall i.$$

This model is executed using the Multi-Objective Ant Colony Optimization (MOACO) for 500 iterations, each involving 60 ants, implemented in MATLAB 9 software. The resulting Pareto solutions are presented in Table 12.3. Several of these Pareto solutions are illustrated in Figs. 12.13, 12.14, 12.15, 12.16 and 12.17. Additionally, Fig. 12.12 displays the Pareto solutions represented in three-dimensional (cubic) space, derived from the three objective functions.

In Fig. 12.12, the X-axis shows cost, and axes Y and Z are the symmetry of benefit and customers' total value, respectively.

Table 12.3 The Pareto solutions resulted from the three objective functions

The number of the Pareto solution	Cost	Benefit	Customers total value
1	1957	3143	5.8
2	1890	2800	1.2
3	1987	3253	6.5
4	2187	899	9.3
5	2187	3515	9.2
6	1987	2307	6.6
7	2067	3273	7.8
8	2097	2167	8.6
9	2157	3405	8.5
10	2097	3383	8.5
11	2077	3385	7.2
12	2047	3275	6.5
13	1920	2910	1.8
14	1953	1902	3.9
15	1953	3118	3.7
16	1923	3008	3.1

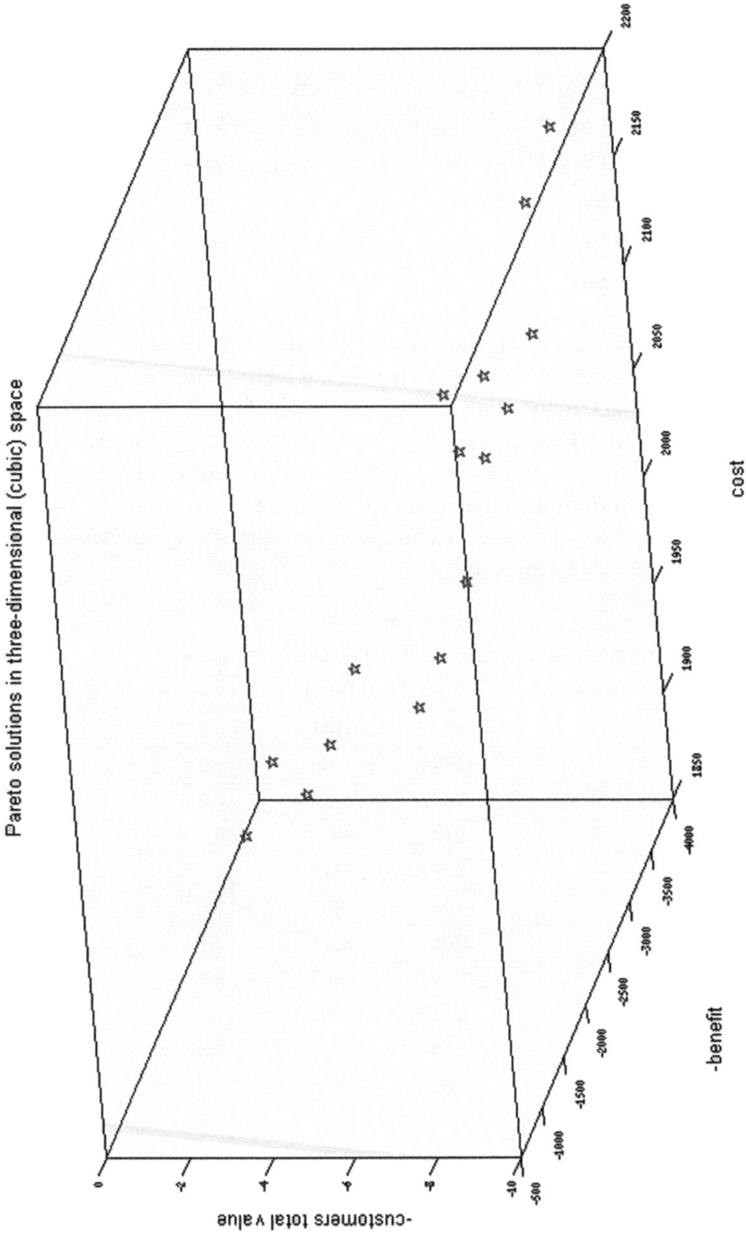

Fig. 12.12 The Pareto solutions in three-dimensional (cubic) space resulted from the three objective functions

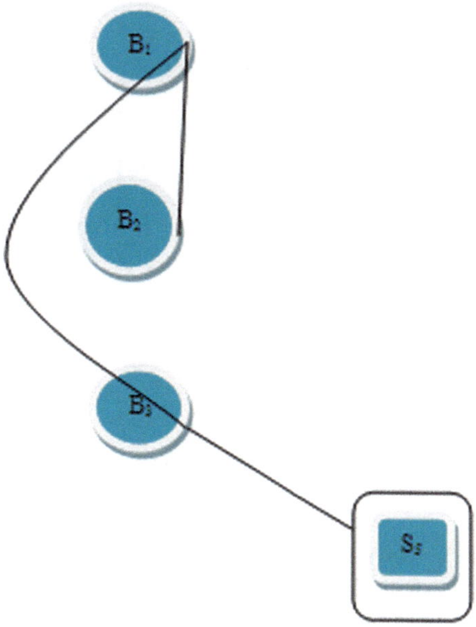

Fig. 12.13 The first Pareto optimal solution

(1) The first Pareto optimal solution is $Y^* = 1920$, $X^* = 2910$ and $X'^* = 1.8$, with the optimal network as shown in Fig. 12.13.
(2) The second Pareto optimal solution is $Y^* = 1923$, $X^* = 3008$ and $X'^* = 3.11$, with the optimal network as shown in Fig. 12.14.
(3) The third Pareto optimal solution is $Y^* = 1957$, $X^* = 3143$ and $X'^* = 5.83$, with the optimal network as shown in Fig. 12.15.

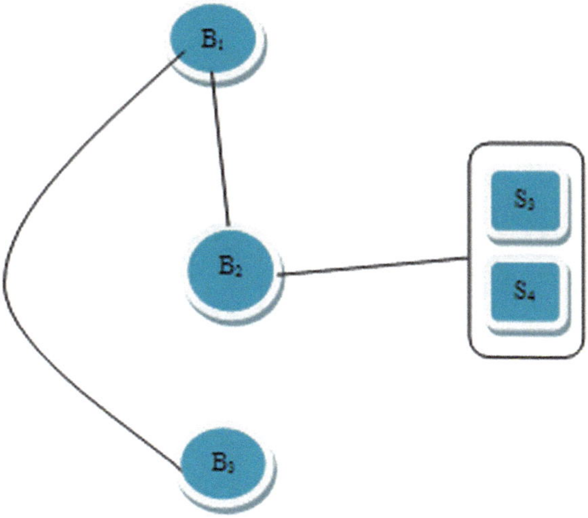

Fig. 12.14 The second Pareto optimal solution

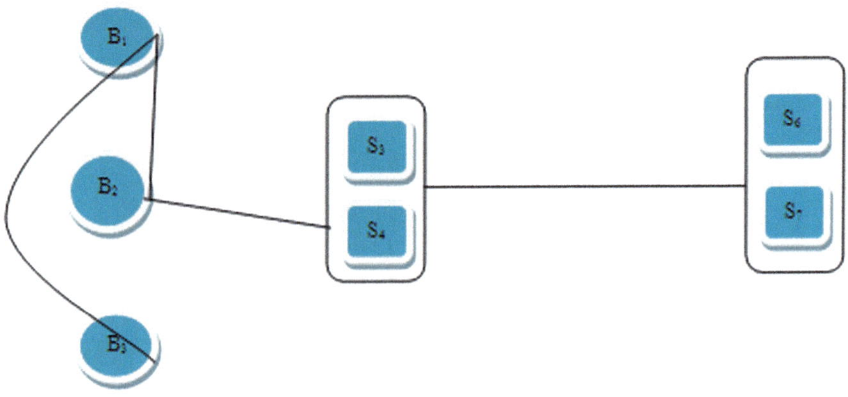

Fig. 12.15 The third Pareto optimal solution

(4) The fourth Pareto optimal solution is $Y^* = 2067$, $X^* = 3273$ and $X'^* = 7.82$, with the optimal network as shown in Fig. 12.16.

(5) The final Pareto optimal solution is $Y^* = 2187$, $X^* = 899$ and $X'^* = 9.32$, with the optimal network as shown in Fig. 12.17.

The proposed methodology generates diverse product configurations, as demonstrated in Figs. 12.13, 12.14, 12.15, 12.16 and 12.17, each characterized by unique cost-benefit-value trade-offs. These solutions emerge from a structured optimization process that incorporates two critical inputs: (1) customer preferences collected

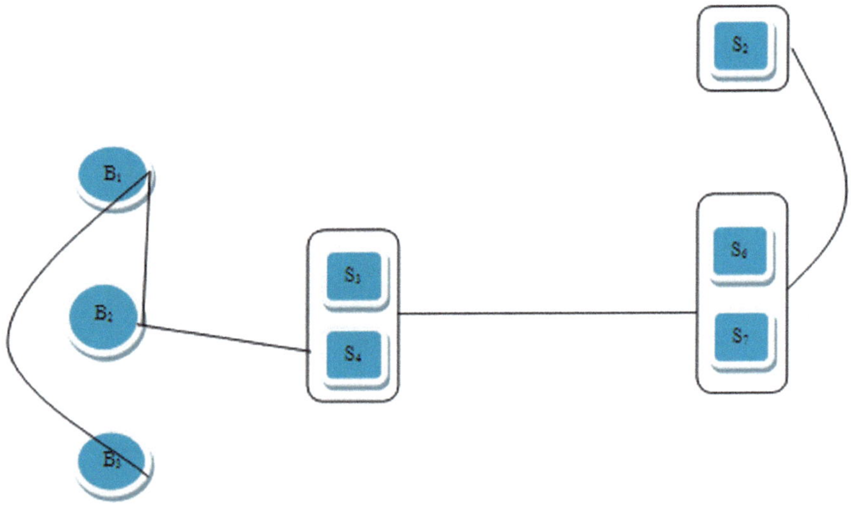

Fig. 12.16 The fourth Pareto optimal solution

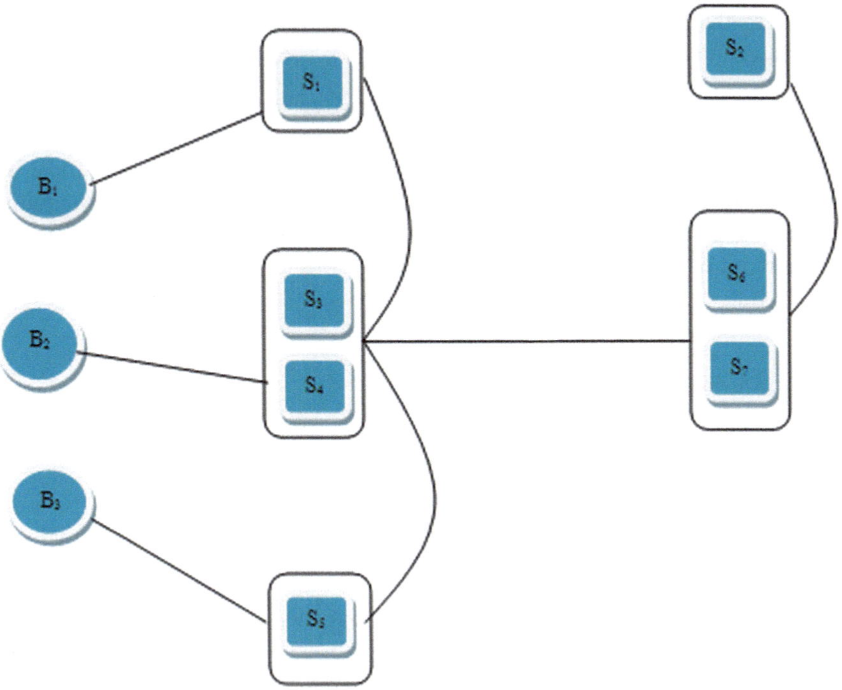

Fig. 12.17 The final Pareto optimal solution

through the web-based classification system, and (2) producer requirements for economically viable configurations. The numerical results reveal several key insights:

1. Customer-Centric Adaptation

- Product variants reflect differentiated customer valuations of functionalities.
- Web-collected preference data directly informs feature inclusion/exclusion decisions.

2. Multi-objective Optimization

- Solutions represent Pareto optimal balances between production cost minimization, functional benefit maximization, and customer value enhancement.
- The three-dimensional trade-off space is navigated through the Steiner tree framework.

3. Stakeholder-Aligned Outcomes

- Producers gain benefit-maximized configurations.
- Consumers receive value-optimized product variants.

- The methodology automatically resolves conflicts through weighted objective optimization.

This systematic approach transforms subjective customer preferences into mathematically optimizable parameters while maintaining producer profitability—a crucial advancement for data-driven product development in convergent industries. The methodology's effectiveness is particularly evident in its ability to generate multiple viable configurations rather than single-point solutions, providing decision-makers with a range of business-optimal alternatives.

12.4 Fuzzy Model

12.4.1 Definitions and Preliminaries

Here, we provide some basic definitions of fuzzy sets and fuzzy numbers, adopted from (Kaufmann & Gupta, 1991).

Definition 1 A fuzzy-set $\tilde{A}$ in a universe of discourse X is characterized by a membership function $\mu_{\tilde{A}}(x)$ which associates with each element x in X a real number in the interval [0, 1]. The function value $\mu_{\tilde{A}}(x)$ is termed the grade of membership of x in $\tilde{A}$.

Definition 2 A fuzzy number $\tilde{A}$ is a fuzzy convex subset of the real line satisfying the following conditions:

(a) $\mu_{\tilde{A}}(x)$ is piecewise continuous.
(b) $\mu_{\tilde{A}}(x)$ is normalized; that is, there exists $m \in \Re$ with $\mu_{\tilde{A}}(m) = 1$, where m is called the mean value of $\tilde{A}$.

Definition 3 A triangular fuzzy number $\tilde{a}$ is defined by a triplet (a_1, a_2, a_3), where $a_i \in R, 1 \leq i \leq 3$, and its conceptual schema and mathematical form are shown by (12.12) below:

$$\mu_{\tilde{a}}(x) = \begin{cases} 0, & x \leq a_1 \\ \frac{x-a_1}{a_2-a_1}, & a_1 \leq x \leq a_2 \\ \frac{a_3-x}{a_3-a_2}, & a_2 \leq x \leq a_3 \\ 0, & a_3 \leq x \end{cases} \tag{12.12}$$

A triangular fuzzy number $\tilde{a}$ in the universe of discourse X that conforms to this definition is shown in Fig. 12.18.

Definition 4 A trapezoidal fuzzy number $\tilde{a}$ can be defined by a quadruplet (a_1, a_2, a_3, a_4), where $a_i \in R, 1 \leq i \leq 4$, and its conceptual schema and mathematical form are shown by (12.13) below:

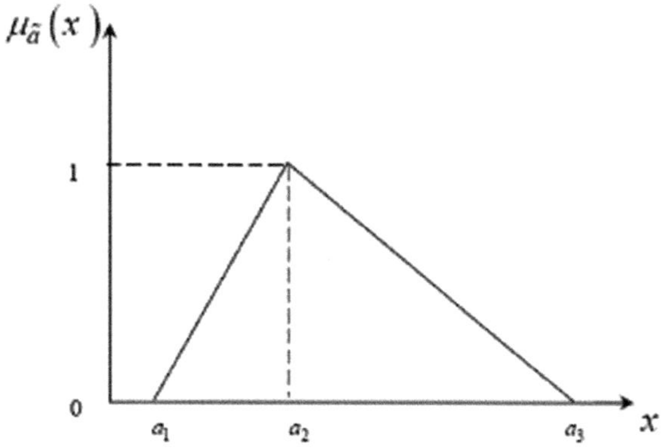

Fig. 12.18 A triangular fuzzy number $\tilde{a}$

$$\mu_{\tilde{a}}(x) = \begin{cases} 0, & x \leq a_1 \\ \frac{x-a_1}{a_2-a_1}, & a_1 \leq x \leq a_2 \\ 1, & a_2 \leq x \leq a_3 \\ \frac{a_4-x}{a_4-a_3}, & a_3 \leq x \leq a_4 \\ 0, & a_4 \leq x. \end{cases} \tag{12.13}$$

A trapezoidal fuzzy number $\tilde{a}$ in the universe of discourse X that conforms to this definition is shown in Fig. 12.19.

Note that one can consider a triangular fuzzy number as a special case of a trapezoidal fuzzy number by having $b_2 = b_3$.

Definition 5 For $\tilde{a} = (a_1, a_2, a_3)$ and $\tilde{b} = (b_1, b_2, b_3)$, triangular fuzzy numbers, the fuzzy sum is given by:

$$\tilde{a} + \tilde{b} = (a_1 + b_1, a_2 + b_2, a_3 + b_3) \tag{12.14}$$

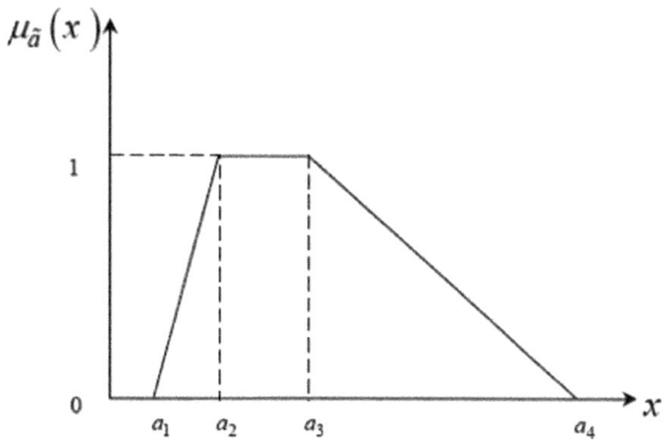

Fig. 12.19 A trapezoidal fuzzy number $\tilde{a}$

and for $\tilde{a} = (a_1, a_2, a_3, a_4)$ and $\tilde{b} = (b_1, b_2, b_3, b_4)$, trapezoidal fuzzy numbers, the fuzzy sum of the two numbers is given by:

$$\tilde{a} + \tilde{b} = (a_1 + b_1, a_2 + b_2, a_3 + b_3, a_4 + b_4).$$

Definition 6 The α-cut and strong α-cut for a fuzzy-set $\tilde{A}$ are shown by $\tilde{A}_\alpha$ and $\tilde{A}_\alpha^+$, respectively, and for $\alpha \in [0,\ 1]$ are defined to be:

$$\tilde{A}_\alpha = \left\{x \,|\, \mu_{\tilde{A}}(x) \geq \alpha\,, x \in X\right\},$$
$$\tilde{A}_\alpha^+ = \left\{\ x \ |\mu_{\tilde{A}}(x) > \alpha\,, x \in X\right\},$$

where X is the universal set.

Upper and lower bounds for any α-cut $\tilde{A}_\alpha$ are shown by $\sup \tilde{A}_\alpha$ and $\inf \tilde{A}_\alpha$, respectively. Here, we assume that the upper and lower bounds of α-cuts are finite values, and for simplicity, we denote $\sup \tilde{A}_\alpha$ by $\tilde{A}_\alpha^R$ and $\inf \tilde{A}_\alpha$ by $\tilde{A}_\alpha^L$.

To effectively compare and identify minimal values among fuzzy numbers in our optimization framework, we introduce a novel ranking method that addresses the limitations of existing approaches while maintaining consistency with fuzzy arithmetic principles. Drawing parallels to fuzzy addition operations, we define the fuzzy minimum operation through a systematic ranking procedure that evaluates the complete characteristics of each fuzzy number—including its membership function shape, support, and core—rather than relying on simplistic defuzzification. This approach enables precise pairwise comparison of fuzzy-valued solutions in our multi-objective Steiner tree model, where critical parameters (costs, benefits, and customer values) may exhibit inherent uncertainty. The proposed method not only preserves the informational richness of fuzzy numbers during comparison but also ensures computational efficiency for optimization purposes, allowing for (1) accurate identification of non-dominated solutions in fuzzy objective space, (2) consistent trade-off analysis under uncertainty, and (3) effective implementation of the fuzzy Steiner tree algorithm. By developing this specialized ranking and minimum operation framework, we establish a rigorous mathematical foundation for handling uncertain parameters throughout the optimization process while maintaining the methodological coherence between fuzzy arithmetic operations and solution selection criteria. The method's adaptability to various fuzzy number types (triangular, trapezoidal, etc.) and its alignment with the principles of fuzzy arithmetic make it particularly suitable for complex, real-world product configuration problems where precise uncertainty management is crucial. To define the fuzzy min operation, we draw parallels to the fuzzy addition as follows:

$$\begin{aligned} \text{MV} &= \text{MV}(\tilde{a}\ ,\ \tilde{b}) = \min(\tilde{a}, \tilde{b}) \\ &= (\min(a_1, b_1), \min(a_2, b_2), \min(a_3, b_3), \min(a_4, b_4)). \end{aligned} \tag{12.15}$$

It is evident that, for non-comparable fuzzy numbers $\tilde{a}$ and $\tilde{b}$, this fuzzy min operation results in a fuzzy number different from both $\tilde{a}$ and $\tilde{b}$. For example, for

$\tilde{a} = (5, 10, 13, 19)$ and $\tilde{b} = (6, 9, 15, 20)$, using (12.4), we get $M\tilde{V} = \min(\tilde{a}, \tilde{b}) = (5, 9, 13, 19)$, which is different from both $\tilde{a}$ and $\tilde{b}$.

Definition 7 The $D_{p,q}$-distance, indexed by parameters p, $1 < p < \infty$, and q, $0 < q < 1$, between two fuzzy numbers $\tilde{a}$ and $\tilde{b}$ is a non-negative function given as follows:

$$D_{p,q}(\tilde{a}, \tilde{b}) = \begin{cases} \left[(1-q) \int\limits_0^1 \left| a_\alpha^- - b_\alpha^- \right|^p d\alpha + q \int\limits_0^1 \left| a_\alpha^+ - b_\alpha^+ \right|^p d\alpha \right]^{\frac{1}{p}}, & p < \infty, \\ (1-q) \sup\limits_{0<\alpha\leq 1} \left(\left| a_\alpha^- - b_\alpha^- \right| \right) + q \inf\limits_{0<\alpha\leq 1} \left(\left| a_\alpha^+ - b_\alpha^+ \right| \right), & p = \infty. \end{cases} \tag{12.16}$$

The analytical properties of $D_{p,q}$ depend on the first parameter p, while the second parameter q of $D_{p,q}$ characterizes the subjective weight attributed to the endpoints of the support of the fuzzy numbers. If there is no reason for distinguishing any side of the fuzzy numbers, $D_{2,\frac{1}{2}}$ is recommended.

For triangular fuzzy numbers $\tilde{a} = (a_1, a_2, a_3)$ and $\tilde{b} = (b_1, b_2, b_3)$, the above distance with $p = 2$ and $q = \frac{1}{2}$ is calculated to be:

$$D_{2,\frac{1}{2}}(\tilde{a}, \tilde{b}) = \sqrt{\frac{1}{6}\left[\sum_{i=1}^{3} (b_i - a_i)^2 + (b_2 - a_2)^2 + \sum_{i\in\{1,2\}} (b_i - a_i)(b_{i+1} - a_{i+1}) \right]}. \tag{12.17}$$

And if $\tilde{a} = (a_1, a_2, a_3, a_4)$ and $\tilde{b} = (b_1, b_2, b_3, b_4)$ are trapezoidal fuzzy numbers, then the distance is calculated as:

$$D_{2,\frac{1}{2}}(\tilde{a}, \tilde{b}) = \sqrt{\frac{1}{6}\left[\sum_{i=1}^{4} (b_i - a_i)^2 + \sum_{i\in\{1,3\}} (b_i - a_i)(b_{i+1} - a_{i+1}) \right]}. \tag{12.18}$$

Now, we make use of the distance function (12.18) to acquire the distances of $\tilde{a}$ and $\tilde{b}$ to MV, as obtained by (12.15). We then compute $D_{2,\frac{1}{2}}(\tilde{a}, \text{MV}) = 0.1667$ and $D_{2,\frac{1}{2}}(\tilde{b}, \text{MV}) = 1.33$. Thus, $\tilde{a}$, having a lower distance from MV, is decided to be smaller than $\tilde{b}$. Algorithm 4 shows to find the minimal values among fuzzy numbers.

Algorithm 4 Find the minimal values among fuzzy numbers $\tilde{L}_i$, $i = 1, 2, \ldots, m$.
Step 1. Set $\tilde{L}_{\min} = \tilde{L}_1$.
Step 2. **For** $i = 2$ till m **do**
Calculate $\text{MV} = \text{Min}(\tilde{L}_{\min}, \tilde{L}_i)$ by Eq. (12.4);

$$D_1 = D_{p,q}(MV, \tilde{L}_{\min});$$
$$D_2 = D_{p,q}(MV, \tilde{L}_i);$$
$$\tilde{L}_{\min} = \arg\min(D_1, D_2);$$

End for.

Step 3. Stop.

Building upon the clustered network structure, we develop an advanced Steiner tree formulation that preserves all base functions while simultaneously optimizing two fuzzy objectives: (1) cost minimization and (2) profit maximization in the convergent product value chain. This dual-objective approach extends the conventional Steiner tree paradigm, which traditionally focuses solely on crisp cost minimization, by incorporating fuzzy-set representations of both financial parameters. Each node i in the network (representing either a function or sub-function) is characterized by two distinct fuzzy cost components: [specify cost types here]. The model's architecture maintains strong theoretical alignment with classical Steiner tree problems while introducing three critical innovations:

1. Fuzzy parameter integration:
   - dual fuzzy cost attributes per node;
   - profit potential as a fuzzy maximization objective;
   - preservation of base functionality constraints.
2. Multi-objective optimization:
   - simultaneous cost-profit trade-off analysis;
   - fuzzy ranking system for solution comparison;
   - threshold-based solution filtering.
3. Computational efficiency:
   - cluster-based network simplification;
   - adaptive ant colony optimization;
   - parallel processing implementation.

The mathematical formulation leverages this structure to identify optimal tree configurations that balance producer economics (cost/profit) with product integrity (base function preservation), while accounting for the inherent uncertainties in product valuation and cost estimation through its fuzzy-set representations. This approach proves particularly valuable in convergent product development, where traditional crisp optimization models often fail to capture the nuanced trade-offs between functionality expansion and economic viability.

$\tilde{c}_{i1}$: software cost,

$\tilde{c}_{i2}$: hardware cost.

Our proposed methodology addresses the NP-hard Steiner tree problem in fuzzy-valued convergent product networks through a structured optimization framework. Each arc (i, j) in the network is associated with a fuzzy benefit measure $\tilde{p}_{ii'}$ derived from its terminal nodes i and i', while nodes carry multiple fuzzy cost components.

To manage computational complexity, we implement a hierarchical decomposition approach consisting of: (1) a leveling process that establishes functional hierarchies from base functions (level 0) to sub-functions, and (2) a clustering phase that groups related nodes at each level l while enforcing strict single-cluster membership through a zero-one fuzzy integer program. This mathematical formulation minimizes total fuzzy assignment costs subject to topological constraints, with a subsequent purification procedure ensuring functional coherence within clusters. The preprocessing stage effectively reduces the original complex problem into tractable subproblems while preserving the network's essential characteristics for subsequent Steiner tree construction. By incorporating fuzzy arithmetic operations, our approach maintains robust handling of uncertain cost/benefit parameters and value estimations throughout the optimization process. The resulting framework provides a computationally efficient solution to the multi-objective challenge of simultaneously minimizing costs and maximizing benefits in convergent product design, while ensuring preservation of core functionalities and proper management of the inherent trade-offs between solution quality and computational tractability in fuzzy-valued networks.

The Zero/One Fuzzy Mathematical Program for Level l

$$\min T = \sum_{i \in n_l} \sum_{j \in m_{li}} (\alpha_{ij}\tilde{c}_{ij1} + \beta_{ij}\tilde{c}_{ij2}) z_{ij}$$

$$\text{s.t.} \quad \sum_{j \in m_{li}} z_{ij} = 1, \quad i = 1, \ldots, n_l$$

$$z_{ij} = \begin{cases} 1, & \text{if node } i \text{ is in cluster } j \\ 0, & \text{otherwise,} \end{cases}$$

where $\beta_{ij}, \alpha_{ij} \in [0, 1], \ \forall i, j$, and

m_{li} :set of indices of clusters at level l where node i is included,

n_l: set of indices of different nodes at level l,

$\tilde{c}_{ij2}$: hardware fuzzy cost of node i in cluster j ($\tilde{c}_{ij2} = \tilde{c}_{i2}$, for all i and j),

$\tilde{c}_{ij1}$: software fuzzy cost of node i in cluster j ($\tilde{c}_{ij1} = \tilde{c}_{i1}$, for all i and j),

α_{ij} : the software reduction cost coefficient of node i in cluster j,

β_{ij} : the hardware reduction cost coefficient of node i in cluster j.

Purifying Fuzzy Benefits and Fuzzy Costs at Level l

To determine the cost for each cluster at level l, we use

$$\tilde{c}_{jl} = \sum_{i \in n_l} (\alpha_{ij}\tilde{c}_{ij1} + \beta_{ij}\tilde{c}_{ij2}) - \sum_{\forall i, i'} \tilde{p}_{ii'},$$

with $\tilde{c}_{ij1}$, $\tilde{c}_{ij2}$, and n_l as defined above, $\beta_{ij}, \ \alpha_{ij} \in [0, 1], \ \forall i, j$, and

$\tilde{c}_{jl}$: fuzzy cost of cluster j at level l,

α_{ij} : the software reduction cost coefficient of node i in cluster j,

β_{ij} : the hardware reduction cost coefficient of node i in cluster j,

$\tilde{p}_{ii'}$: the benefit of an arc connecting node i in cluster j to node i' in cluster j.

Also, to adjust the combined arc fuzzy benefits in clusters, the following equation is used:

$$\tilde{p}_{jj'} = (1 + \gamma_{jj'}) \sum_{\forall i\,,\, i'} \tilde{p}_{ii'}$$

where $\forall i, i', j\ , j'$, and

$\tilde{p}_{jj'}$: the adjusted arc fuzzy benefit connecting cluster j to cluster j',

$\tilde{p}_{ii'}$: the fuzzy benefit of an arc connecting node i in cluster j to node i' in cluster j',

$\gamma_{jj'}$: the added value configured from nodes in clusters j and j'.

Additionally, each node must belong to only one cluster at level l. The node with the minimal cost (according to the ranking defined in Definition 7) is selected for level l. Instead of employing the zero/one mathematical program for this level, we can utilize step 4 of Algorithm 3. This approach significantly reduces computations by eliminating the need for zero/one programming.

Algorithm 5 Leveling and clustering in the network.

Step 1: Set the base functions at level 0. Let $l = 0$.

Step 2: **While** sub-functions exist for processing **do**

Find sub-functions with a link to a function (or sub-function) at level l and place them at level $l + 1$. Let $l = l + 1$;

End while.

$\{l$ is the number of levels$\}$

Step 3: Set each node at level 0 to be a cluster.

Step 4: **For** $i = 1$ till l **do**

$\{$Form clusters at level $i\}$

Cluster all sub-functions at level i linked to a single parent at level $i - 1$;

While $|n_i| > 0$ **do**

Select $k \in n_i$ such that $\alpha_{kp}\tilde{c}_{kp1} + \beta_{kp}\tilde{c}_{kp2} = \min\limits_{j \in m_{ik}} \left\{\alpha_{kj}\tilde{c}_{kj1} + \beta_{kj}\tilde{c}_{kj2}\right\}$.

Set $z_{kp} = 1$,

and $z_{kj} = 0\ ,\ \forall j \in m_{ik}\ ,\ j \neq p$(using Algorithm 0);

set $n_i \leftarrow n_i - \{k\}$.

End while;

For $j = 1$ till q_i **do** $\{q_i$ is the number of clusters at level $i\}$

Set $\tilde{c}_{ji} = \sum\limits_{i \in n_l} (\alpha_{ij}\tilde{c}_{ij1} + \beta_{ij}\tilde{c}_{ij2}) - \sum\limits_{\forall i, i'} \tilde{p}_{ii'}$;

End for;

For $j = 1$ till q_i **do**
  For $j' = 1$ till q_i **do**
    Set $\tilde{p}_{jj'} = (1 + \gamma_{jj'}) \sum_{\forall i,\, i'} \tilde{p}_{ii'}$;
  End for;
End for;
End for.
Step 5: **Stop**.

12.4.2 Fuzzy Mathematical Formulation

Building upon the Dantzig-Fulkerson-Johnson framework (Costa et al., 2006), we present a binary integer programming formulation for the Convergent Product Steiner Tree Problem (CPSTP) that integrates both topological and economic considerations. The model utilizes two key decision variables: $x_{ij} \in \{0, 1\}$ indicating the activation of link (i,j) and $y_i \in \{0, 1\}$ representing the selection of cluster i, with mandatory inclusion of terminal nodes (base functions) through the constraint $y_i = 1 \; \forall \; i \in N$. The objective function minimizes total implementation costs, combining both cluster costs ($\Sigma c_i y_i$) and connection costs ($\Sigma c_{ij} x_{ij}$), while enforcing critical network properties via three constraint classes: (1) connectivity constraints ($\Sigma x_{ij} \geq 1 \; \forall \; S \subset V$ with $S \cap N \neq \emptyset$) ensuring all terminals form a unified network, (2) tree structure constraints ($\Sigma x_{ij} = \Sigma y_i - 1$) preventing cyclic configurations, and (3) consistency constraints ($x_{ij} \leq y_i, x_{ij} \leq y_i$) maintaining logical relationships between cluster selection and link activation. This formulation preserves the theoretical foundations of Steiner tree problems while specifically addressing convergent product design requirements through its terminal set specification and cost-sensitive optimization framework, providing both a rigorous mathematical foundation and practical basis for subsequent algorithmic developments.

The mathematical model can then be expressed as follows:

$$\text{Maximize} \quad \sum_{(i,j)\in E} \tilde{p}_{ij}.x_{ij}, \tag{12.19}$$

$$\text{Minimize} \quad \sum_{i\in V} \tilde{c}_i.y_i, \tag{12.20}$$

Such that

$$\sum_{(i,j)\in E} x_{ij} = \sum_{i\in V} y_i - 1, \tag{12.21}$$

$$\sum_{(i,j)\in E(S)} x_{ij} \leq \sum_{i\in S-\{k\}} y_i, \quad \forall k \in S \subseteq V, \; \forall S : |S| \geq 2, \tag{12.22}$$

$$y_h = 1, \quad \forall h \in N, \tag{12.23}$$

$$x_{ij} \in \{0, 1\}, \quad \forall i, j \in E, \tag{12.24}$$

$$y_i \in \{0, 1\}, \quad \forall i \in V. \tag{12.25}$$

Our mathematical model simultaneously addresses two competing objectives: (1) maximizing the total benefits $\sum b_n x_n$ from selected links and (2) minimizing the total costs $\sum \left(c_n^x + c_n y\right)$ of both cluster activation and connection establishment. The constraint system enforces critical solution properties through four key components: Constraint (12.22) establishes the fundamental tree structure requirement by equating the number of selected clusters ($\sum \mathrm{y_i}$) to one less than the number of activated links ($\sum x_{ij} - 1$), while constraints (12.11) ensure network connectivity through exponentially growing cutset conditions ($2^{|V|} - |V| - 1$). Constraints (12.23) mandate the inclusion of all terminal clusters ($y_i = 1 \ \forall \ i \in N$), and relations (12.24)–(12.25) formally define the binary nature of all decision variables ($x_{ij}, y_i \in \{0, 1\}$). The model's computational complexity manifests in two dimensions: the variable set expands linearly with cluster count, while the constraint system grows exponentially, reflecting the inherent NP-hard nature of the problem. This dual expansion characteristic underscores the trade-off between solution quality and computational tractability in convergent product network optimization, motivating our subsequent development of heuristic approaches to handle practical-scale instances while maintaining near-optimal performance guarantees.

12.4.3 The Extended Bi-objective Particle Swarm Optimization Approach

Since the Steiner Tree Problem (STP) is NP-hard, solving medium to large-sized instances efficiently requires metaheuristic approaches. A bi-objective particle swarm optimization (BOPSO) algorithm is proposed (Abedi et al., 2015) for this purpose.

Classic Particle Swarm Optimization (PSO):

PSO is inspired by artificial life (A-life) concepts, particularly bird flocking, fish schooling, and swarming behavior. It also has connections to evolutionary computing, sharing similarities with genetic algorithms (GA) and evolutionary programming (Kennedy & Eberhart, 1995). A swarm consists of multiple particles (e.g., birds, fish, bees), each moving through the solution space with a velocity vector. Every particle tracks two key values:

pbest (personal best): The best position found by the particle itself.

gbest (global best): The best position found by the entire swarm or its best informant.

Each particle updates its position based on three components:

- its own current velocity,
- a pull toward its personal best (pbest) position,
- a pull toward the global best (gbest) position of the swarm.

These three vectors are combined linearly using confidence coefficients, forming the foundation of the classic PSO update rule.

The Proposed BOPSO

Due to the increased complexity of the discrete particle swarm optimization algorithm for the Steiner Tree Problem compared to standard PSO, we present its flowchart in Fig. 12.20. The proposed algorithm comprises several key components, including preprocessing operations, representation, update operations, improvement strategies, and the updating of velocity and position values.

A. **The Preprocessing Operations**

The solution to the Steiner Tree Problem (STP) is always a connected tree. However, since the graph G is not necessarily a complete graph, certain binary strings may correspond to unconnected nodes. Consequently, it may not be possible to construct a maximum spanning tree (MST) in the graph based on these binary strings.

Another preprocessing step involves determining the longest distance from each node to all terminal nodes. Following this, the nodes are sorted based on their longest distances in ascending order. This sorting serves as a preparatory step for Modification 7 in the update operations.

Remark 1 In all cases needing to compute $\tilde{a} = (a_1, a_2, a_3)$, we use the distance measure $D_{2,\frac{1}{2}}(\tilde{0}, \tilde{a})$ in Eq. (12.6), so that $\tilde{0} = (0, 0, 0)$.

B. **The Representation**

The core challenge in solving the Steiner Tree Problem (STP) lies in identifying optimal intermediate nodes that minimize total connection costs while ensuring connectivity to all terminals. To address this efficiently, we encode potential solutions as binary strings $s_k^i \in \{0, 1\}$, where:

k denotes the generation index in the optimization process,

i represents the ith node in the graph,

$s_k^i = 1$ indicates node selection (either as an intermediate node or terminal).

This encoding enables polynomial-time conversion to a Maximum Spanning Tree (MST) through the following steps:

1. String Decoding: The binary string directly maps to node selection in the graph.
2. MST Construction: Applies Kruskal's or Prim's algorithm to the subgraph induced by selected nodes.

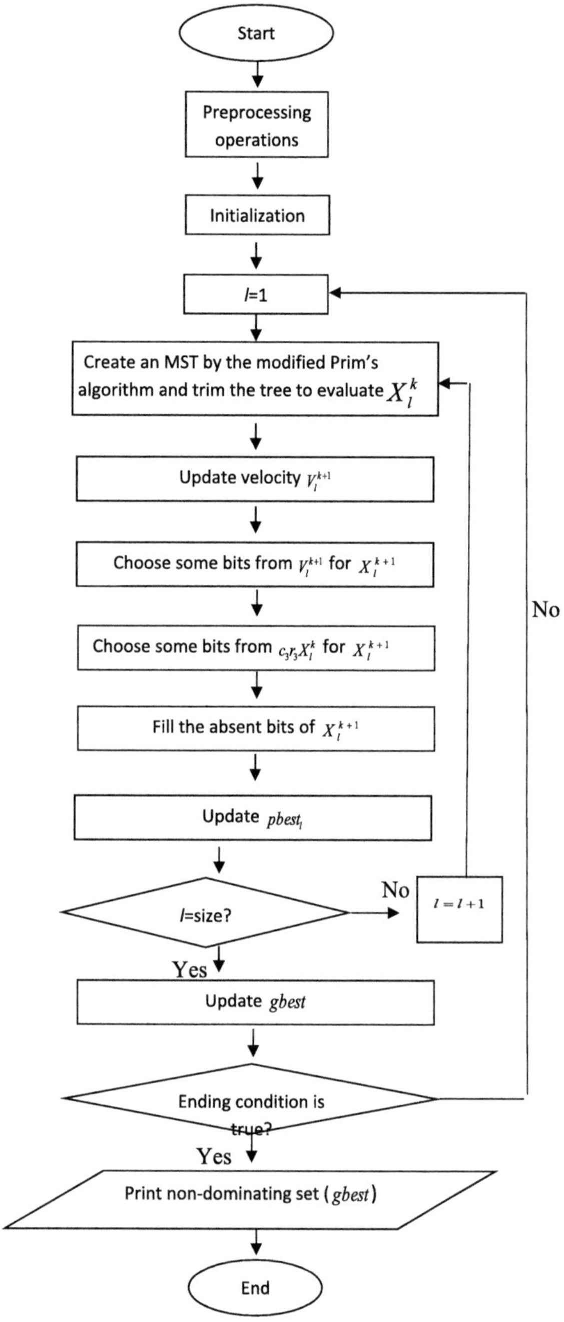

Fig. 12.20 Flow chart of the bi-objective PSO algorithm

3. Solution Validation: Ensures all terminals are connected in the resulting tree.

Remark 2 The bits standing for the terminal nodes are always set to 1.

The velocities are represented as

$$V = \begin{pmatrix} v_1^0, & v_2^0, & \ldots, & v_n^0 \\ v_1^1, & v_2^1, & \ldots, & v_n^1 \end{pmatrix}$$

such that $v_i^0 = \frac{\lambda}{D_{2,\frac{1}{2}}(\tilde{0}, \tilde{c}_i)}$ and $v_i^1 = \frac{\gamma(D_{2,\frac{1}{2}}(\tilde{0}, \tilde{c}_i)-1)}{D_{2,\frac{1}{2}}(\tilde{0}, \tilde{c}_i)}$,

where v_i^0 and v_i^1 are real numbers in the internal [0, 1] and denote the probabilities of the ith bit to be 0 or 1. Also, λ and γ are real numbers in the internal [0, 1]. According to *Modification* 5 below, the sum of v_i^0 and v_i^1 is not necessarily equal to 1.

C. The Update Equations

The proposed algorithm implements an evolutionary optimization process where candidate solutions (particles) undergo iterative refinement until meeting termination criteria. During initialization, each particle receives randomized binary-encoded positions representing potential intermediate node selections and associated velocity vectors. These positions are evaluated against dual objectives: (1) minimizing connection and cluster activation costs ($f_1 = \Sigma\left(c_{ij}x_{ij} + c_i y_i\right)$) and (2) maximizing benefit accumulation ($f_2 = \Sigma b_{ij} x_{ij}$). The solution space is explored through generation-wise updates governed by the velocity equation $v_{k+1}^i = w \cdot v_k^i + c_1 \cdot \text{rand}() \cdot \left(\text{pbest}^i - s_k^i\right) + c_2 \cdot \text{rand}() \cdot \left(\text{gbest} - s_k^i\right)$ and position update $s_{k+1}^i = s_k^i + v_{k+1}^i$, where w maintains momentum while c_1 and c_2 weight the attraction toward personal best (*pbest*) and global best (*gbest*) solutions, respectively. Following each update, a repair mechanism enforces solution feasibility by preserving tree topology constraints and terminal connectivity requirements. The algorithm maintains and progressively refines two solution sets: a global best (*gbest*) for single-objective guidance and a Pareto front tracking non-dominated solutions for multi-objective optimization. This dual approach balances intensive local search around promising solutions (through *pbest* attraction) with global exploration (via *gbest* guidance), efficiently navigating the exponential solution space characteristic of Steiner tree problems while addressing the fundamental cost-benefit trade-offs in convergent product design.

$$V_l^{k+1} = \omega V_l^k + c_1 r_1 (\text{pbest}_l^k - X_l^k) + c_2 r_2 (\text{gbest}^k - X_l^k), \tag{12.26}$$

$$X_l^{k+1} = X_l^k + V_l^{k+1}, \tag{12.27}$$

where k denotes the kth generation. The velocity update equation $v_{k+1}^i = \omega \cdot v_k^i + c_1 \cdot r_1 \cdot \left(\text{pbest}^i - s_k^i\right) + c_2 \cdot r_2 \cdot \left(\text{gbest} - s_k^i\right)$ in generation k consists of three distinct components: (1) an inertial term ($\omega \cdot v_k^i$) preserving previous search momentum through weight parameter $\omega \in (0, 1)$, (2) a self-cognitive component $\left(c_1 \cdot r_1 \cdot \left(\text{pbest}^i - s_k^i\right)\right)$ directing particles toward their historical best solutions, and (3) a social-cognitive component $\left(c_2 \cdot r_2 \cdot \left(\text{gbest} - s_k^i\right)\right)$ steering particles toward the swarm's global optimum, where $c_1, c_2 \in \mathbb{R}^+$ are acceleration constants and $r_1, r_2 \sim U[0, 1]$ introduce stochastic exploration. Following position updates via $s_{k+1}^i = s_k^i + v_{k+1}^i$, solutions undergo reevaluation with potential updates to pbest and gbest based on dominance relationships.

For effective adaptation to the discrete Steiner Tree Problem, we implement seven key modifications to standard continuous PSO:

Binary Position Encoding: Solutions represented as bitstrings indicating node inclusion,
Discrete Velocity Interpretation: Velocities map to probabilities of bit flips,
Topology-Aware Repair: Invalid solutions adjusted to maintain tree connectivity,
Terminal Node Preservation: Mandatory inclusion of base functions enforced,
Elitist Archive Maintenance: Non-dominated solutions preserved across generations,
Adaptive Parameter Tuning: Dynamic adjustment of ω, c_1, c_2 during search,
Hybrid Local Search: Combined with deterministic refinement heuristics.

These adaptations address the combinatorial nature of STP while retaining PSO's global exploration capabilities, enabling effective navigation of the solution space despite the problem's NP-hard complexity. The modified algorithm maintains polynomial-time complexity per iteration while substantially improving solution quality over conventional approaches through its balanced exploitation-exploration mechanism and problem-specific solution handling.

Modification 1 The result of subtraction operator "–" between two positions (binary strings) X_1 and X_2 in (12.26) is defined as a velocity, in which v_i^b is set to 1 if the jth bit in X_1 is b while it is not in X_2. Otherwise, v_i^b is set to 0.

Modification 2 In Eq. (12.15), the result V of subtraction is multiplied by $c_j r_j$ $(j = 1, 2)$, where c_j is a constant real number and r_j is a random real number belonging to [0, 1]. We do the same in our proposed algorithm. As a result, each element is probably multiplied by a unique real number belonging to $[0,\ c_j]$, because the r_j are spawn randomly.

Remark 3 If v_i^b is greater than 1 after the multiplication, it will be set 1.

Remark 4 As to ωV, the result is that each element v_i^b in V is multiplied by the same ω.

Modification 3 The result of the operator "+" in Eq. (12.26) between two velocities is a new velocity, that is $V = V_1 + V_2$. The v_i^b in V is the greater one between v_{1i}^b and v_{2i}^b.

Following the three modifications above, the first equation is totally redefined for STP. An example is given as follows.

Assume $\omega = 0.5$, $r_1 = 2$, $r_2 = 2$ and $\text{pbest}_l^k = (1, 1, 0, 0, 1, 1, 1, 0)$, $\text{gbest}^k = (1, 1, 1, 0, 1, 0, 1, 0), V_l^k = \begin{pmatrix} 0.4, 0, 0, 0.2, 0, 0, 1, 0 \\ 0.6, 0, 0.2, 0.8, 0, 0, 0, 1 \end{pmatrix}$.

So $V_{l1}^{k+1} = c_1 r_1(\text{pbest}_l^k - X_l^k) = \begin{pmatrix} 0, 0, 0.8, 0, 0, 0, 0, 0 \\ 0, 0, 0, 0, 0, 0, 1, 0 \end{pmatrix}$. Suppose that $c_1 r_1 =$ 0.8 and 1.5 for each v_i^b ($v_i^b \neq 0$).

Let, $V_{l2}^{k+1} = c_2 r_2(\text{gbest}^k - X_l^k) = \begin{pmatrix} 0, 0, 0, 0, 0, 0.7, 0, 0 \\ 0, 0, 0, 0, 0, 0, 0.6, 0 \end{pmatrix}$. Suppose that $c_2 r_2 =$ 0.7 and 0.6 for each v_i^b ($v_l^b \neq 0$).

Then, $V_l^{k+1} = \omega V_l^k + V_{l1}^{k+1} + V_{l2}^{k+1} = \begin{pmatrix} 0.2, 0, 0.8, 0.1, 0, 0.7, 0.5, 0 \\ 0.3, 0, 0.1, 0.4, 0, 0, 1, 0.5 \end{pmatrix}$.

The second update equation is composed of four modifications.

Modification 4 The next position X_l^{k+1} is initialized to be an empty string.

Modification 5 Generate a random number $\alpha \in [0,\ 1]$, and if v_i^b in V_l^{k+1} is greater than α, then the ith bit of X_l^{k+1} is set to b. If both v_i^0 and v_i^1 are greater than α, then the ith bit is set to 0 or 1 randomly. And, if both v_i^0 and v_i^1 are smaller than α, then the ith bit is set to "–", which stands for the absent bits. After these two steps, X_l^{k+1} is a string with several absent bit.

Modification 6 A new parameter c_3 is added to Eq. (12.27) to keep a balance between exploitation and exploration, that is, the second update equation is modified as:

$$X_l^{k+1} = V_l^{k+1} \otimes c_3 r_3 X_l^k.$$

For each absent bit in X_l^k, a random real number r_3 belonging to [0, 1] is spawned, and if $c_3 r_3 > \alpha$, then the corresponding bit in X_l^k is copied to X_l^{k+1}.

However, the number of $v_i^b = 0$ in V_l^{k+1} increases rapidly. So, if we choose as many as possible bits in X_l^k, then almost all elements in X_l^{k+1} will come from X_l^k, and the algorithm stagnates.

With c_3, some absent bits in X_l^{k+1} are not filled until the operation in *Modification* 7, like mutation. So, c_3 is called the mutation factor. What the algorithm benefits from is shown by experiments in the next section.

Modification 7 Because several bits are perhaps absent in X_l^{k+1}, a strategy to fill the blanks is adopted. For the ith absent bit, a random real number t belonging to [0, 1] is generated, if $t < \text{rand}_i/n$, where rand_i is the rank of the ith node figured out in the preprocessing operations, and n is the number of nodes. Otherwise, the ith bit is set to 0. So, for each node, the farer to all the terminals it is, the more opportunity to be selected it will have.

Assume $\alpha = 0.5$, $c_3 = 2$, and V_l^{k+1} is given as in the example after *Modification* 3. So, a possible case is $X_l^{k+1} = (-, -, 0, -, -, 0, 0, 1)$.

Assume $c_3 r_3 = 1,\ 0.5,\ 0.4,\ 0.7$ for the first, second, fourth, and fifth bits, respectively. Then, the first, second, and fifth bits in X_l^k are copied to X_l^{k+1}, and so $X_l^{k+1} = (1, 1, 0, -, 1, 0, 0, 1)$.

Suppose $t < \text{rand}_4/n$. Then $X_l^{k+1} = (1, 1, 0, 0, 1, 0, 0, 1)$.

D. **The Modified Prim's Algorithm and the Trimming Strategy**

Based on the seven modifications, X_l^{k+1} a new n-bit binary string is generated. Since the solution to the Convergent Product Steiner Tree Problem (CPSTP) is a maximum spanning tree (MST), it is essential to devise a method for converting a binary string into the corresponding MST. The benefits of the edges and the costs of the nodes in the MST will indicate the fitness of the particle. Prim's algorithm is an effective approach for constructing an MST in a connected graph. However, it remains uncertain whether all selected nodes X_l^{k+1} are directly connected.

It is important to note that upon termination, the MST generated will include all selected nodes X_l^{k+1}. Due to the restoration operation, additional nodes may also be incorporated into the MST.

E. **Updating Velocity and Position Values**

The selection mechanism for global best (gbest) and personal best (pbest) solutions represents a critical component when extending particle swarm optimization (PSO) to multi-objective problems. Drawing inspiration from evolutionary algorithms, our approach implements a Pareto ranking scheme where:

1. Elite Solution Archiving:
   - Each particle maintains a pbest archive containing all non-dominated solutions from its search history.
   - A global gbest archive stores Pareto optimal solutions identified by the entire swarm, updated using fast non-dominated sorting.
   - Both archives follow strict dominance principles:
     - New solutions that dominate existing members trigger replacement.
     - Non-dominated additions expand the archive.
     - Dominated solutions are immediately removed.
2. Diversity-Preserving Selection Process:
   (a) Calculate the selection quota:
   - Quotient $Q = \lfloor \text{total particles/gbest archive size} \rfloor$.
   - Remainder $R = $ total particles mod gbest archive size.

   (b) For $R \neq 0$:
   - Compute crowding distances for all gbest members.
   - Sort members by descending crowding distance.

- Top R members receive $(Q + 1)$ selections.
- Remaining members receive Q selections.

(c) Particle-specific assignments:

- gbest selection: Choose the nearest available gbest by Euclidean distance in objective space.
- pbest selection: Select the pbest with maximum Euclidean distance from the assigned gbest.

3. Numerical Example Implementation:

- With 90 particles and 7 gbest archive members:
  - $Q = 12, R = 6$ (since $90 \div 7 = 12$ with remainder 6).
  - Top 6 members by crowding distance get 13 selections each.
  - Remaining member gets 12 selections.
- Ensures comprehensive coverage of the Pareto front.

4. Termination and Output:

- Final gbest archive contains the approximated Pareto optimal set.
- Solutions represent optimal trade-offs between:
  - Minimizing total costs: $\Sigma\left(c_{ij}x_{ij} + c_i y_i\right)$.
  - Maximizing total benefits: $\Sigma b_{ij}x_{ij}$.
- All solutions maintain:
  - Tree topology constraints $(\Sigma x_{ij} = \Sigma y_i - 1)$.
  - Terminal connectivity requirements $(y_i = 1 \ \forall \ i \in N)$.

This approach provides three key advantages:

1. Theoretical Soundness: Maintains the mathematical rigor of Pareto optimality.
2. Computational Efficiency: Uses $O(MN^2)$ non-dominated sorting.
3. Solution Diversity: Crowding distance metric prevents clustering.

The complete procedure continues iterating until reaching predetermined termination criteria (maximum generations or solution stability), at which point the gbest archive contains the final set of non-dominated solutions for the Convergent Product Steiner Tree Problem. The implementation includes adaptive mechanisms for archive size control and objective normalization to ensure balanced optimization across both cost and benefit dimensions.

12.4.4 Experimental Study

To demonstrate the applicability and effectiveness of our proposed multi-objective optimization process, we conducted an experiment. We consider an undirected graph $G = (i, j)$ with a set of clusters $V = \{1, ..., n\}$ and a set of links $E = \{e = (i, j) : i, j \in V,\ i < j\}$, where each link has non-negative profits p_e, associated with it, and each cluster has non-negative costs c_i.

In this Steiner tree problem, the objective is to identify a tree that maximizes revenue, which is defined as the total profits from the links p_e included in the solution, while simultaneously minimizing the total costs of the clusters. On one hand, we aim to span all links to avoid losing profit; however, this may lead to excessive costs for the tree-structured network that serves all clusters. Thus, a trade-off exists between the costs associated with the clusters included in the solution and the profits derived from the links.

The proposed methodology employs a structured three-dimensional matrix (Table 12.1) that systematically captures the relationships between core functions, sub-functions, and performance attributes. Through comprehensive data aggregation, we compute arithmetic means across all dimensions to generate a normalized comparison matrix. Based on analytical evaluation of the complete dataset (Table 12.4), we establish a threshold value of 0.561—corresponding to the arithmetic mean—which serves as a critical filter for identifying significant functional relationships. Application of this threshold yields a refined matrix (Table 12.5) that retains only the most substantive connections, effectively reducing solution space complexity while preserving essential product configurations. The resulting network topology, visualized in Fig. 12.21, demonstrates the optimized product architecture derived from this thresholding process. This integrated approach achieves an optimal balance between maintaining critical functional relationships and ensuring computational tractability, while providing a clear visualization of the core product structure for subsequent optimization stages in the convergent product design framework.

The proposed methodology implements a two-phase network configuration process through Algorithm 3, comprising sequential leveling and clustering operations. In the initial leveling phase (Steps 1–2), the network establishes hierarchical relationships by assigning base functions to level 0 and progressively organizing

Table 12.4 The three-dimensional comparison matrix for all the attributes for fuzzy network

Attributes	B_1	B_2	B_3	S_1	S_2	S_3	S_4	S_5	S_6	S_7
B_1	0	0.59	0.58	0.6	0.53	0.66	0.46	0.52	0.48	0.46
B_2	–	0	0.46	0.36	0.46	0.63	0.86	0.58	0.43	0.43
B_3	–	–	0	0.59	0.56	0.53	0.9	0.63	0.33	0.53
S_1	–	–	–	0	0.58	0.59	0.6	0.53	0.6	0.6
S_2	–	–	–	–	0	0.36	0.43	0.43	0.7	0.83
S_3	–	–	–	–	–	0	0.63	0.6	0.63	0.58
S_4	–	–	–	–	–	–	0	0.46	0.6	0.43
S_5	–	–	–	–	–	–	–	0	0.65	0.63
S_6	–	–	–	–	–	–	–	–	0	0.63
S_7	–	–	–	–	–	–	–	–	–	0

Table 12.5 The thresholded comparison matrix for all the attributes for fuzzy network

Attributes	B_1	B_2	B_3	S_1	S_2	S_3	S_4	S_5	S_6	S_7
B_1	0	1	1	1	0	1	0	0	0	0
B_2	–	0	0	0	0	1	1	1	0	0
B_3	–	–	0	1	0	0	1	1	0	0
S_1	–	–	–	0	1	1	1	0	1	1
S_2	–	–	–	–	0	0	0	0	1	1
S_3	–	–	–	–	–	0	1	1	1	1
S_4	–	–	–	–	–	–	0	0	1	0
S_5	–	–	–	–	–	–	–	0	1	1
S_6	–	–	–	–	–	–	–	–	0	1
S_7	–	–	–	–	–	–	–	–	–	0

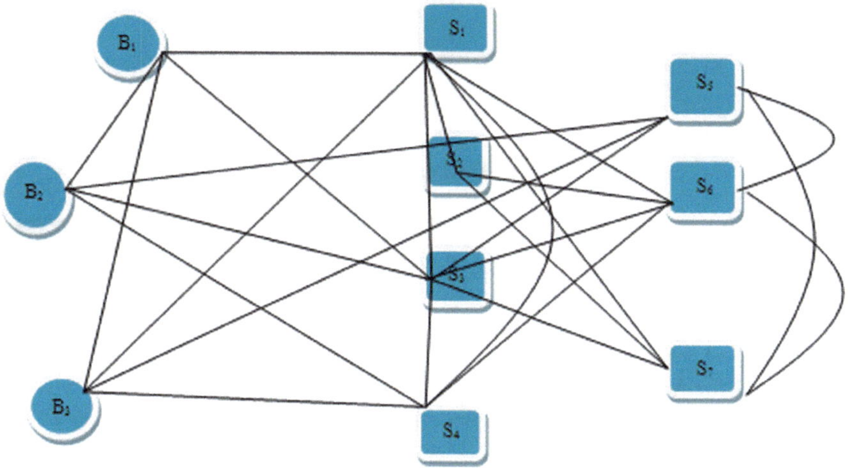

Fig. 12.21 The configured thresholded network for fuzzy model

sub-functions into subsequent levels, with the resulting tiered structure visualized in Fig. 12.22. The subsequent clustering phase (Steps 3–4) then groups functionally related nodes at each level while enforcing strict single-cluster membership, producing the optimized topology shown in Fig. 12.23. This structured transformation progressively refines the network architecture by first creating logical dependencies through leveling and then improving computational efficiency via clustering, all while maintaining essential connectivity requirements and preserving all terminal nodes. The visual progression from Figs. 12.22 to 12.23 demonstrates the methodology's effectiveness in converting raw functional relationships into an optimized network structure ready for Steiner tree implementation, while maintaining complete

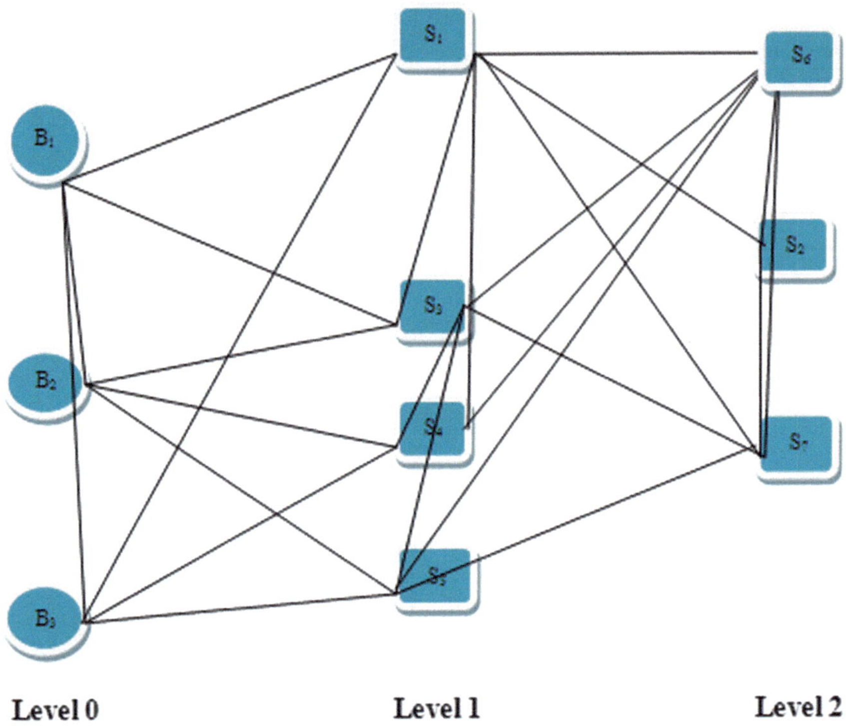

Fig. 12.22 The configured leveled network for fuzzy model

traceability of all functional dependencies and architectural decisions throughout both transformation stages.

The fuzzy cost vectors, the fuzzy benefit matrix, and the matrices $\alpha = [\alpha_{ij}]$, $\beta = [\beta_{ij}]$ are:

$$C_1 = ((120,\ 150,\ 170),\ (200,\ 210,\ 240),\ (160,\ 180,\ 200),\ (15,\ 20,\ 30),\ (23,\ 30,\ 35),\ (25,\ 30,\ 32),\ (40,\ 50,\ 60),\ (8,\ 10,\ 11),\ (17,\ 20,\ 22),\ (15,\ 20,\ 20))$$

$$C_2 = ((280,\ 300,\ 330),\ (400,\ 450,\ 480),\ (560,\ 600,\ 650),\ (65,\ 70,\ 80),\ (73,\ 80,\ 85),\ (45,\ 50,\ 52),\ (30,\ 40,\ 50),\ (18,\ 20,\ 21),\ (17,\ 20,\ 22),\ (15,\ 20,\ 20))$$

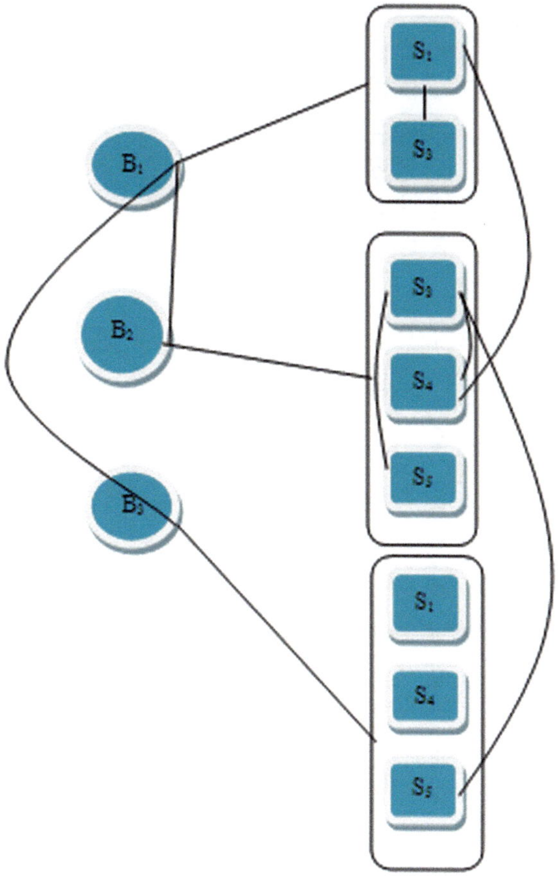

Fig. 12.23 The configured clustered network (the third and the fourth steps of Algorithm 3)

$$
\alpha = \begin{bmatrix} - & - & - \\ - & - & - \\ - & - & - \\ 0.65 & - & 0.7 \\ - & - & - \\ 1 & 0.8 & - \\ - & 0.6 & 0.9 \\ - & 0.7 & 0.65 \\ - & - & - \\ - & - & - \end{bmatrix} \quad \beta = \begin{bmatrix} - & - & - \\ - & - & - \\ - & - & - \\ 0.9 & - & 0.9 \\ - & - & - \\ 0.9 & 0.7 & - \\ - & 0.6 & 0.9 \\ - & 0.7 & 0.6 \\ - & - & - \\ - & - & - \end{bmatrix}
$$

$$p=\begin{bmatrix} - & (1260,\ 1500,\ 1880) & (1000,\ 1300,\ 1550) & (80,\ 100,\ 130) & - & (60,\ 80,\ 110) & - & - & - & - \\ - & - & - & - & - & (50,\ 70,\ 80) & (80,\ 90,\ 105) & (100,\ 120,\ 130) & - & - \\ - & - & - & (55,\ 60,\ 70) & - & - & (90,\ 100,\ 115) & (100,\ 110,\ 118) & - & - \\ - & - & - & - & (40,\ 50,\ 55) & (80,\ 90,\ 100) & (105,\ 110,\ 120) & - & (50,\ 60,\ 70) & (35,\ 40,\ 43) \\ - & - & - & - & - & - & - & - & (65,\ 70,\ 73) & (27,\ 30,\ 35) \\ - & - & - & - & - & - & (70,\ 80,\ 90) & (68,\ 70,\ 73) & (33,\ 40,\ 45) & (17,\ 20,\ 25) \\ - & - & - & - & - & - & - & - & (27,\ 30,\ 35) & - \\ - & - & - & - & - & - & - & - & (17,\ 20,\ 25) & (6,\ 10,\ 15) \\ - & - & - & - & - & - & - & - & - & (15,\ 20,\ 22) \\ - & - & - & - & - & - & - & - & - & - \end{bmatrix}.$$

For level 1, using iteration 1 of the while loop in Step 4 of Algorithm 3, we obtain:

$$\begin{aligned}
&(\alpha_{41}c_{411} + \beta_{41}c_{412}) = ((0.65)(15,\ 20,\ 30) + (0.9)(65,\ 70,\ 80)) = (68.25,\ 76.5,\ 91.5)\\
&(\alpha_{43}c_{431} + \beta_{43}c_{432}) = ((0.7)(15,\ 20,\ 30) + (0.9)(65,\ 70,\ 80)) = (69,\ 77,\ 93)\\
&MV = (68.25,\ 76.5,\ 91.5)
\end{aligned}$$

$$\begin{aligned}
&(\alpha_{61}c_{611} + \beta_{61}c_{612}) = ((1)\,(25,\ 30,\ 32) + (0.9)(45,\ 50,\ 52)) = (65.5,\ 75,\ 78.8)\\
&(\alpha_{62}c_{621} + \beta_{62}c_{622}) = ((0.8)\,(25,\ 30,\ 32) + (0.7)(45,\ 50,\ 52)) = (51.5,\ 59,\ 62)\\
&MV = (51.5,\ 59,\ 62)
\end{aligned}$$

$$\begin{aligned}
&(\alpha_{72}c_{721} + \beta_{72}c_{722}) = ((0.6)(40,\ 50,\ 60) + (0.6)(30,\ 40,\ 50)) = (42,\ 54,\ 66)\\
&(\alpha_{73}c_{731} + \beta_{73}c_{732}) = ((0.9)(40,\ 50,\ 60) + (0.9)(30,\ 40,\ 50)) = (63,\ 81,\ 99)\\
&MV = (42,\ 54,\ 66)
\end{aligned}$$

$$\begin{aligned}
&(\alpha_{82}c_{821} + \beta_{82}c_{822}) = ((0.7)\,(8,\ 10,\ 11) + (0.7)(18,\ 20\,,\ 21)) = (18.2,\ 21,\ 22.4)\\
&(\alpha_{83}c_{831} + \beta_{83}c_{832}) = ((0.65)\,(8,\ 10,\ 11) + (0.6)(18,\ 20\,,\ 21)) = (16,\ 18.5,\ 19.75)\\
&MV = (16,\ 18.5,\ 19.75).
\end{aligned}$$

Therefore, $z_{41} = 1,\ z_{62} = 1,\ z_{72} = 1$ and $z_{83} = 1$ with other variables being zero.

After purifying the benefits and costs for level 2, the final cost and benefit matrices are formed as follows:

$$\tilde{p} = \begin{bmatrix}
- & (1260,\ 1500,\ 1880) & (1000,\ 1300,\ 1550) & (80,\ 100,\ 130) & - & - & - & -\\
- & - & - & - & - & (169,\ 208,\ 240.5) & - & -\\
- & - & - & - & - & - & (100,\ 110,\ 118) & -\\
- & - & - & - & (40,\ 50,\ 55) & (126,\ 132,\ 144) & - & -\\
- & - & - & - & - & - & - & (119.6,\ 130,\ 140.4)\\
- & - & - & - & - & - & (81.6,\ 84,\ 87.6) & (115.5,\ 135,\ 157.5)\\
- & - & - & - & - & - & - & -\\
- & - & - & - & - & - & - & -
\end{bmatrix}$$

$$\begin{aligned}
\tilde{c} = (&(400,\ 450,\ 500)\quad (600, 660, 720)\quad (720,\ 780,\ 850)\quad (80,\ 90,\ 110)\\
&(96,\ 110,\ 120)\quad (23.5,\ 33,\ 38)\quad (26,\ 30,\ 32)\quad (28.3,\ 34,\ 34.8))
\end{aligned}$$

With respect to these matrices, the Steiner tree model is:

$$\begin{aligned}
\max\ u = {} & (1260,\ 1500,\ 1880)x_{12} + (1000,\ 1300,\ 1550)x_{13} + (80,\ 100,\ 130)x_{14}\\
& + (169,\ 208,\ 240.5)\,x_{26} + (100,\ 110,\ 118)x_{37} + (40,\ 50,\ 55)x_{45}\\
& + (126,\ 132,\ 144)\,x_{46} + (119.6,\ 130,\ 140.4)\,x_{58} + (81.6,\ 84,\ 87.6)\,x_{67}\\
& + (115.5,\ 135,\ 157.5)\,x_{68}\\
\min\ z = {} & (400,\ 450,\ 500)y_1 + (600,\ 660,\ 720)y_2 + (720,\ 780,\ 850)y_3
\end{aligned}$$

$$+ (80,\ 90,\ 110)y_4 + (96,\ 110,\ 120)y_5 + (23.5,\ 33,\ 38)y_6$$
$$+ (26,\ 30,\ 32)y_7 + (28.3,\ 34,\ 34.8)y_8$$

s.t.

$$x_{12} + x_{13} + x_{14} + x_{26} + x_{37} + x_{45} + x_{46} + x_{58} + x_{67} + x_{68}$$
$$= y_1 + y_2 + y_3 + y_4 + y_5 + y_6 + y_7 + y_8 - 1$$
$$x_{14} \le y_1$$
$$x_{14} \le y_4$$
$$x_{26} \le y_2$$
$$x_{26} \le y_6$$
$$x_{37} \le y_3$$
$$x_{37} \le y_7$$
$$x_{46} \le y_4$$
$$x_{46} \le y_6$$
$$x_{58} \le y_5$$
$$x_{58} \le y_8$$
$$x_{12} \le y_1$$
$$x_{12} \le y_2$$
$$x_{67} \le y_6$$
$$x_{67} \le y_7$$
$$x_{13} \le y_1$$
$$x_{13} \le y_3$$
$$x_{45} \le y_5$$
$$x_{45} \le y_4$$
$$x_{68} \le y_6$$
$$x_{68} \le y_8$$
$$x_{14} + x_{26} + x_{12} + x_{46} \le y_1 + y_2 + y_4$$
$$x_{14} + x_{26} + x_{12} + x_{46} \le y_1 + y_2 + y_6$$
$$x_{14} + x_{26} + x_{12} + x_{46} \le y_1 + y_6 + y_4$$
$$x_{14} + x_{26} + x_{12} + x_{46} \le y_6 + y_2 + y_4$$
$$x_{45} + x_{58} + x_{68} + x_{46} \le y_4 + y_5 + y_6$$
$$x_{45} + x_{58} + x_{68} + x_{46} \le y_4 + y_5 + y_8$$
$$x_{45} + x_{58} + x_{68} + x_{46} \le y_4 + y_8 + y_6$$
$$x_{45} + x_{58} + x_{68} + x_{46} \le y_8 + y_5 + y_6$$
$$x_{12} + x_{26} + x_{67} + x_{37} + x_{13} \le y_1 + y_2 + y_3 + y_6$$
$$x_{12} + x_{26} + x_{67} + x_{37} + x_{13} \le y_1 + y_2 + y_3 + y_7$$
$$x_{12} + x_{26} + x_{67} + x_{37} + x_{13} \le y_1 + y_2 + y_7 + y_6$$

$$x_{12} + x_{26} + x_{67} + x_{37} + x_{13} \leq y_1 + y_7 + y_3 + y_6$$
$$x_{12} + x_{26} + x_{67} + x_{37} + x_{13} \leq y_7 + y_2 + y_3 + y_6$$
$$x_{14} + x_{46} + x_{67} + x_{37} + x_{13} \leq y_1 + y_4 + y_3 + y_6$$
$$x_{14} + x_{46} + x_{67} + x_{37} + x_{13} \leq y_1 + y_4 + y_3 + y_7$$
$$x_{14} + x_{46} + x_{67} + x_{37} + x_{13} \leq y_1 + y_4 + y_7 + y_6$$
$$x_{14} + x_{46} + x_{67} + x_{37} + x_{13} \leq y_1 + y_7 + y_3 + y_6$$
$$x_{14} + x_{46} + x_{67} + x_{37} + x_{13} \leq y_7 + y_4 + y_3 + y_6$$
$$x_{14} + x_{45} + x_{58} + x_{26} + x_{12} + x_{68} \leq y_1 + y_2 + y_4 + y_5 + y_6$$
$$x_{14} + x_{45} + x_{58} + x_{26} + x_{12} + x_{68} \leq y_1 + y_2 + y_4 + y_5 + y_8$$
$$x_{14} + x_{45} + x_{58} + x_{26} + x_{12} + x_{68} \leq y_1 + y_2 + y_4 + y_8 + y_6$$
$$x_{14} + x_{45} + x_{58} + x_{26} + x_{12} + x_{68} \leq y_1 + y_2 + y_8 + y_5 + y_6$$
$$x_{14} + x_{45} + x_{58} + x_{26} + x_{12} + x_{68} \leq y_1 + y_8 + y_4 + y_5 + y_6$$
$$x_{14} + x_{45} + x_{58} + x_{26} + x_{12} + x_{68} \leq y_8 + y_2 + y_4 + y_5 + y_6$$
$$x_{14} + x_{45} + x_{58} + x_{68} + x_{37} + x_{67} + x_{13} \leq y_1 + y_3 + y_4 + y_5 + y_6 + y_7$$
$$x_{14} + x_{45} + x_{58} + x_{68} + x_{37} + x_{67} + x_{13} \leq y_1 + y_3 + y_4 + y_5 + y_6 + y_8$$
$$x_{14} + x_{45} + x_{58} + x_{68} + x_{37} + x_{67} + x_{13} \leq y_1 + y_3 + y_4 + y_5 + y_8 + y_7$$
$$x_{14} + x_{45} + x_{58} + x_{68} + x_{37} + x_{67} + x_{13} \leq y_1 + y_3 + y_4 + y_8 + y_6 + y_7$$
$$x_{14} + x_{45} + x_{58} + x_{68} + x_{37} + x_{67} + x_{13} \leq y_1 + y_3 + y_8 + y_5 + y_6 + y_7$$
$$x_{14} + x_{45} + x_{58} + x_{68} + x_{37} + x_{67} + x_{13} \leq y_1 + y_8 + y_4 + y_5 + y_6 + y_7$$
$$x_{14} + x_{45} + x_{58} + x_{68} + x_{37} + x_{67} + x_{13} \leq y_8 + y_3 + y_4 + y_5 + y_6 + y_7$$
$$y_h = 1, \qquad \forall h \in \{1, 2, 3\}$$
$$x_{ij} \in \{0, 1\}, \qquad \forall i, j$$
$$y_i \in \{0, 1\}, \qquad \forall i.$$

The proposed model was successfully solved using a binary particle swarm optimization (BOPSO) approach implemented in MATLAB R2019b (Version 9.0), executing 150 iterations with a swarm size of 20 particles per generation. The optimization process generated a comprehensive set of Pareto optimal solutions that effectively capture the fundamental trade-off between the two competing objectives: (1) minimizing total system costs and (2) maximizing overall benefits. These non-dominated solutions are systematically presented in Table 12.6, while Fig. 12.24 provides a two-dimensional visualization of the resulting Pareto frontier, with the x-axis representing cost minimization and the y-axis representing benefit maximization. The solution set demonstrates several key characteristics: first, it reveals the inherent inverse relationship between cost and benefit objectives; second, it provides decision-makers with a diverse range of optimal configurations; and third, all solutions strictly maintain the required problem constraints, including tree topology and terminal node connectivity. The visualization in Fig. 12.24 particularly highlights the algorithm's effectiveness in exploring the objective space, as evidenced by the well-distributed solutions along the Pareto frontier. This implementation confirms the BOPSO approach's capability to handle the problem's combinatorial complexity while delivering high-quality solutions within reasonable computational time, as facilitated by MATLAB's efficient execution environment. The results not only validate the methodology's effectiveness but also provide practical insights for product configuration decisions in real-world applications.

In Fig. 12.24, the X-axis is cost, and the Y-axis is the symmetry of benefit.

The first Pareto optimal solution is $D_{2,\frac{1}{2}}(\tilde{0}, \text{benefit}^*) = 3558.6$ and $D_{2,\frac{1}{2}}(\tilde{0}, \text{cost}^*) = 2191.7$.

The second Pareto optimal solution is $D_{2,\frac{1}{2}}(\tilde{0}, \text{benefit}^*) = 3449.2$ and $D_{2,\frac{1}{2}}(\tilde{0}, \text{cost}^*) = 2162.1$.

The third Pareto optimal solution is $D_{2,\frac{1}{2}}(\tilde{0}, \text{benefit}^*) = 3428.9$ and $D_{2,\frac{1}{2}}(\tilde{0}, \text{cost}^*) = 2082.5$.

The fourth Pareto optimal solution is $D_{2,\frac{1}{2}}(\tilde{0}, \text{benefit}^*) = 3319.5$ and $D_{2,\frac{1}{2}}(\tilde{0}, \text{cost}^*) = 2052.9$.

The fifth Pareto optimal solution is $D_{2,\frac{1}{2}}(\tilde{0}, \text{benefit}^*) = 3186.3$ and $D_{2,\frac{1}{2}}(\tilde{0}, \text{cost}^*) = 1960.1$.

Table 12.6 The Pareto solutions resulted from the two objective functions

The number of the Pareto solution	$D_{2,\frac{1}{2}}(\tilde{0}, \text{cost})$	$D_{2,\frac{1}{2}}(\tilde{0}, \text{benefit})$	Cost	Benefit
1	2191.7	3558.6	(1973.8, 2187, 2404.8)	(2889.5, 3515, 4230.4)
2	2162.1	3449.2	(1947.8, 2157, 2372.8)	(2789.5, 3405, 4112.4)
3	2082.5	3428.9	(1877.8, 2077, 2284.8)	(2770.5, 3385, 4090)
4	2052.9	3319.5	(1851.8, 2047, 2252.8)	(2670.5, 3275, 3972)
5	1989.6	3295.6	(1797.8, 1987, 2174.8)	(2644.5, 3253, 3946)
6	1960.1	3186.3	(1771.8, 1957, 2142.8)	(2544.5, 3143, 3828)
7	1956.8	3159.4	(1769.5, 1953, 2140)	(2529, 3118, 3788.5)
8	1927.2	3050	(1743.5, 1923, 2108)	(2429, 3008, 3670.5)
9	1924.7	2952	(1746, 1920, 2102)	(2360, 2910, 3548)
10	1881.9	2842.6	(1720, 1890, 2070)	(2260, 2800, 3430)

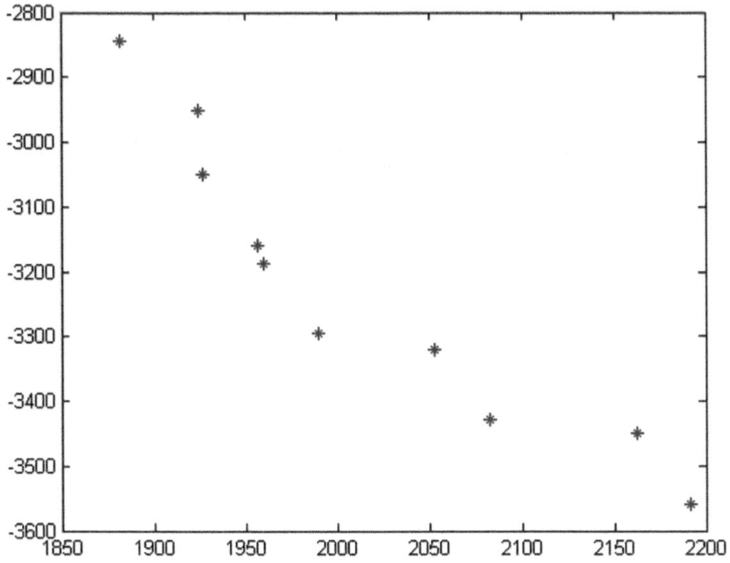

Fig. 12.24 The Pareto solutions in two-dimensional space resulted from the two objective functions

The sixth Pareto optimal solution is $D_{2,\frac{1}{2}}(\tilde{0}, \text{benefit}^*) = 3050$ and $D_{2,\frac{1}{2}}(\tilde{0}, \text{cost}^*) = 1927.2$.

The proposed approach offers distinct products tailored for producers and consumers, each associated with varying benefits and costs. The numerical findings indicate that the configuration of these products, which have different costs, is informed by customer feedback gathered through a web-based system. These products are designed to deliver maximum benefits to the producers.

A key decision within the proposed methodology involves balancing the cost and benefit objectives. This balance is shaped by customer perspectives on the addition of product features, as well as producers' insights into the configuration of advantageous features.

12.5 Conclusions

We present a dual-method optimization framework that combines ant colony optimization (ACO) and particle swarm optimization (PSO) approaches to identify optimal value-adding functionalities for convergent products. The methodology begins by modeling product features as a digital network, where nodes represent base functions and sub-functions, and arcs denote their interconnections. Through an initial purification process, we establish fundamental relationships between these components, subsequently applying leveling and clustering techniques to assign quantitative benefit values to arcs and cost values to nodes. The ACO approach adapts Steiner tree methodology to create a deterministic multi-objective model that identifies optimal functionality trees maximizing product value. Complementing this, we developed a fuzzy PSO variant that handles parameter uncertainty through bi-objective optimization of cost and benefit metrics. Together, these methods generate diverse product configurations with varying cost-benefit profiles, revealing critical trade-offs between implementation costs, functional benefits, and customer-perceived value. The framework's web-based implementation enables continuous incorporation of customer feedback and producer constraints, yielding dynamic solutions that balance technical feasibility with market demands. Our results demonstrate how this integrated approach can guide product configuration decisions by providing both deterministic and fuzzy-optimized solutions tailored to different business scenarios and uncertainty levels.

References and Further Reading

Abbasian, M., Nosratabadi, H., & Fazlollahtabar, H. (2015). Applying an intelligent dynamic genetic algorithm for solving a multi-objective flexible job shop scheduling problem with maintenance considerations. *Journal of Applied & Computational Mathematics, 4*(4).

Abedi, M., Seidgar, H., Fazlollahtabar, H., & Bijani, R. (2015). Bi-objective optimisation for scheduling the identical parallel batch-processing machines with arbitrary job sizes, unequal job release times and capacity limits. *International Journal of Production Research, 53*(6), 1680–1711.

Bellman, R. E. and Zadeh, L. A. (1970). Decision-making in a fuzzy environment, Management Science 171, 41–164, 1970.

Costa, A. M., Cordeau, J. F., & Laporte, G. (2006). Steiner tree problems with profits. *INFOR, 44*(2), 99–115.

Dreyfuss, S. E., & Wagner, R. A. (1971). The Steiner problem in graphs. *Networks, 1*, 195–207.

Duin, C. W., & Volgenant, A. (1989). Reduction tests for the Steiner problem in graphs. *Networks, 19*, 549–567.

Fazlollahtabar, H., & Tehranchian, A. M. (2015). Integration between mathematical programming and fuzzy logic to optimize consumers behavior. In *Research Methods: Concepts, Methodologies, Tools, and Applications* (pp. 1600–1615). IGI Global.

Fazlollahtabar, H., & Aghasi, E. (2012). A two-stage mechanism as a decision support for marketing via customer satisfaction measures. *Journal of the Chinese Institute of Industrial Engineers, 29*(4), 270–281.

Fazlollahtabar, H., & Ehsani, M. A. (2010). Integration between regression model and fuzzy logic approach for analyzing various electronic commerce effects on economic growth in organizations. *Journal of Electronic Commerce in Organizations (JECO), 8*(4), 17–31.

Fazlollahtabar, H., & Gholizadeh, H. (2020). Fuzzy possibility regression integrated with fuzzy adaptive neural network for predicting and optimizing electrical discharge machining parameters. *Computers and Industrial Engineering, 140*, Article 106225.

Fazlollahtabar, H., Hassanzadeh, R., Mahdavi, I., & Mahdavi-Amiri, N. (2012). A genetic optimization algorithm and perceptron learning rules for a bi-criteria parallel machine scheduling. *Journal of the Chinese Institute of Industrial Engineers, 29*(3), 206–218.

Fazlollahtabar, H. H., & Mohseninejad, M. (2015). A neural network decision support system for analyzing markets. *Journal of Advances in Management Sciences and Information Systems, 1*, 38–46.

Gholizadeh, H., & Fazlollahtabar, H. (2019). Production control process using integrated robust data envelopment analysis and fuzzy neural network. *International Journal of Mathematical, Engineering and Management Sciences, 4*(3), 580.

Golneshini, F. P., & Fazlollahtabar, H. (2019). Meta-heuristic algorithms for a clustering-based fuzzy bi-criteria hybrid flow shop scheduling problem. *Soft Computing, 23*(22), 12103–12122.

Hassanzadeh, R., Mahdavi I., Fazlollahtabar, & Mahdavi-Amiri, N. (2012). A web-based multi-optimization approach for convergent product networks using Steiner tree. In *Proceedings of the 2012 International Conference on Industrial Engineering and Operations Management*, Istanbul, Turkey, July 3–6 (pp. 2057–2062).

Hwang, F. K., & Richards, D. S. (1992). Steiner tree problems. *Networks, 22*, 55–89.

Johnson, D. S. (1985). The NP-completeness column: An ongoing guide. *Journal of Algorithms, 6*, 434–451.

Kaufmann, A., & Gupta, M. M. (1991). *Introduction to Fuzzy Arithmetic: Theory and Applications*. Van Nostrand-Reinhold.

Kennedy, J., & Eberhart, R. (1995). Particle Swarm optimization. In *The 1995 IEEE International Conference on Neural Networks* (Vol. 4, pp. 1942–1948).

Lawler, E. L. (1976). *Combinatorial Optimization Networks and Matroids*. Holt.

Mahdavi, I., Fazlollahtabar, H., Heidarzade, A., Mahdavi-Am, N., & Rooshan, Y. I. (2008). A heuristic methodology for multi-criteria evaluation of web-based e-learning systems based on user satisfaction. *Journal of Applied Sciences, 8*(24), 4603–4609.

Mahdavi, I., Fazlollahtabar, H., Mahdavi-Amiri, N., & Kakouei, M. (2010). Designing integrated models as a decision aid in web-based marketing. *Journal of Information Technology Applications and Management, 17*(1), 1–10.

Robins, G., & Zelikovsky, A. (2000). Improved Steiner tree approximation in graphs. In *Proceedings of the 11th Annual ACM-SIAM Symposium on Discrete Algorithms* (pp. 770–779). SIAM, ACM.

Robins, G., & Zelikovsky, A. (2005). Tighter bounds for graph Steiner tree approximation. *SIAM Journal on Discrete Mathematics, 19*(1), 122–134.

Shirazi, B., Fazlollahtabar, H., & Mahdavi, I. (2010). A six sigma based multi-objective optimization for machine grouping control in flexible cellular manufacturing systems with guide-path flexibility. *Advances in Engineering Software, 41*(6), 865–873.

Takahashi, H., & Matsuyama, A. (1980). An approximate solution for the Steiner problem in graphs. *Mathematica Japonica, 24*(6), 573–577.

Thomas, H. C., Charles, E. L., Ronald, L. R., & Clifford, S. (1990). *Introduction to Algorithms*. MIT Press.

Winter, P., & MacGregor Smith, J. (1992). Path-distance heuristics for the Steiner problem in undirected networks. *Algorithmica, 7*, 309–327.

Bibliography

Abbasian, M., Nosratabadi, H., & Fazlollahtabar, H. (2015). Applying an intelligent dynamic genetic algorithm for solving a multi-objective flexible job shop scheduling problem with maintenance considerations. *Journal of Applied & Computational Mathematics, 4*(4).

Abbasian, P., Mahdavi-Amiri, N., & Fazllolahtabar, H. (2015b). Modeling and optimizing of a non-constrained multi-objective problem having multiple utility functions using the Bayesian theory. *Journal of Information Sciences and Computing Technologies, 4*(3), 332–342.

Abbasian, P., Mahdavi-Amiri, N., & Fazlollahtabar, H. (2018). Multiple utility constrained multi-objective programs using Bayesian theory. *Journal of Industrial Engineering International, 14*, 111–118.

Abbaspour, M., Fazlollahtabar, H., & Stevic, Z. (2022). Multi-objective rough best-worst method to evaluate sustainability of a biofuel energy supply chain. *International Journal of Industrial Engineering & Production Research, 33*(1), 34–50.

Abedi, M., Seidgar, H., Fazlollahtabar, H., & Bijani, R. (2015). Bi-objective optimisation for scheduling the identical parallel batch-processing machines with arbitrary job sizes, unequal job release times and capacity limits. *International Journal of Production Research, 53*(6), 1680–1711.

Akbari, F., Fazlollahtabar, H., & Mahdavi, I. (2013). An uncertain decision making process considering customers and services in evaluating banks: A case study. *International Journal of Strategic Decision Sciences (IJSDS), 4*(2), 48–78.

Asadi, M., Fazlollahtabar, H., & Shirazi, B. (2015a). An enterprise management decision making system based on possibility theory. *International Journal of Enterprise Information Systems (IJEIS), 11*(4), 1–27.

Asadi, M., Shirazi, B., & Fazlollahtabar, H. (2015b). A sales management decision making system based on possibility theory. *Journal of Information Technology & Software Engineering, 5*(3), 1000159.

Bagheri, H., Ghavareshki, M. H. K., Abbasi, M., & Fazlollahtabar, H. (2025). Integrated smart industry 4.0-based decision support for optimizing tactical-operational decisions: Case study. *Journal of Industrial and Management Optimization, 21*(1), 783–808.

Chaleshigar Kordasiabi, M., Gholizadeh, H., Fazlollahtabar, H., & Javadian, N. (2020). Analysis of cost model with queuing system under uncertainty. *Journal of Industrial and Production Engineering, 37*(6), 292–304.

Ebadi, S., Tajdin, A., Mahdavi, I., & Fazlollahtabar, H. (2024). Rough multi-period network data envelopment analysis for evaluation of supply chain: A case study of skill training in Iran. *Facta Universitatis, Series: Mechanical Engineering*.

H. Fazlollahtabar, *Uncertain Multi-objective Decision Making: Methods and Industrial Models*, https://doi.org/10.1007/978-981-95-6035-6

Esmaeilpour, R., Fazlollahtabar, H., & Aghasi, E. (2013). Multi-objective mathematical model as a decision support for customer service marketing. *International Journal of Applied Operational Research, 3*(2), 27–41.

Fazlollahtabar, H. (2023). Rough DT2R2ML for renewable energy supplier selection in the presence of big data. *Concurrency and Computation Practice and Experience, 35*(23), Article e7773. https://doi.org/10.1002/cpe.7773

Fazlollahtabar, H. (2008). Applying multiple-criteria decision making methods for developing information technology industry. *International Journal of Information and Decision Sciences, 1*(1), 115–131.

Fazlollahtabar, H. (2016). An integration between fuzzy PROMETHEE and fuzzy linear program for supplier selection problem: Case study. *Journal of Applied Mathematical Modelling and Computing, 1*(1).

Fazlollahtabar, H. (2022). Intelligent marketing decision model based on customer behavior using integrated possibility theory and K-means algorithm. *Journal of Intelligent Management Decision, 1*(2), 88–96.

Fazlollahtabar, H. H., & Mohseninejad, M. (2015). A neural network decision support system for analyzing markets. *Journal of Advances in Management Sciences & Information Systems, 1*, 38–46.

Fazlollahtabar, H., & Aghasi, E. (2012). A two-stage mechanism as a decision support for marketing via customer satisfaction measures. *Journal of the Chinese Institute of Industrial Engineers, 29*(4), 270–281.

Fazlollahtabar, H., & Aghasi, E. (2014). An integrated stochastic multi-criteria acceptability analysis and mathematical optimisation for service marketing. *International Journal of Services and Operations Management, 17*(1), 38–49.

Fazlollahtabar, H., & Ashena, M. (2013). A Bayesian network decision support system for order management in new product development. *Computer Science and Information Technology, 1*, 82–89.

Fazlollahtabar, H., & Ebadi, S. (2023). Skill training supply chain performance evaluation using network data envelopment analysis. *International Journal of Industrial Engineering & Production Research, 34*(1), 1–14.

Fazlollahtabar, H., & Ehsani, M. A. (2010). Integration between regression model and fuzzy logic approach for analyzing various electronic commerce effects on economic growth in organizations. *Journal of Electronic Commerce in Organizations (JECO), 8*(4), 17–31.

Fazlollahtabar, H., & Gholizadeh, H. (2019). Economic analysis of the M/M/1/N queuing system cost model in a vague environment. *International Journal of Fuzzy Logic and Intelligent Systems, 19*(3), 192–203.

Fazlollahtabar, H., & Gholizadeh, H. (2020). Fuzzy possibility regression integrated with fuzzy adaptive neural network for predicting and optimizing electrical discharge machining parameters. *Computers & Industrial Engineering, 140*, Article 106225.

Fazlollahtabar, H., & Gholizadeh. (2020). Application of queuing theory in quality control of multi-stage flexible flow shop. *Yugoslav Journal of Operations Research, 30*(1), 101–119.

Fazlollahtabar, H., & Hassanli, S. (2018). Hybrid cost and time path planning for multiple autonomous guided vehicles. *Applied Intelligence, 48*, 482–498.

Fazlollahtabar, H., & Saidi-Mehrabad, M. (2013). Optimising a multi-objective reliability assessment in multiple AGV manufacturing system. *International Journal of Services and Operations Management, 16*(3), 352–372.

Fazlollahtabar, H., & Saidi-Mehrabad, M. (2015). Optimizing multi-objective decision making having qualitative evaluation. *Journal of Industrial & Management Optimization, 11*(3).

Fazlollahtabar, H., & Tehranchian, A. M. (2015). Integration between mathematical programming and fuzzy logic to optimize consumers behavior. In *Research methods: Concepts, methodologies, tools, and applications* (pp. 1600–1615). IGI Global.

Fazlollahtabar, H., Aghasi, E., & Forte, P. (2012a). Bi-objective two-stage decision-making process for service marketing. *International Journal of Strategic Decision Sciences (IJSDS), 3*(3), 24–39.

Fazlollahtabar, H., Akbari, F., & Mahdavi, I. (2013a). A fuzzy goal programming for optimizing service industry market using virtual intelligent agent. *Journal of Industrial and Production Engineering, 30*(1), 20–29.

Fazlollahtabar, H., Eslami, H., & Salmani, H. (2010). Designing a fuzzy expert system to evaluate alternatives in fuzzy analytic hierarchy process. *Journal of Software Engineering and Applications, 3*(04), 409.

Fazlollahtabar, H., Hassanzadeh, R., Mahdavi, I., & Mahdavi-Amiri, N. (2012b). A genetic optimization algorithm and perceptron learning rules for a bi-criteria parallel machine scheduling. *Journal of the Chinese Institute of Industrial Engineers, 29*(3), 206–218.

Fazlollahtabar, H., Mahdavi, I., & Mohajeri, A. (2013b). Applying fuzzy mathematical programming approach to optimize a multiple supply network in uncertain condition with comparative analysis. *Applied Soft Computing, 13*(1), 550–562.

Fazlollahtabar, H., Mahdavi, I., Ashoori, M. T., Kaviani, S., & Mahdavi-Amiri, N. (2011). A multi-objective decision-making process of supplier selection and order allocation for multi-period scheduling in an electronic market. *The International Journal of Advanced Manufacturing Technology, 52*, 1039–1052.

Fazlollahtabar, H., Mahdavi-Amiri, N., & Muhammadzadeh, A. (2015). A genetic optimization algorithm for nonlinear stochastic programs in an automated manufacturing system. *Journal of Intelligent & Fuzzy Systems, 28*(3), 1461–1475.

Fazlollahtabar, H., Vasiljević, M., Stević, Ž., & Vesković, S. (2017, September). Evaluation of supplier criteria in automotive industry using rough AHP. In *The 1st International Conference on Management, Engineering and Environment ICMNEE* (pp. 186–197).

Gholizadeh, H., & Fazlollahtabar, H. (2019). Production control process using integrated robust data envelopment analysis and fuzzy neural network. *International Journal of Mathematical, Engineering and Management Sciences, 4*(3), 580.

Gholizadeh, H., Fazlollahtabar, H., & Gholizadeh, R. (2019). A modified branch and bound algorithm for a vague flow-shop scheduling problem. *Iranian Journal of Fuzzy Systems, 16*(4), 55–64.

Gholizadeh, H., Fazlollahtabar, H., & Khalilzadeh, M. (2020). A robust fuzzy stochastic programming for sustainable procurement and logistics under hybrid uncertainty using big data. *Journal of Cleaner Production, 258*, Article 120640.

Golneshini, F. P., & Fazlollahtabar, H. (2019). Meta-heuristic algorithms for a clustering-based fuzzy bi-criteria hybrid flow shop scheduling problem. *Soft Computing, 23*(22), 12103–12122.

Gunantara, N., & Hendrantoro, G. (2013, September 23). Multi-objective cross-layer optimization for selection of cooperative path pairs in multihop wireless ad hoc networks. *Journal of Communications Software and Systems, 9*(3), 170–177.

Katiraee, N., Shirazi, B., & Fazlollahtabar, H. (2017). Fuzzy multi-objective supplier selection considering production requirements in resilient supply chain: Case study in steel industry. *International Journal of Information Systems and Supply Chain Management (IJISSCM), 10*(3), 65–83.

Kazemirash, N., & Fazlollahtabar, H. (2022). A multidimensional decision making for supplier selection in the presence of information systems. *Croatian Operational Research Review, 13*(1), 31–47.

Kazemitash, N., Fazlollahtabar, H., & Abbaspour, M. (2021). Rough best-worst method for supplier selection in biofuel companies based on green criteria. *Operational Research in Engineering Sciences: Theory and Applications, 4*(2), 1–12.

Komleh, A. A., & Fazlollahtabar, H. (2019). Stochastic multi-criteria acceptability analysis for technology transfer evaluation: A case study in construction digging. *International Journal of Mathematical, Engineering and Management Sciences, 4*(4), 1031.

Mahdavi, I., Fazlollahtabar, H., Heidarzade, A., Mahdavi-Am, N., & Rooshan, Y. I. (2008a). A heuristic methodology for multi-criteria evaluation of web-based e-learning systems based on user satisfaction. *Journal of Applied Sciences, 8*(24), 4603–4609.

Mahdavi, I., Fazlollahtabar, H., Mahdavi-Amiri, N., & Kakouei, M. (2010). Designing integrated models as a decision aid in web-based marketing. *Journal of Information Technology Applications & Management, 17*(1), 1–10.

Mahdavi, I., Fazlollahtabar, H., Mozaffari, E., Heidari, M., & Mahdavi-Amiri, N. (2008b). Data envelopment analysis based comparison of two hybrid multi-criteria decision-making approaches for mobile phone selection: A case study in Iranian telecommunication environment. *International Journal of Information and Decision Sciences, 1*(2), 194–220.

Mojaveri, H. S., & Fazlollahtabar, H. (2012). Designing an integrated AHP based fuzzy expert system and SWOT Analysis to prioritize development strategies of Iran agriculture. *Revista De Management Comparat International, 13*(1), 117.

Rahimikelarijani, B., Fazlollahtabar, H., & Nayeri, S. (2020). Multi-objective multi-load tandem autonomous guided vehicle for robust workload balance and material handling optimization. *SN Applied Sciences, 2*(7), 1200.

Salarian, R., & Fazlollahtabar, H. (2015). Multioptimization in a cellular manufacturing system having stochastic parameters considering pricing. *Journal for Manufacturing Science and Production, 15*(3), 257–265.

Salarian, R., Fazlollahtabar, H., & Mahdavi, I. (2014). Inter-cell movement minimisation in a cellular manufacturing system having stochastic parameters. *International Journal of Services and Operations Management, 17*(1), 67–87.

Shirazi, B., Fazlollahtabar, H., & Mahdavi, I. (2010). A six sigma based multi-objective optimization for machine grouping control in flexible cellular manufacturing systems with guide-path flexibility. *Advances in Engineering Software, 41*(6), 865–873.

Tavana, M., Fazlollahtabar, H., & Hassanzadeh, R. (2014). A bi-objective stochastic programming model for optimising automated material handling systems with reliability considerations. *International Journal of Production Research, 52*(19), 5597–5610.

Torkjazi, M., & Fazlollahtabar, H. (2015a). Applying a utility based fuzzy probabilistic α-cut method to optimize a constrained multi objective model. *Creative Mathematics and Informatics, 24*(2), 221–230.

Torkjazi, M., & Fazlollahtabar, H. (2015b). A fuzzy probabilistic maximum technique to optimize an unconstrained utility based multi objective model. *Industrial Engineering and Management, 4*(147), 2169–2316.

Torkjazi, M., Fazlollahtabar, H., & Mahdavi, I. (2014). Optimizing an unconstrained multi objective model using a utility based fuzzy probabilistic α-cut method. *Journal of Intelligent & Fuzzy Systems, 26*(6), 2927–2936.

Vasiljević, M., Fazlollahtabar, H., Stević, Ž, & Vesković, S. (2018). A rough multicriteria approach for evaluation of the supplier criteria in automotive industry. *Decision Making: Applications in Management and Engineering, 1*(1), 82–96.

GPSR Compliance

The European Union's (EU) General Product Safety Regulation (GPSR) is a set of rules that requires consumer products to be safe and our obligations to ensure this.

If you have any concerns about our products, you can contact us on ProductSafety@springernature.com

In case Publisher is established outside the EU, the EU authorized representative is:

Springer Nature Customer Service Center GmbH
Europaplatz 3
69115 Heidelberg, Germany

Batch number: 10725823

Printed by Printforce, the Netherlands